Recent Advances In

Stochastic Operations Research

Recent Advances In Stochastic Operations Research

Editors

Tadashi Dohi
Hiroshima University, Japan

Shunji Osaki
Katsushige Sawaki
Nanzan University, Japan

World Scientific

NEW JERSEY • LONDON • SINGAPORE • BEIJING • SHANGHAI • HONG KONG • TAIPEI • CHENNAI

Published by

World Scientific Publishing Co. Pte. Ltd.
5 Toh Tuck Link, Singapore 596224
USA office: 27 Warren Street, Suite 401-402, Hackensack, NJ 07601
UK office: 57 Shelton Street, Covent Garden, London WC2H 9HE

British Library Cataloguing-in-Publication Data
A catalogue record for this book is available from the British Library.

RECENT ADVANCES IN STOCHASTIC OPERATIONS RESEARCH

ISBN-13 978-981-256-704-8
ISBN-10 981-256-704-6

Printed in Singapore by World Scientific Printers (S) Pte Ltd

PREFACE

Operations Research uses quantitative models to analyze and predict the behavior of systems, and to provide information for decision makers. Two key concepts in Operations Research are Optimization and Uncertainty. Uncertainty is emphasized in Operations Research that could be called "Stochastic Operations Research" in which uncertainty is described by stochastic models. The typical models in Stochastic Operations Research are queuing models, inventory models, financial engineering models, reliability models, and simulation models.

The International Workshop on Recent Advances in Stochastic Operations Research (2005 RASOR Canmore) was held in Canmore, Alberta, Canada, on August 25–26, 2005. Based on the 40 papers presented, participants exchanged ideas, discussed common problems, and found new ideas and problems during the conference. Fruitful and keen discussions occurred among the participants during the two days of the conference.

After the conference, we asked all the authors to submit their papers for the proceedings, and as a result, almost all papers that had been presented were submitted. After a careful peer-review, 20 papers were chosen for the Proceedings. All in all, it took one year to edit the Proceedings. We believe that the Proceedings will be useful for the researchers interested in Stochastic Operations Research.

This conference was sponsored by the Research Center for Mathematical Sciences and Information Engineering, Nanzan University, 27 Seirei-cho, Seto-shi, Aichi 489-0863, JAPAN, to whom we would like to express our appreciation for their financial support. We also appreciated the financial support we received in the form of Grant-in-Aid for Scientific Research from the Ministry of Education, Sports, Science and Culture of Japan under Grant Nos. 16201035 and 16510128. Our special thanks are due to Professor Hiroyuki Okamura and Dr. Koichiro Rinsaka, Hiroshima University, Japan, for their continual support from the initial planning of the conference to the final stage of editing the proceedings. Finally, we would like to thank Chelsea Chin, World Scientific Publishing Co., Singapore, for her warm help and patience.

Tadashi Dohi — Hiroshima University
Shunji Osaki — Nanzan University
Katsushige Sawaki — Nanzan University

September 2006

LIST OF CONTRIBUTORS

M. Arafuka	– Kinjo Gakuin University, Japan
M. Arai	– Tokyo Metropolitan University, Japan
R. Arnold	– Victoria University of Wellington, New Zealand
S. Chukova	– Victoria University of Wellington, New Zealand
T. Dohi	– Hiroshima University, Japan
E. A. Elsayed	– Rutgers, The State University of New Jersey, USA
F. Ferreira	– University of Trás-os-Montes e Alto Douro, Portugal
S. Fukumoto	– Tokyo Metropolitan University, Japan
H. Goko	– Bank of Japan, Japan
G. Hardy	– LaRIA, France
Y. Hayakawa	– Waseda University, Japan
T. Hino	– Tokyo Metropolitan University, Japan
N. Hirotsu	– Japan Institute of Sports Sciences, Japan
H. Hohjo	– Osaka Prefecture University, Japan
M. Imaizumi	– Aichi Gakusen University, Japan
S. Inoue	– Tottori University, Japan
K. Ito	– Mitsubishi Heavy Industries, LTD., Japan
K. Iwasaki	– Tokyo Metropolitan University, Japan
N. Kaio	– Hiroshima Shudo University, Japan
H. Kawai	– Tottori University, Japan
M. Kimura	– Gifu City Women's College, Japan
J. Koyanagi	– Tottori University, Japan
N. Limnios	– LMAC, UTC, France
C. Lucet	– LaRIA, France

S. Nakagawa	– Kinjo Gakuin University, Japan
T. Nakagawa	– Aichi Institute of Technology, Japan
T. Nakai	– Kyushu University, Japan
S. Nakamura	– Kinjo Gakuin University, Japan
K. Naruse	– Aichi Institute of Technology, Japan
M. Ohnishi	– Osaka University, Japan
Y. Okuda	– Aichi Institute of Technology, Japan
S. Osaki	– Nanzan University, Japan
A. Pacheco	– Technical University of Lisbon, Portugal
H. Ribeiro	– Polytechnic Institute of Leiria, Portugal
K. Rinsaka	– Hiroshima University, Japan
K. Sawaki	– Nanzan University, Japan
M. Suzaki	– Nanzan University, Japan
A. Suzuki	– Nanzan University, Japan
Y. Teraoka	– Osaka Prefecture University, Japan
M. Tsujimura	– Ryukoku University, Japan
K. Yagi	– Nanzan University, Japan
S. Yamada	– Tottori University, Japan
K. Yasui	– Aichi Institute of Technology, Japan
H. Zhang	– Rutgers, The State University of New Jersey, USA

CONTENTS

PART A

Reliability

WARRANTY ANALYSIS: ESTIMATION OF THE DEGREE OF IMPERFECT REPAIR VIA A BAYESIAN APPROACH

S. CHUKOVA

School of Mathematics, Statistics and Computer Science, Victoria University of Wellington, PO Box 600, Wellington, New Zealand
Stefanka.Chukova@mcs.vuw.ac.nz

Y. HAYAKAWA

School of International Liberal Studies, Waseda University, 1-21-1 Nishi-Waseda, Shinjuku-ku, Tokyo 169-0051, Japan
yu.hayakawa@waseda.jp

R. ARNOLD

School of Mathematics, Statistics and Computer Science, Victoria University of Wellington, PO Box 600, Wellington, New Zealand
Richard.Arnold@mcs.vuw.ac.nz

An approach to modeling imperfect repairs under warranty settings is presented in Chukova, Arnold and Wang[12]. They model the imperfect repairs using the concepts of delayed and accelerated distribution functions. As an extension of their approach, we design a procedure for estimating the degree of repair as well as other modeling parameters by Markov chain Monte Carlo (McMC) methods.

1. Introduction

The growth of using the product warranty as a strategic tool has increased quite significantly over the past decade. For example, in automobile industry warranty is considered as an attribute of the the products and it is used as a valuable selling point. A few of the many reasons closely tied to the usage of warranties by vendors are the following: customers are more concerned with quality issues; many consumers have neither the time nor the inclination to deal with products' failures or repairs; due to the increasing complexity of the products, consumers are often unable to judge quality before buying a product, and so on.

A product warranty is an agreement offered by a producer to a consumer to repair or replace a faulty item, or to partially or fully reimburse the consumer in the event of a failure. The form of reimbursement of the customer on failure of the product or dissatisfaction with service, is one of the most important characteristics of warranty. The most common forms are (see Blischke and Murthy[3,4]):

- A lump-sum rebate (e.g., "money-back guarantee"), which is usually assigned for relatively small time interval immediately after the purchase of the product or service. Usually a "money back guarantee" offer promotes the trust between the seller and the buyer. It addresses the risk of information asymmetry, i.e. lack of information on the buyer's part, which can cause a wrong purchase decision.
- A free repair of the failed item. The associated warranty coverage is called a free repair warranty (FRW).
- A repair provided at reduced cost to the buyer. The cost reduction is usually a decreasing function of the time to failure. The corresponding warranty is called pro-rata warranty (PRW).
- A combination of the preceding terms. Usually combination warranty starts with a FRW up to a specified time and switches to a repair at pro-rated cost for the remainder of the warranty period. This is called FRW/PRW.

Regarding the mechanism of the warranty coverage, there are two types of warranty policies used in the marketplace and studied in the literature:

- Non-renewing warranty (NR): A newly sold item is covered by a warranty for some calendar time of duration W, called warranty period, which usually starts at the time of the purchase of the product. During the warranty period, the warranter assumes all (NRFRW), or a portion of the expenses (NRPRW) associated with the failure of the product.

Most of the domestic appliances, such as vacuum cleaners, refrigerators, washing machines and dryers, TV's, are covered by non-renewing warranty.

- Renewing warranty (R): The warranter repairs any faulty item from the time of the purchase up to time W, the length of the warranty period. At the time of each repair within an existing warranty the item is warranted anew for a period of length W. The warranty coverage expires when the lifetime of the item (the original one or its repaired version) exceeds W. During the warranty coverage, the warranter assumes all (RFRW), or a portion of the expenses (RPRW) associated with the failure of the product.

For example, light bulbs are covered by renewing free repair warranty (RFRW). A light bulb has an initial warranty period of thirty days and if it fails during this period it is replaced by a new bulb and the warranty starts anew. The replacement can be considered as a particular type of repair, namely, complete, or perfect repair (see Section 2). Usually renewing warranty is assigned to inexpensive products.

In a renewing warranty scenario, the warranty coverage is a random variable, whereas under a non-renewing warranty, the warranty period is a constant, which might be predetermined or a decision variable.

Despite the fact that warranties are so commonly used, the accurate pricing of warranties in many situations remains an unsolved problem. This may seem surprising since the fulfillment of warranty claims may cost companies large amount of money. Underestimating true warranty cost results in losses for a company. On the other hand, overestimating them will lead to uncompetitive product prices. As a result the amount of product sales will decrease. The data relevant to the modeling of warranty costs in a particular industry are usually highly confidential, since they are commercially sensitive. Therefore, much warranty analysis takes place in internal research divisions in large companies.

The main objective in product warranty analysis is to model and estimate the warranty cost. The expected warranty cost over the warranty period (or coverage) or the expected warranty cost per unit time over the warranty period (or coverage), as well as the expected warranty cost over the life-cycle of a product are of particular interest. Of course, corresponding standard deviations are also important. These quantities summarize the financial risk or burden carried by buyers, sellers and decision makers.

The evaluation of the parameters (e.g., warranty period or price) of the warranty contract can be obtained, by using appropriate models, from the producer's, seller's, buyer's as well as decision maker's point of view. Often these parameters are solutions of an appropriate optimisation problem and their values result from the application of analytical or numerical methods. Due to the complexity of the models, it is almost always necessary to resort to numerical methods, since analytical solutions exist only in the simplest situations. A general treatment of warranty analysis is given in Blischke and Murthy[3,4] and Chukova, Dimitrov and Rykov[11]. For recent literature review see Murthy and Djamaludin[13].

The outline of this paper is as follows. Section 2 provides a review on classification of repairs based on the degree of repair. In Section 3, several types of repairs are compared by the failure rate functions, in particular, the worse than new, better than used (WTNBTU) case is studied. Section 4 deals with the model assumptions. In Section 5, the Gibbs sampler is described in the setting of WTNBTU repair in order to make inference on the degree of repair as well as the other model parameters. In Section 6, using an example, we illustrate the ideas and compare our findings with the findings in [12]. The last section concludes our study.

2. Types of Repair

The evaluation of the warranty cost or any other parameter of interest in modeling warranties depends on the failure and repair processes and on the assigned preventive warranty maintenance of the items. Assuming that the failure rate function of the product's lifetime distribution is an increasing function of time, repairs can be classified according to the degree to which they restore the ability of the item to function (see Brown and Proschan[10], Pham and Wang[14]). The post-failure repairs affect repairable products in one of the following ways:

(1) **Improved Repair**: A repair brings the product to a state *better* than when it was initially purchased. This is equivalent to the replacement of the faulty item by a new and improved item.
(2) **Complete Repair**: A repair completely resets the performance of the product so that upon restart the product operates as a new one. This type of repair is equivalent to a replacement of the faulty item by a new one, identical to the original.
(3) **Imperfect Repair**: A repair contributes to some noticeable improvement of the product. It effectively sets back the clock for the repaired item. After the repairs the performance and expected lifetime of the item are as they were at an earlier age.
(4) **Minimal Repair**: A repair has no impact on the performance of the item. The repair brings the product from a 'down' to and 'up' state without affecting its performance.
(5) **Worse Repair**: A repair contributes to some noticeable worsening of the product. It effectively sets forward the clock for the repaired item. After the repairs, the performance of the item is as it would have been at a later age.

(6) **Worst Repair**: A repair accidentally leads to the product's destruction.

3. Comparison of Lifetime Distributions by Failure Rate Functions

Chukova, Arnold and Wang in [12] consider the first 5 types of repairs listed in the previous section (excluding the worst repair) and compare them in terms of the expected lifetimes and the failure rates of the first two interfailure times. In what follows we focus on the second type of comparison based on the post-repair failure rate functions.

3.1. *Classification of the repair based on the post-repair failure rate functions*

We denote by:

- X_1 - the initial lifetime of the product and X_i - the lifetime after the $(i-1)^{th}$ repair, $i = 2, 3, \ldots$;
- x_i - a realisation of X_i;
- $F_i(x),\ \bar{F}_i(x),\ f_i(x),\ \lambda_i(x),\ x \geq 0$ - the distribution function, the reliability function, the probability density function and the failure rate function of X_i. We recall (see Barlow and Proshan[1]) that

$$\bar{F}_i(x) = \exp(-\int_0^x \lambda_i(u)du) \quad \text{and} \quad \lambda_i(x) = f_i(x)/\bar{F}_i(x). \tag{1}$$

We rank the types of repairs corresponding to those introduced in the previous section in the same way as Chukova, Arnold and Wang[12] by introducing an index parameter, τ, which reflects the degree of repair. For any particular time $x > 0$, the following classification of the repairs is considered:

(1) **Better than new (BTN)** $\lambda_{i+1}(x) \leq \lambda_i(x)$,
i.e., $\lambda_{i+1}(x) = \frac{1}{\tau}\lambda_i(x)$, for $\tau > 1$
(2) **Good as new (GAN)** $\lambda_{i+1}(x) = \lambda_i(x)$
(3) **Worse than new, better than used (WTNBTU)**
$\lambda_i(x) < \lambda_{i+1}(x) < \lambda_i(x_i + x)$,
i.e., $\lambda_{i+1}(x) = \tau\lambda_i(x) + (1-\tau)\lambda_i(x_i + x)$ for $0 < \tau < 1$.
(4) **Minimal repair (MIN)** $\lambda_{i+1}(x) = \lambda_i(x_i + x)$
(5) **Worse than used (WTU)** $\lambda_{i+1}(x) > \lambda_i(x_i + x)$,
i.e., $\lambda_{i+1}(x) = \frac{1}{\tau+1}\lambda_i(x_i + x)$ for $-1 < \tau < 0$.

3.2. *Worse Than New Better Than Used (WTNBTU) repair*

From now onwards our study will focus only on worse than new, better than used (WTNBTU) repairs. Next, we discuss the relationship between the characteristics of the initial lifetime of the product X_1 and its lifetime after the $(i-1)^{st}$ WTNBTU repair X_i. By our definition of WTNBTU we have

$$\lambda_2(x) = \tau\lambda_1(x) + (1-\tau)\lambda_1(x_1 + x).$$

and for $i = 3$, it follows that

$$\begin{aligned}\lambda_3(x) &= \tau\lambda_2(x) + (1-\tau)\lambda_2(x_2 + x) \\ &= \tau^2\lambda_1(x) + \tau(1-\tau)[\lambda_1(x_1+x) + \lambda_1(x_2+x)] + (1-\tau)^2\lambda_1(x_1+x_2+x).\end{aligned}$$

Using (1) and mathematical induction the following result can be derived:

Theorem 1. The failure rate of the ith operational time $\lambda_i(x)$ can be expressed in terms of $\lambda_1(\cdot)$ and τ as follows:

$$\lambda_i(x) = \sum_{k=0}^{i-1} \tau^{(i-1)-k}(1-\tau)^k \sum_{S\in M_i^k} \lambda_1\big(x + \sum_{j\in S} x_j\big), \tag{2}$$

where the collections of sets $M_i^k,\ k = 1, 2, \cdots, i-1$ are defined as follows

$$M_i^k = \{\{j_1, \ldots, j_k\} \mid 1 \le j_1 < j_2 < \ldots < j_k < i\} \tag{3}$$

and M_i^0 contains the empty set.

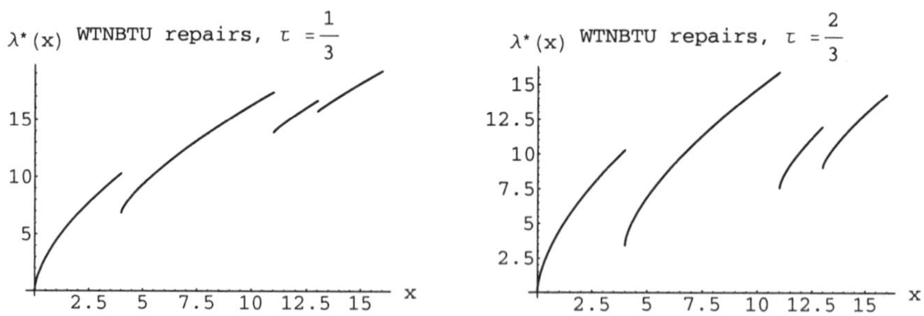

Figure 1. The failure rate $\lambda^*(x)$ for two different WTNBTU repairs

If we denote by $\lambda^*(x)$ the failure rate function of a product maintained by WTNBTU repairs with identical degrees with a Weibull underlying initial failure distribution, then Figure 1 represents $\lambda^*(x)$ for two different degrees of repair, $\tau = \frac{1}{3}$ and $\tau = \frac{2}{3}$ with $x_1 = 4, x_2 = 7$, and $x_3 = 2$.

Theorem 2. The conditional reliability function of the ith operational time $\bar{F}_i(x|x_1, \cdots, x_{i-1})$ can be expressed in terms of $\bar{F}_1(\cdot)$ and τ as follows:

$$\bar{F}_i(x|x_1, \cdots, x_{i-1}) = \left[\bar{F}_1(x)\right]^{\tau^{i-1}} \prod_{k=1}^{i-1} \prod_{S \in M_i^k} \left[\frac{\bar{F}_1\left(x + \sum_{j\in S} x_j\right)}{\bar{F}_1\left(\sum_{j\in S} x_j\right)}\right]^{\tau^{(i-1)-k}(1-\tau)^k}, \tag{4}$$

where the collections of sets M_i^k, $k = 1, 2, \cdots, i-1$ are given in (3).

4. The Model

In Chukova, Arnold and Wang[12], the indexing parameter τ and the other parameters of the product lifetime distribution are estimated using the maximum likelihood approach. Here the proposed estimations are from a Bayesian perspective. To simplify the matter, we assume that the first two lifetimes (x_1, x_2) are observed and the repair type is known to be WTNBTU.

Three lifetime distributions for X_1 are considered:

(1) Weibull(α, β) with $\bar{F}_1(x) = e^{-\beta x^\alpha}, f_1(x) = \alpha\beta x^{\alpha-1} e^{-\beta x^\alpha}$

(2) Gamma(α, β) with $\bar{F}_1(x) = \bar{\gamma}(\beta x, \alpha), f_1(x) = \frac{\beta^\alpha}{\Gamma(\alpha)} x^{\alpha-1} e^{-\beta x}$,
where $\bar{\gamma}(x, a) = 1 - \frac{1}{\Gamma(\alpha)} \int_0^x u^{a-1} e^{-u} du$ and $\gamma(x, a) = 1 - \bar{\gamma}(x, a)$

(3) Exp(β) with $\bar{F}_1(x) = e^{-\beta x}, f_1(x) = \beta e^{-\beta x}$

For $\alpha = 1$, Weibull(α, β) and Gamma(α, β) both reduce to Exp(β).

Using Theorem 1, the respective likelihood functions for Cases 1, 2, 3 given (x_1, x_2) are:

(1) Weibull case:

$$L(\tau, \alpha, \beta \,|\, x_1, x_2) = (\alpha\beta)^2 x_1^{\alpha-1}[(1-\tau)(x_1+x_2)^{\alpha-1} + \tau x_2^{\alpha-1}] \times e^{-(1-\tau)\beta(x_1+x_2)^\alpha - \tau\beta(x_1^\alpha + x_2^\alpha)};$$

(2) Gamma case:

$$L(\tau, \alpha, \beta \,|\, x_1, x_2) = \left[\frac{\beta^\alpha}{\Gamma(\alpha)}\right]^2 x_1^{\alpha-1} e^{-\beta(x_1+x_2)} \left\{\frac{\bar{\gamma}(\beta(x_1+x_2), \alpha)}{\bar{\gamma}(\beta x_1, \alpha)\bar{\gamma}(\beta x_2, \alpha)}\right\}^{1-\tau} \times \left\{\tau x_2^{\alpha-1} + (1-\tau)(x_1+x_2)^{\alpha-1} e^{-\beta x_1} \frac{\bar{\gamma}(\beta x_2, \alpha)}{\bar{\gamma}(\beta(x_1+x_2), \alpha)}\right\}.$$

(3) Exponential case: $L(\tau, \beta \,|\, x_1, x_2) = \beta^2 e^{-\beta(x_1+x_2)}$.

Due to the memoryless property of the exponential distribution corresponding likelihood function does not depend on the degree of repair τ. Hence, in the exponential case, the statistical inference for β can be done via a standard Bayesian method by updating the posterior for β.

For strictly Weibull and gamma cases, a set of possible standard prior distributions for the parameters α, β and τ are listed below. The hyperparameter values such as α_0, β_0 are not necessarily the same for Weibull and gamma scenarios.

- $\alpha \sim \text{Unif}(\alpha_0, \beta_0)$ with density function $\pi(\alpha) = 1/(\beta_0 - \alpha_0)$ for $\alpha \in (\alpha_0, \beta_0)$.
- $\beta \sim \text{Gamma}(s, w)$.
- $\tau \sim \text{Beta}(a, b)$ with density function $\pi(\tau) = \frac{\Gamma(a+b)}{\Gamma(a)\Gamma(b)} \tau^{a-1}(1 - \tau)^{b-1}$ for $0 < \tau < 1$.

Another possible choice of a prior distribution for the parameter α is a gamma distribution.

5. The Markov Chain Monte Carlo Method

In this section, we present the utilisation of Markov chain Monte Carlo method (see Chen, Shao and Ibrahim[6]) for estimating τ and the parameters of the initial lifetime distribution. To simplify the matter, we assume that the first two lifetimes are observed and the repair type is known to be WTNBTU.

The McMC methods enable us to simulate a Markov chain whose stationary distribution is our "target" distribution, for instance, the posterior distribution of the parameter of interest. We will use the Gibbs sampler which is one of the McMC methods. Under our model, the Gibbs sampler can be described as follows:

(1) Start with an arbitrary initial vector $\theta^{(0)} = (\tau^{(0)}, \alpha^{(0)}, \beta^{(0)})$ and set $k = 0$.
(2) Sample $\tau^{(k+1)}$ from $\pi(\tau \,|\, \alpha^{(k)}, \beta^{(k)}, x_1, x_2)$.
(3) Sample $\alpha^{(k+1)}$ from $\pi(\alpha \,|\, \tau^{(k+1)}, \beta^{(k)}, x_1, x_2)$.
(4) Sample $\beta^{(k+1)}$ from $\pi(\beta \,|\, \tau^{(k+1)}, \alpha^{(k+1)}, x_1, x_2)$.
(5) Set $\theta^{(k+1)} = (\tau^{(k+1)}, \alpha^{(k+1)}, \beta^{(k+1)})$ and $k = k + 1$. Go back to the second step.

Here, $\pi(\cdot \,|\, \cdot)$ denotes a full conditional density function. For instance,

$$\pi(\tau \,|\, \alpha^{(k)}, \beta^{(k)}, x_1, x_2)$$

denotes the conditional density function of τ given current values of all the other parameters as well as two failure times. One can obtain each full conditional density up to a proportionality constant as follows by, firstly, obtaining the joint density function of all the parameters and failure times, i.e.,

$$L(\tau, \alpha, \beta \mid x_1, x_2)\pi(\tau)\pi(\alpha)\pi(\beta), \tag{5}$$

and then viewing (5) as a function of the parameter of interest. With our model assumptions, direct sampling from each full conditional cannot be easily done due to the indexing parameter τ (see Gilks[5]). Metropolis-Hastings algorithms (see Metropolis et al.[8] and Hastings[9]) can be used to draw samples from the full conditionals. The Metropolis-Hastings method is an McMC methods that includes the Gibbs sampler as a special case (see Chib and Greenberg[7]). Under certain regularity conditions, for sufficiently large k, $\{\theta^{(m)} : k \leq m \leq (k+n-1)\}$ is approximately an i.i.d. sample of size n from the posterior distribution of τ, α, β given (x_1, x_2). Numerical examples will illustrate our approach.

6. Example

In this section we will reconsider the example discussed in[12] and summarize the findings on the estimation of τ using maximum likelihood approach. Further, we will estimate τ from Bayesian prospective and compare the results of the two approaches. As in [12], we will restrict our attention on the comparison of two consecutive lifetime distributions by using corresponding failure rate functions.

We will begin with the following assumptions:

- The item, S, subject to failures/repairs, has a complex structure comprising m subsystems, i.e., $S = \{S_1, S_2, ...S_m\}$.
- A failure of a particular subsystem requires a type of repair which is known in advance.

For example, let us consider a car. If the failure affects the tires of a car (say subsystem S_1), usually a complete repair is required. On the other hand, if the charging system of the car (say subsystem S_2) fails, worse than new, better than used (WTNBTU) repair is performed. The information on warranty failures and repairs is usually strictly confidential and it is very difficult to obtain real warranty data even for research purposes. For this reason we demonstrate how one might estimate the parameters of our model using simulated data.

In general, the information on the performance of the system S under NRFRW with warranty period W can be depicted as follows:

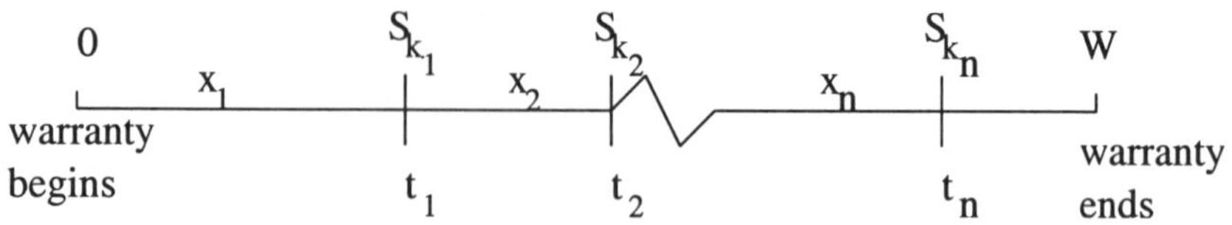

where

- t_i is the instant of the ith failure,
- x_i is the lifetime between the $(i-1)^{th}$ and i^{th} failure, and
- S_{k_i} is the subsystem affected by the ith failure.

In this very general scenario, each repair can be of different degree (τ_{k_i}). For the purposes of the current study and easier comparison between the two estimation approaches, we consider only one transition, i.e., there is a failure and and degree τ repair of the subsystem S_{k_1}, followed by a failure of subsystem S_{k_2}.

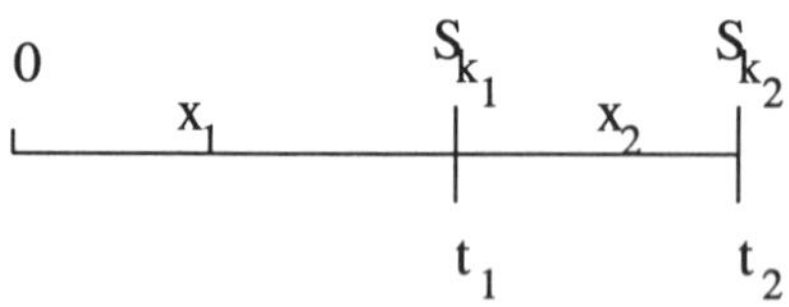

The data in this case are pairs of observations (X_1, X_2) of the first $(X_1 \sim \bar{F}_1(x_1))$ and the second $(X_2 \sim \bar{F}_2(x_2|X_1))$ lifetimes of the system, and the labels of the subsystems (k_1, k_2) which failed. Then, assuming that failure in S_{k_1} requires WTNBTU repair $(0 < \tau < 1)$ and given the data, we estimate τ along with the other parameters of the lifetime distributions, firstly using the maximum likelihood approach, see [12], followed by McMC approach.

6.1. *Maximum likelihood approach for estimating* τ

As in [12], we assume that the distribution of the first lifetime is Weibull, with distribution functions for the first and second lifetimes as given in Section 3.2. Thus, the likelihood of a single observation is

$$\begin{aligned} f(x_1, x_2) = (\alpha\beta)^2 x_1^{\alpha-1}[(1-\tau)(x_1+x_2)^{\alpha-1} + \tau x_2^{\alpha-1}] \\ \times \exp\{-(1-\tau)\beta(x_1+x_2)^{\alpha} - \tau\beta(x_1^{\alpha} + x_2^{\alpha})\}. \end{aligned}$$

We estimate the values of α, β and τ by maximum likelihood from a sample of size n by solving the following system of equations

$$\frac{2n}{\beta} - (1-\tau)\sum_{i=1}^{n}(x_{1i}+x_{2i})^{\alpha} - \tau\sum_{i=1}^{n}(x_{1i}^{\alpha}+x_{2i}^{\alpha}) = 0$$

$$\begin{aligned}\frac{2n}{\alpha} + \sum_{i=1}^{n}\log(x_{1i}) - (1-\tau)\beta\sum_{i=1}^{n}(x_{1i}+x_{2i})^{\alpha}\log(x_{1i}+x_{2i})& \\ -\tau\beta\sum_{i=1}^{n}(x_{1i}^{\alpha}\log(x_{1i}) + x_{2i}^{\alpha}\log(x_{2i}))& \\ +\sum_{i=1}^{n}\frac{(1-\tau)(x_{1i}+x_{2i})^{\alpha-1}\log(x_{1i}+x_{2i}) + \tau x_{2i}^{\alpha-1}\log(x_{2i})}{(1-\tau)(x_{1i}+x_{2i})^{\alpha-1}+\tau x_{2i}^{\alpha-1}} &= 0\end{aligned}$$

$$\begin{aligned}\sum_{i=1}^{n}\frac{x_{2i}^{\alpha-1} - (x_{1i}+x_{2i})^{\alpha-1}}{(x_{1i}+x_{2i})^{\alpha-1} + \tau[x_{2i}^{\alpha-1} - (x_{1i}+x_{2i})^{\alpha-1}]}& \\ -\beta\sum_{i=1}^{n}[(x_{1i}^{\alpha}+x_{2i}^{\alpha}) - (x_{1i}+x_{2i})^{\alpha}] &= 0.\end{aligned}$$

Samples of 100 observations of X_1 and X_2 were generated from $f(x_1, x_2)$. The values of parameters α, β were determined by appropriate choice of the expectation and the standard deviation of X_1. The mean was set to 7 months and the standard deviation to 4 months. For a range of τ, the results of the maximum likelihood fitting are given in Table 1.

Table 1. Maximum Likelihood estimates of α, β and τ from simulations of $n = 100$ observations.

	α=1.81		β=0.024			
τ	$\hat{\alpha}$	SE($\hat{\alpha}$)	$\hat{\beta}$	SE($\hat{\beta}$)	$\hat{\tau}$	SE($\hat{\tau}$)
0.1	1.75	0.09	0.031	0.006	0.16	0.15
0.2	1.82	0.12	0.021	0.007	0.00	0.14
0.3	1.92	0.11	0.016	0.006	0.32	0.17
0.4	1.71	0.10	0.036	0.006	0.50	0.15
0.5	1.87	0.13	0.020	0.008	0.41	0.14
0.6	1.78	0.12	0.025	0.008	0.57	0.11
0.7	1.95	0.11	0.017	0.007	0.72	0.10
0.8	1.67	0.12	0.034	0.008	0.66	0.10
0.9	1.69	0.10	0.029	0.005	0.85	0.07

In general the maximum likelihood estimates of α, β and τ estimates found in Table 1 are close to the true values that were used to simulate the data. However we note that the likelihood surface is very flat for τ, especially where τ is close to zero, leading to large standard errors. In some cases (most notably the $\tau = 0.2$ case) the maximum occurs on the boundary of the region (i.e. at $\tau = 0$). This behaviour meant that standard errors, at least for this small sample size, could not be estimated using the curvature of the likelihood function at the maximum. Instead the standard errors in Table 1 are the standard deviations of the maximum likelihood parameter estimates from repeated simulations of sample size 100.

Larger sample sizes (such as $n = 1000$ used by Chukova, Arnold and Wang[12]) lead to likelihoods with more clearly defined maxima, and improved maximum likelihood estimation.

6.2. *Bayesian approach for estimating τ*

We assume that the distribution of the first lifetime is Weibull(α, β), as in Section 6.1. The conditional reliability function for the second lifetimes is as given in Theorem 2, Section 3.2. Also, we considered the following set of prior distributions for the parameters α and β of the Weibull distribution:

- $\alpha \sim$ Gamma(t, v) with shape parameter t where $t = 0.0324, v = 0.018$.
- $\beta \sim$ Gamma(s, w) with shape parameter s where $s = 0.00000625, w = 0.00025$.

Table 2. Case 2 (Informative prior): Posterior means, standard deviations and credible intervals

	Posterior means and standard deviations					
	$\alpha = 1.81$		$\beta = 0.024$		τ	
τ	Mean	sd	Mean	sd	Mean	sd
0.1	1.73	0.12	0.033	0.009	0.10	0.08
0.2	1.85	0.11	0.021	0.006	0.16	0.06
0.3	1.91	0.13	0.017	0.005	0.30	0.07
0.4	1.70	0.12	0.037	0.010	0.42	0.08
0.5	1.87	0.12	0.021	0.006	0.48	0.07
0.6	1.77	0.11	0.027	0.008	0.59	0.07
0.7	1.94	0.12	0.018	0.006	0.71	0.06
0.8	1.66	0.10	0.038	0.010	0.75	0.07
0.9	1.67	0.10	0.032	0.008	0.89	0.07
	95% credible intervals					
	α		β		τ	
0.1	1.52	1.95	0.018	0.054	0.005	0.290
0.2	1.63	2.07	0.011	0.036	0.052	0.295
0.3	1.69	2.15	0.009	0.030	0.164	0.454
0.4	1.49	1.92	0.021	0.060	0.269	0.569
0.5	1.65	2.10	0.011	0.035	0.335	0.614
0.6	1.56	1.99	0.014	0.044	0.444	0.725
0.7	1.71	2.18	0.009	0.031	0.577	0.822
0.8	1.46	1.86	0.022	0.059	0.593	0.884
0.9	1.48	1.87	0.018	0.050	0.735	0.991

We study the influence of the prior of τ on the posterior analysis of the parameters by considering two priors of τ and compare the results.

Case 1 (Uninformative prior): The first choice of prior of τ is $\tau \sim \text{Beta}(1,1)$, which is equivalent to $\tau \sim \text{Unif}(0,1)$ for all true τ values.

Case 2 (Informative prior): The second choice of prior of τ is $\tau \sim \text{Beta}(a,b)$ with density

$$\pi(\tau) = \frac{\Gamma(a+b)}{\Gamma(a)\Gamma(b)}\tau^{a-1}(1-\tau)^{b-1} \text{ for } 0 < \tau < 1.$$

We set the prior mean of τ equal to the true value of τ. The following table summarises the values of the hyperparameters of the prior for τ.

τ	0.1	0.2	0.3	0.4	0.5	0.6	0.7	0.8	0.9
a	1.18571	4.37143	8.7	13.3143	17.3571	19.9714	20.3	17.4857	10.6714
b	10.6714	17.4857	20.3	19.9714	17.3571	13.3143	8.7	4.37143	1.18571

The full conditional of β is also a gamma distribution with updated values of the hyperparameters and β was sampled directly from its full conditional. Both α and τ were sampled from their respective full conditionals via the Metropolis-Hastings algorithm within the Gibbs sampler.

We have refitted the data used in the maximum likelihood section above. Table 2 summarises the results of our Bayesian estimating procedure for Case 2, whereas Table 3 reflects our finding for Case 1.

Table 3. Case 1 (Uninformative prior): Posterior means, standard deviations and credible intervals

	Posterior means and standard deviations					
	$\alpha = 1.81$		$\beta = 0.024$		τ	
τ	Mean	sd	Mean	sd	Mean	sd
0.1	1.75	0.11	0.032	0.009	0.21	0.13
0.2	1.84	0.11	0.022	0.006	0.12	0.09
0.3	1.91	0.12	0.017	0.005	0.30	0.14
0.4	1.69	0.11	0.039	0.010	0.46	0.16
0.5	1.85	0.12	0.021	0.006	0.38	0.14
0.6	1.76	0.11	0.027	0.008	0.53	0.15
0.7	1.94	0.12	0.018	0.006	0.69	0.10
0.8	1.65	0.10	0.036	0.009	0.61	0.15
0.9	1.68	0.10	0.030	0.008	0.82	0.10
	95% credible intervals					
	α		β		τ	
0.1	1.54	1.98	0.018	0.053	0.012	0.495
0.2	1.63	2.06	0.011	0.036	0.005	0.341
0.3	1.68	2.15	0.009	0.030	0.037	0.569
0.4	1.49	1.91	0.022	0.062	0.116	0.742
0.5	1.63	2.08	0.011	0.036	0.077	0.644
0.6	1.55	1.98	0.014	0.044	0.209	0.779
0.7	1.70	2.19	0.009	0.032	0.473	0.858
0.8	1.46	1.86	0.021	0.058	0.280	0.861
0.9	1.48	1.88	0.017	0.048	0.578	0.982

In Figure 2 the traces and the posterior density estimates of the parameters of interest are depicted for $\tau = 0.2$ under Case 2. These plots are obtained as a standard output of CODA (Convergence Diagnosis and Output Analysis[2]) which provides convergence diagnostics of the McMC methods as well as statistical summaries of outputs of the methods.

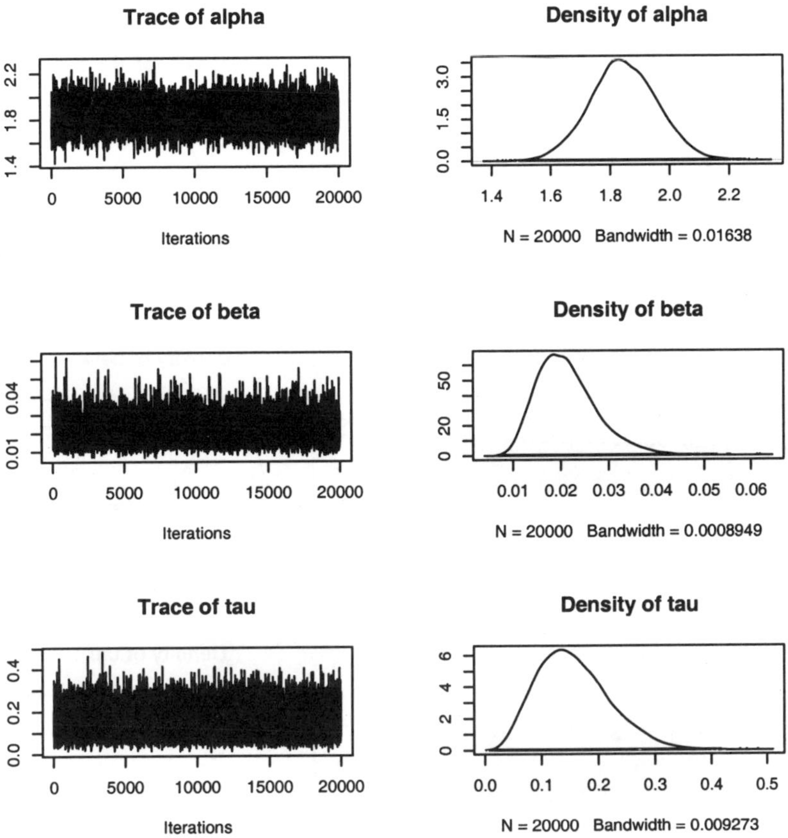

Figure 2. Case 2 (Informative prior): CODA Output Analysis for $\tau = 0.2$

Figure 3 depicts the traces and posterior densities estimates of the parameter for Case 1 for $\tau = 0.2$. This is the case where the prior for τ is uniform, and the posterior is similar to the likelihood. Here it can be seen, as noted above, that the Bayesian maximum *a posteriori* estimate, like the maximum likelihood estimate, lies at $\tau = 0$. The advantage of the Bayesian approach over maximum likelihood estimation is that in cases such as this we can report posterior medians or (as in our case) posterior means which are better estimates of the true parameter values.

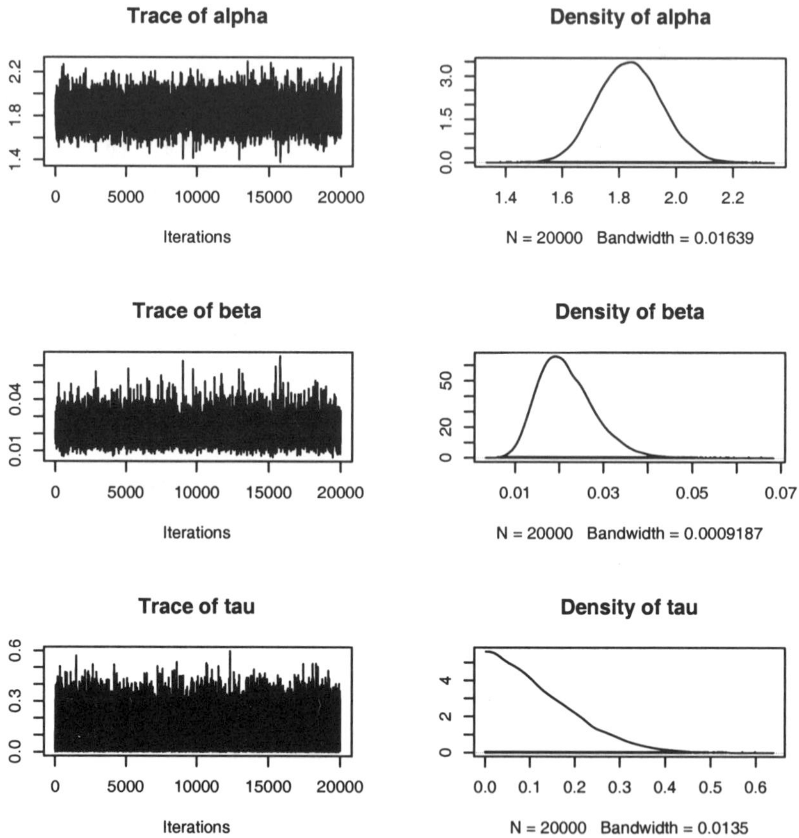

Figure 3. Case 1 (Uninformative prior): CODA Output Analysis for $\tau = 0.2$

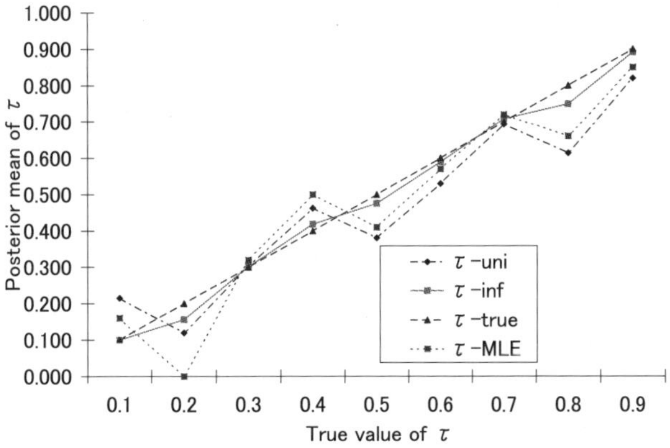

Figure 4. Comparison of the estimates for τ

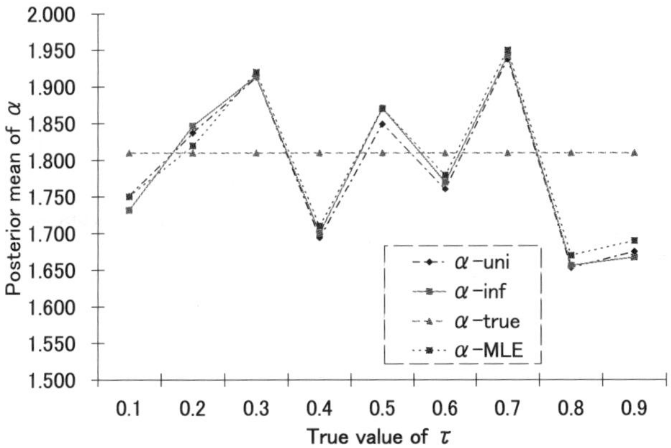

Figure 5. Comparison of the estimates for α

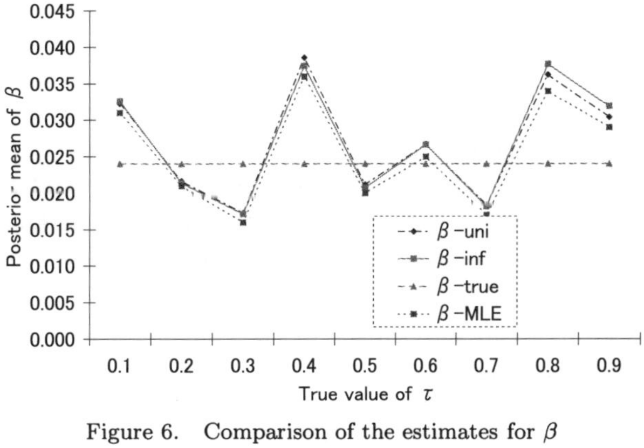

Figure 6. Comparison of the estimates for β

6.3. *Comparison of the two approaches*

Next, we will summarize and compare the two approaches discussed in 6.2 and 6.1 for estimating the degree of repair τ. As we can see from Table 1, if the sample size is relatively large (as in 6.1), the maximum likelihood approach gives very accurate results for τ and for the other parameters of the model. If a large set of data is available maximum likelihood approach and Bayesian approaches are similar. However for the small datasets (sample size $n = 100$) used here the Bayesian approach is preferable since it provides the full posterior distribution of the parameters from which better point estimates can be calculated. In particular the posterior allows direct calculation of standard errors and credible intervals. Pictorial comparison of the results shown in Tables 1–3 is given in Figures 4–6.

Unfortunately, in most real situations where an estimation of the degree of repair is of interest, the sample size we expect to work with is relatively small. It is well known that the maximum likelihood estimations can change dramatically with small sample sizes. That is why, even though the Bayesian approach (Gibbs sampling with Metropolis-Hastings component) is not as accurate as the maximum likelihood approach, it is preferable when dealing with small samples. Moreover, the most natural choice of the initial value τ_0 is as shown in Table 3, because the true value of τ is unknown.

7. Conclusion

In this paper we have focused on the lifetime of a product which undergoes multiple Worse than New Better than Used (WTNBTU) repairs. Our models allow for each repair to affect the lifetime of the repaired unit. This approach to modeling imperfect repairs is based on the concepts of the delayed distribution functions and related failure rate functions (see Chukova, Arnold and Wang[12]). Using a Markov chain Monte Carlo (McMC) method, in particular Gibbs sampler with Metropolis-Hasting component, we propose an approach for estimating the degree of repair and the parameters of the product lifetime distribution.

Our example shows the basic principle of how these models should be fitted. It is clear that more research is needed into the modelling of more complex real life situations, which take into account consecutive failures of different subsystems and their different degrees of repair. Our future research will address the application of these models and method to real data from warranty repairs.

References

1. R.E. Barlow and F. Proshan, *Statistical Theory of Reliability and Life Testing*, McArdle Press, Inc. (1981).
2. N.G. Best, M.K. Cowles, and S.K. Vines, *CODA manual version 0.30.* Cambridge, UK: MRC Biostatistics Unit. (1995).
3. W. Blischke and D.N.P. Murthy, *Warranty Cost Analysis*, Marcel Dekker, Inc. (1993).
4. W. Blischke and D.N.P. Murthy, *Product Warranty Handbook*, Marcel Dekker, Inc. (1996).
5. W.R. Gilks, Full conditional distributions, In *Markov Chain Monte Carlo in Practice*, (Edited by W.R. Gilks, S. Richardson and D.J. Spiegelhalter), 75-88, Chapman & Hall (1996).
6. M.-H. Chen, Q.-M. Shao and J.G. Ibrahim, *Monte Carlo Methods in Bayesian Computation*, Springer (2000).
7. S. Chib and E. Greenberg, Understanding the Metropolis-Hastings algorithm, *The American Statistician*, **49**, 327-335 (1995).
8. N. Metropolis, A.W. Rosenbluth, M.N. Rosenbluth, A.H. Teller and E. Teller, *Journal of Chemical Physics*, **21**, 1087-1092 (1953).
9. W.K. Hastings, Monte Carlo sampling methods using Markov chains and their applications, *Biometrika*, **57**, 97-109 (1970)
10. M. Brown and F. Proschan, Imperfect repair, *Journal of Applied Probability*, **20**, 851 - 859 (1983).
11. S. Chukova, B. Dimitrov and V. Rykov, Warranty Analysis. A survey, *Journal of Soviet Mathematics*, **67** (6), 3486-3508 (1993).
12. S. Chukova, R. Arnold and D. Wang, Warranty Analysis: An Approach to modelling Imperfect Repairs, *International Journal of Production Economics*, **89** (1), 57-68 (2004).
13. D.N.P. Murthy and I. Djamaludin, New product warranty: A literature review, *International Journal of Production Economics*, **79**, 231 - 260 (2000).
14. H. Pham and H. Wang H., Imperfect maintenance, *European Journal of Operational Research*, **94**, 425 - 438 (1996).

DESIGN OF OPTIMUM SIMPLE STEP-STRESS ACCELERATED LIFE TESTING PLANS

ELSAYED A. ELSAYED[†] and HAO ZHANG

Department of Industrial and Systems Engineering,
Rutgers, The State University of New Jersey
96 Frelinghuysen Road, Piscataway, NJ 08854-8018, USA
E-mail: elsayed@rci.rutgers.edu

The mission time of today's products is extended so much that it is difficult to observe failures under normal operating conditions. Therefore, accelerated life testing (ALT) is widely conducted to obtain failure time data in a much shorter time and to make inference about reliability at normal conditions. The accuracy of the reliability prediction is dependent on well-designed ALT plans. A step-stress ALT allows the test conditions to change at a given time or upon the occurrence of a specified number of failures. Step-stress accelerated life testing, an important type of ALT, is more difficult to model compared with constant stress ALT, however it yields failures more quickly. A test unit starts at a specified low stress. If the unit does not fail in a specified time, the stress is increased and held constant for another specified time. Stress is then repeatedly increased and held constant until the test unit fails. In this paper, we propose a procedure to determine the parameters of the optimum simple step-stress testing plan so that the reliability prediction at normal conditions is accurately determined. The parameters of the process are the lower stress level, the number of failures at the lower stress level, the duration of test at the lower stress level (change time to higher stress level), the higher stress level, and the number of failures at the higher stress level and the duration of test at the higher stress level. In many cases, most of these parameters are predetermined based on experience and field failures. We intend to investigate efficient procedures to estimate most, if not all of these parameters under different operating conditions. The resultant optimum plan is verified through numerical example and sensitivity analysis.

1. Introduction

Accelerated life testing (ALT) is used to quickly obtain reliability-related information on products' life and/or degradation data, and often promoted as a solution to save test time and costs. Inference about the reliability of products at normal operating conditions can be obtained using data obtained from the accelerated conditions. The accuracy of the inference procedure profoundly affects the reliability estimates at normal conditions and the subsequent decisions regarding system configuration, warranties and preventive

[†] Corresponding author.

maintenance schedules. The accuracy of the reliability estimates mainly depends on two factors: the ALT models and the experimental design of the ALT plans. Optimum plans yield the most accurate reliability estimates of products' life at the normal conditions, or the design stress conditions. ALT is usually conducted by subjecting the product to severer conditions than normal design conditions (accelerated stress) or by using the product more intensively than in normal use without changing the normal operating conditions (accelerated failure time). Conducting an accelerated life testing requires the development of a proper reliability model, which relates the failure data and reliability estimates at accelerated stresses with those at design stress conditions. An ALT plan is also required to obtain appropriate and sufficient information on accurate reliability estimates about products' performance at design conditions. A test plan determines the type of stresses to be applied, stress levels, methods of stress loading, number of units at each stress level, minimum number of failures at each stress level, optimum test duration and other parameters.

In recent years, studies of ALT plans have attracted many researchers. Chernoff (1962) considers optimum plans for an exponential distribution data at design stress level. Nelson and Meeker (1978) provide optimum test plans to estimate percentiles of Weibull and smallest extreme-value distribution at design stress conditions. Nelson (1990) provides guidelines for planning ALTs. Yang (1994) proposes an optimum design of 4-level constant-stress ALT plans with various censoring times. Tang (1999) considers the optimal plans for both constant stress and step-stress ALTs with Weibull exponential failure time distributions. Alhadeed and Yang (2002) give optimal times of changing stress-level for the simple step-stress plans under Khamis-Higgins model (1996, 1998).

In this paper, we present an optimum simple step-stress ALT plan based on Cox's proportional hazards (PH) model to obtain the most accurate reliability estimates at design conditions. The remainder of the paper is organized as follows. Section 2 presents the well-known Cox's PH model as well as cumulative exposure model for step-stress ALT. In this section, we also propose the PH-based optimum ALT plans with simple step-stress to obtain the optimal reliability function estimates, and formulate the nonlinear programming problem used to minimize the asymptotic variance of the reliability prediction at design stress conditions over a pre-specified period of time. We verify the proposed optimum ALT plans with a numerical example and perform sensitivity analysis in section 3. The concluding remarks are given in the last section.

2. PH-Based Optimum ALT Plans with Simple Step-Stress

2.1. *Nomenclature*

F	Fisher information matrix
Σ	variance-covariance matrix
ln	natural logarithm
MLE	maximum likelihood estimate
n	number of test units placed on test
τ_1	stress changing time
τ_2	censoring time
z_1, z_2	low and high stress levels respectively
z_H	specified maximum stress
z_D	specified design stress
T	pre-specified period of time over which the reliability estimate is of interest
$\lambda_0(t)$	unspecified baseline hazard function at time t
$\lambda(t;z)$	hazard function at time t, for given z
$R(t;z)$	reliability at time t, for given z
$f(t;z)$	PDF at time t, for given z
$\Lambda(t;z)$	cumulative hazard function at time t, for given z
$\lambda(t;z_1,z_2)$	hazard function at time t under step-stress
$R(t;z_1,z_2)$	reliability function at time t under step-stress
$f(t;z_1,z_2)$	PDF at time t under step-stress
$\Lambda(t;z_1,z_2)$	cumulative hazard function at time t under step-stress

2.2. *Proportional hazards models*

Cox's proportional hazards (PH) model is a semi-parametric multiple regression approach for reliability estimation, in which the baseline hazard function is affected multiplicatively by the covariates (i.e. applied stresses). The PH model is distribution-free requiring that the ratio of hazard rates between two stress levels be constant with time.

The proportional hazards model has the following form,

$$\lambda(t;z) = \lambda_0(t)\exp(\boldsymbol{\beta} z)\,. \tag{1}$$

We assume the baseline hazard function $\lambda_0(t)$ to be linear with time:

$$\lambda_0(t) = \gamma_0 + \gamma_1 t\,. \tag{2}$$

Substituting $\lambda_0(t)$ in Eq. (2) into the PH model, we obtain:

$$\lambda(t;z) = (\gamma_0 + \gamma_1 t)\exp(\boldsymbol{\beta} z), \tag{3}$$

where $z = (z_1, z_2, \ldots z_p)^T$ is a column vector of the covariates (or applied stresses); and $\boldsymbol{\beta} = (\beta_1, \beta_2, \ldots \beta_p)$ is a row vector of the unknown coefficients.

Unlike standard regression models, the PH models assume that the applied stresses act multiplicatively, rather than additively, on the hazard rate. The PH model is a class of models that have the property that testing units under different stress levels have hazard functions that are proportional to each other, that is, the ratio of the hazard rates for two devices tested at two different stress levels z_1 and z_2 does not vary with time.

2.3. *Step-stress ALT and model*

A step-stress ALT allows test conditions to change during testing. In step-stress, stress applied on each unit is not constant but is increased by planned steps at specified times. A test unit starts at a specified low stress. If the unit does not fail in a specified time period, the stress is increased and held constant for another specified time period. Stress is repeatedly increased and held constant until the test unit fails. The step-stress pattern is chosen to assure that failures occur quickly.

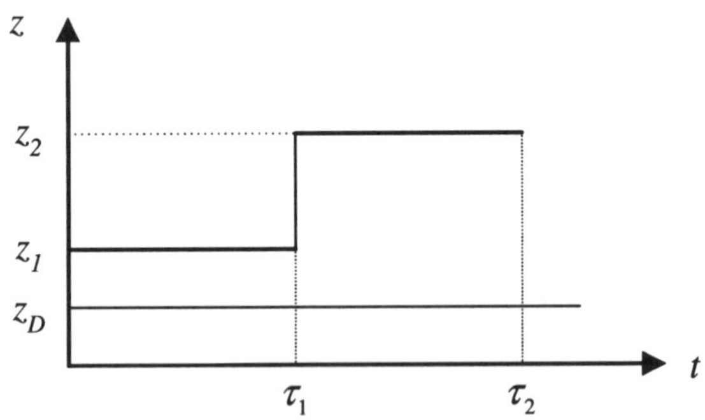

Figure 1. Stress application in simple step-stress test.

Simple step-stress tests use only two stress levels as shown in Figure 1. In a simple step-stress test, units are initially placed on test at a low stress level z_1 and run until a specified time τ_1. The stress is changed to the high stress level z_2; and the test is continued until censoring time τ_2. The objective of is to design the optimum simple step-stress ALT test plan by determining the optimal low stress level z_1 and the optimal stress changing point τ_1 employing the proportional hazards model.

The test procedure of simple step-stress ALT is as follows: n test units are initially placed at low stress z_1 and run until the stress changing time τ_1. Surviving units at time τ_1 are then subjected to a higher stress z_2 until a predetermined censoring time τ_2. We observe n_i failure times corresponding to stresses z_i, $i=1,2$ respectively.

The assumptions of simple step-stress ALT are:

1. There is a single accelerating stress type z.
2. Two stress levels z_1 and z_2 ($z_1 < z_2$) are used in the simple step-stress test.
3. The assumption of PH model is satisfied under different stress levels:

$$\lambda(t;z) = \lambda_0(t)\exp(\beta z) .$$

4. The baseline hazard function $\lambda_0(t)$ is linear with time:

$$\lambda_0(t) = \gamma_0 + \gamma_1 t .$$

5. The lifetimes of the test units are s-independent.

To analyze data from a step-stress ALT test, one needs a model that relates the life distribution under step-stress to the distribution under a constant stress. In this paper we adopt the most widely used cumulative exposure model to derive the cumulative distribution function of the failure time.

The cumulative exposure model assumes that the remaining life of a test unit depends only on the "exposure" it has seen, and the unit does not remember how the exposure was accumulated. Figure 2 shows the relationship between constant-stress and step-stress distributions.

At step 1 units are tested at stress level z_1 until stress changing time τ_1. Let $F_i(t)$ denote the CDF of time to failure for units tested at constant stress z_i, $i=1,2$. The population CDF of units failed by time t in step 1 is: $F_0(t) = F_1(t)$. Step 2 has an equivalent starting time s, which would have produced the same population cumulative failures at the stress level z_1 as the amount cumulated throughout step 1, which ends at the stress changing time τ_1. Thus, s is the solution of $F_2(s) = F_1(\tau_1)$, or equivalently

$$\exp[-(\gamma_0 s + \frac{\gamma_1}{2} s^2)e^{\beta z_2}] = \exp[-(\gamma_0 \tau_1 + \frac{\gamma_1}{2}\tau_1^2)e^{\beta z_1}] .$$

The population CDF of units failing by time $t > \tau_1$ is $F_0(t) = F_2(t - \tau_1 + s)$.

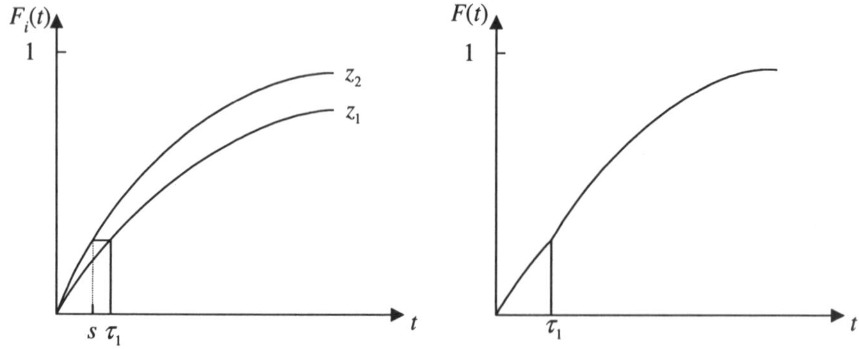

Figure 2. Relationship between constant-stress and step-stress distributions.

A test unit may experience two types of failing patterns: (a) it either fails under stress level z_1 before the stress is changed at time τ_1, (b) or does not fail by time τ_1 and continues to run until either its failure or censoring time τ_2 at stress level z_2. The following provides the log likelihood of an observation t (time to failure) of a single test unit. First, we define the indicator function $I_1 = I_1(t \le \tau_1)$ in terms of the stress changing point τ_1 by:

$$I_1 = I_1(t \le \tau_1) = \begin{cases} 1 & \text{if } t \le \tau_1, \text{ failure observed before time } \tau_1, \\ 0 & \text{if } t > \tau_1, \text{ failure observed after time } \tau_1. \end{cases} \tag{4}$$

and the indicator function $I_2 = I_2(t \le \tau_2)$ in terms of τ_2 by:

$$I_2 = I_2(t \le \tau_2) = \begin{cases} 1 & \text{if } t \le \tau_2, \text{ failure observed before time } \tau_2, \\ 0 & \text{if } t > \tau_2, \text{ censored at time } \tau_2. \end{cases} \tag{5}$$

Therefore the log likelihood of the observation t is expressed as

$$\begin{aligned} \ln L(t; z_1, z_2) &= I_2\{I_1[\ln f(t; z_1)] + (1-I_1)[\ln f(t'; z_2)]\} + (1-I_2)[-\Lambda(t'; z_2)] \\ &= I_1 I_2[\ln(\gamma_0 + \gamma_1 t) + \beta z_1 - (\gamma_0 t + \frac{\gamma_1 t^2}{2})e^{\beta z_1}] \\ &\quad +(1-I_1)I_2[\ln(\gamma_0 + \gamma_1 t') + \beta z_2 - (\gamma_0 t' + \frac{\gamma_1 t'^2}{2})e^{\beta z_2}] \\ &\quad -(1-I_2)[(\gamma_0 t' + \frac{\gamma_1 t'^2}{2})e^{\beta z_2}] \end{aligned} \tag{6}$$

where, $t' = t - \tau_1 + s$.

The first partial derivatives of the log likelihood with respect to the model parameters are,

$$\frac{\partial \ln L}{\partial \gamma_0}=\frac{I_1 I_2}{\lambda_0(t)}-I_1 I_2 t e^{\beta z_1}+\frac{(1-I_1)I_2}{\lambda_0(t')}-(1-I_1)I_2 t' e^{\beta z_2}-(1-I_2)t' e^{\beta z_2}, \tag{7}$$

$$\frac{\partial \ln L}{\partial \gamma_1}=\frac{I_1 I_2 t}{\lambda_0(t)}-\frac{I_1 I_2 t^2}{2}e^{\beta z_1}+\frac{(1-I_1)I_2 t'}{\lambda_0(t')}-\frac{(1-I_1)I_2 t'^2}{2}e^{\beta z_2}-\frac{(1-I_2)t'^2}{2}e^{\beta z_2}, \tag{8}$$

$$\begin{aligned}\frac{\partial \ln L}{\partial \beta}&=I_1 I_2 z_1-I_1 I_2 z_1 \Lambda(t;z_1)+(1-I_1)I_2 z_2-(1-I_1)I_2 z_2 \Lambda(t';z_2)\\&\quad-(1-I_2)z_2 \Lambda(t';z_2).\end{aligned} \tag{9}$$

The second partial derivatives with respect to the model parameters are,

$$\frac{\partial^2 \ln L}{\partial \gamma_0^2}=-\frac{I_1 I_2}{\lambda_0^2(t)}-\frac{(1-I_1)I_2}{\lambda_0^2(t')}, \tag{10}$$

$$\frac{\partial^2 \ln L}{\partial \gamma_1^2}=-\frac{I_1 I_2 t^2}{\lambda_0^2(t)}-\frac{(1-I_1)I_2 t'^2}{\lambda_0^2(t')}, \tag{11}$$

$$\frac{\partial^2 \ln L}{\partial \gamma_0 \gamma_1}=-\frac{I_1 I_2 t}{\lambda_0^2(t)}-\frac{(1-I_1)I_2 t'}{\lambda_0^2(t')}, \tag{12}$$

$$\frac{\partial^2 \ln L}{\partial \beta^2}=-I_1 I_2 {z_1}^2 \Lambda(t;z_1)-(1-I_1)I_2 {z_2}^2 \Lambda(t';z_2)-(1-I_2){z_2}^2 \Lambda(t';z_2). \tag{13}$$

These are given in terms of the random quantities I_1, I_2 and stress levels z_1, z_2 as well as the model parameters. The elements of the Fisher information matrix for an observation are the negative expectations of the above equations:

$$\begin{aligned}E\left\{-\frac{\partial^2 \ln L}{\partial \gamma_0^2}\right\}&=\int_0^{\tau_1}\frac{1}{\lambda_0^2(t)}f(t;z_1)dt+\int_{\tau_1}^{\tau_2}\frac{1}{\lambda_0^2(t')}f(t';z_2)dt\\&=\int_0^{\tau_1}\frac{1}{\lambda_0(t)}e^{\beta z_1}R(t;z_1)dt+\int_{\tau_1}^{\tau_2}\frac{1}{\lambda_0(t')}e^{\beta z_2}R(t';z_2)dt\end{aligned} \tag{14}$$

$$\begin{aligned}E\left\{-\frac{\partial^2 \ln L}{\partial \gamma_1^2}\right\}&=\int_0^{\tau_1}\frac{t^2}{\lambda_0^2(t)}f(t;z_1)dt+\int_{\tau_1}^{\tau_2}\frac{t'^2}{\lambda_0^2(t')}f(t';z_2)dt\\&=\int_0^{\tau_1}\frac{t^2}{\lambda_0(t)}e^{\beta z_1}R(t;z_1)dt+\int_{\tau_1}^{\tau_2}\frac{t'^2}{\lambda_0(t')}e^{\beta z_2}R(t';z_2)dt\end{aligned} \tag{15}$$

$$
\begin{aligned}
E\left\{-\frac{\partial^2 \ln L}{\partial \gamma_0 \gamma_1}\right\} &= \int_0^{\tau_1} \frac{t}{\lambda_0^2(t)} f(t;z_1)dt + \int_{\tau_1}^{\tau_2} \frac{t'}{\lambda_0^2(t')} f(t';z_2)dt \\
&= \int_0^{\tau_1} \frac{t}{\lambda_0(t)} e^{\beta z_1} R(t;z_1)dt + \int_{\tau_1}^{\tau_2} \frac{t'}{\lambda_0(t')} e^{\beta z_2} R(t';z_2)dt
\end{aligned} \tag{16}
$$

$$
\begin{aligned}
E\left\{-\frac{\partial^2 \ln L}{\partial \beta^2}\right\} &= \int_0^{\tau_1} z_1^{\,2} \Lambda(t;z_1) f(t;z_1)dt + \int_{\tau_1}^{\tau_2} z_2^{\,2} \Lambda(t';z_2) f(t';z_2)dt \\
&\quad + \int_{\tau_2}^{\infty} z_2^{\,2} \Lambda(t';z_2) f(t';z_2)dt
\end{aligned} \tag{17}
$$

The Fisher information matrix for MLE $(\hat{\gamma}_0, \hat{\gamma}_1, \hat{\beta})$ of $(\gamma_0, \gamma_1, \beta)$ can be obtained as the expectations of the negative of the second partial derivatives of the log likelihood with respect to the model parameters $(\gamma_0, \gamma_1, \beta)$ (Nelson, 1990). Equations (14) to (17) show the components of the Fisher information matrix for a single observation. Since all n units placed under the step-stress test experience the same test conditions, the Fisher information matrix for the n samples is expressed as

$$
F = n \begin{bmatrix}
E[-\frac{\partial^2 \ln L(t;z)}{\partial \gamma_0^2}] & E[-\frac{\partial^2 \ln L(t;z)}{\partial \gamma_0 \gamma_1}] & 0 \\
E[-\frac{\partial^2 \ln L(t;z)}{\partial \gamma_0 \gamma_1}] & E[-\frac{\partial^2 \ln L(t;z)}{\partial \gamma_1^2}] & 0 \\
0 & 0 & E[-\frac{\partial^2 \ln L(t;z)}{\partial \beta^2}]
\end{bmatrix}. \tag{18}
$$

The variance-covariance matrix for MLE $(\hat{\gamma}_0, \hat{\gamma}_1, \hat{\beta})$ is defined as the inverse matrix of the Fisher information matrix

$$
\Sigma = \begin{bmatrix}
Var(\hat{\gamma}_0) & Cov(\hat{\gamma}_0, \hat{\gamma}_1) & 0 \\
Cov(\hat{\gamma}_0, \hat{\gamma}_1) & Var(\hat{\gamma}_1) & 0 \\
0 & 0 & Var(\hat{\beta})
\end{bmatrix} = F^{-1}. \tag{19}
$$

2.4. *Optimization criterion*

In order to obtain the most accurate reliability estimate under the limitations of testing conditions (time, cost, test units, etc.), we choose the optimization criterion that minimizes the asymptotic variance of the reliability function estimates over a pre-specified period of time at design stress, i.e., minimize

$$\int_0^T Var[\hat{R}(t;z_D)]dt = \int_0^T Var[\exp(-(\hat{\gamma}_0 t + \hat{\gamma}_1 t^2/2)e^{\hat{\beta}z_D})]dt.$$

The asymptotic variance for the reliability function estimate is derived as follows

$$Var[\hat{R}(t;z_D)] = Var[\exp(-(\hat{\gamma}_0 t + \hat{\gamma}_1 t^2/2)e^{\hat{\beta}z_D})]$$
$$= \left[\frac{\partial \hat{R}}{\partial \hat{\gamma}_0}, \frac{\partial \hat{R}}{\partial \hat{\gamma}_1}, \frac{\partial \hat{R}}{\partial \hat{\beta}}\right] \Sigma \left[\frac{\partial \hat{R}}{\partial \hat{\gamma}_0}, \frac{\partial \hat{R}}{\partial \hat{\gamma}_1}, \frac{\partial \hat{R}}{\partial \hat{\beta}}\right]^{-1} \tag{20}$$

where

$$\frac{\partial \hat{R}}{\partial \hat{\gamma}_0} = -te^{\hat{\beta}z_D}\hat{R}(t,z_D),$$

$$\frac{\partial \hat{R}}{\partial \hat{\gamma}_1} = -(t^2/2)e^{\hat{\beta}z_D}\hat{R}(t,z_D),$$

$$\frac{\partial \hat{R}}{\partial \hat{\beta}} = -z_D(\hat{\gamma}_0 t + \hat{\gamma}_1 t^2/2)e^{\hat{\beta}z_D}\hat{R}(t;z_D)$$

The choice of the optimization criterion has a direct impact on the computational difficulty of solving the nonlinear programming optimization process. In this paper we choose to optimize the accuracy of the reliability function estimates as shown in Eq. (20).

2.5. *Problem formulation*

The problem is to optimally design a simple step-stress ALT with Type I censoring under the constraints of available test units, censoring time and specification of a minimum number of failures at low stress level, such that the asymptotic variance of the reliability function estimate at design stress is minimized over a pre-specified period of time T. The optimal decision variables, low stress level z_1^* and stress change time τ_1^*, are determined by solving the following nonlinear optimization problem.

Objective function

Min $\quad f(\boldsymbol{x}) = \int_0^T Var[\exp(-(\hat{\gamma}_0 t + \hat{\gamma}_1 t^2/2)e^{\hat{\beta}z_D})]dt$

Subject to

$$n\Pr[t \le \tau_1 \mid z_1] \ge MNF,$$

$$\Sigma = \begin{bmatrix} Var(\hat{\gamma}_0) & Cov(\hat{\gamma}_0, \hat{\gamma}_1) & 0 \\ Cov(\hat{\gamma}_0, \hat{\gamma}_1) & Var(\hat{\gamma}_1) & 0 \\ 0 & 0 & Var(\hat{\beta}) \end{bmatrix} = F^{-1},$$

where

$$\boldsymbol{x} = \begin{bmatrix} z_1 \\ \tau_1 \end{bmatrix}$$

and *MNF* is the minimum number of failures at low stress level.

The optimal design depends on model parameters ($\gamma_0, \gamma_1, \beta$). A design using the pre-estimates of the model parameters is called a locally optimal design (Chernoff, 1962) and is commonly adopted (Bai and Kim (1989), Bai and Chun (1991), Nelson (1990), Meeker and Hahn (1985)). Here we also assume that pre-estimates ($\hat{\gamma}_0, \hat{\gamma}_1, \hat{\beta}$) are available through either preliminary baseline experiments or engineering experience obtained prior to the design of the optimal test plan.

The nonlinear optimization problem can be only solved by numerical methods. This is a typical constrained nonlinear optimization problem. Since the derivatives of the unknown parameters have complicated forms, we adopt a direct search algorithm, the **C**onstrained **O**ptimization **BY** **L**inear **A**pproximations (COBYLA) algorithm, by Powell (1992) to avoid the calculation of derivatives. The global optimum solution may be obtained by trying different initial values. Details of the algorithm are given in Powell (1992).

3. Numerical Example

3.1. *Problem formulation and solution*

A simple time-step accelerated life test is to be conducted for MOS capacitors in order to estimate its life distribution at design temperature of 50°C. The test needs to be completed in 300 hours. The total number of test items placed under test is 200 units. To avoid the introduction of failure mechanisms other than those expected at the design temperature, it has been decided, through engineering judgment, that the testing temperature cannot exceed 250°C. The minimum number of failures for low temperature is specified as 40. Furthermore, the experiment should provide the most accurate reliability estimate over a 10-year period of time. The test plan is determined through the following steps:

1. According to the Arrehenius model, we use 1/(Absolute Temperature) as the covariate z in the ALT model, i.e., the design stress level z_D=1/323.16 K, and the highest stress level z_H=1/523.16 K.
2. The PH model is used in conducting reliability data analysis and designing the optimal ALT plan. The model is given by:

$$\lambda(t;z) = \lambda_0(t)\exp(\beta z) = (\gamma_0 + \gamma_1 t)\exp(\beta z),$$

where the stress z is related to the temperature ($Temp$) level by

$$z = \frac{1}{273.16 + Temp}.$$

3. A baseline experiment is conducted to obtain initial values for the model parameters. These values are: $\hat{\gamma}_0 = 0.0001$, $\hat{\gamma}_1 = 0.55$, and $\hat{\beta} = -3800$.
4. The problem is to optimally design a simple step-stress ALT with Type I censoring, under the constraints of available test units, censoring time and specification of minimum number of failures at low stress level, such that the asymptotic variance of the hazard rate estimate at design stress is minimized over a pre-specified period of time T. The optimal decision variables, low stress level z_1^* and stress change time τ_1^*, are determined by solving the following nonlinear optimization problem:

Objective function

Min $$f(\mathbf{x}) = \int_0^T Var[\exp(-(\hat{\gamma}_0 t + \hat{\gamma}_1 t^2/2)e^{\hat{\beta} z_D})]dt$$

Subject to

$$Var[\hat{R}(t;z_D)] = Var[\exp(-(\hat{\gamma}_0 t + \hat{\gamma}_1 t^2/2)e^{\hat{\beta} z_D})]$$
$$= \left[\frac{\partial \hat{R}}{\partial \hat{\gamma}_0}, \frac{\partial \hat{R}}{\partial \hat{\gamma}_1}, \frac{\partial \hat{R}}{\partial \hat{\beta}}\right] \Sigma \left[\frac{\partial \hat{R}}{\partial \hat{\gamma}_0}, \frac{\partial \hat{R}}{\partial \hat{\gamma}_1}, \frac{\partial \hat{R}}{\partial \hat{\beta}}\right]^{-1},$$

$$\Sigma = F^{-1},$$

$$50°C \le 1/z_1 - 273.16 \le 250°C,$$

$$\tau_1 \le \tau_2,$$

$$n\Pr[t \le \tau_1 \mid z_1] \ge MNF.$$

where

$$x = \begin{bmatrix} z_1 \\ \tau_1 \end{bmatrix},$$

$$n = 200.$$

$$T = 87600,$$

$$\tau_2 = 300,$$

$$MNF = 40.$$

5. We use nonlinear programming technique to solve the optimization problem:
6. Input the initial baseline values for the model parameters γ_0, γ_1, and β; the design stress level z_D and highest stress level z_H as well as total test units n, test duration τ_2 and minimum number of failures MNF for low stress level.
7. Solving this nonlinear programming problem yields the following optimum plan that optimizes the objective function and meets the constraints:

$$Temp_1^* = 145^o C, \text{ and } \tau_1^* = 262.5\text{hours}.$$

3.2. *Sensitivity analysis*

To solve the nonlinear optimization problem given in this example, we first obtain estimates of the values of the model parameters $\gamma_0, \gamma_1, \beta$. Since these are point estimates it is important to investigate the sensitivity of the reliability estimates to variations of the parameter estimates. Therefore, we investigate and analyze the sensitivity of the solution of the proposed optimum ALT plan to changes in the model parameters. If a small change in a parameter results in relatively large changes in the solution of the optimum ALT plan, the ALT plan is said to be sensitive to that parameter. This means that this specific parameter needs to be investigated further before we design the optimum ALT plan. Meanwhile some parameters in the nonlinear optimization problem are given arbitrarily or are given based on engineering judgment, e.g. the censoring time τ_2, minimum required failure units MNF, and the total period of time, T, in which we are supposed to estimate the reliability performance of product at the design stress levels. If the solution of the optimum ALT plan is sensitive to any of the above mentioned parameters, then accurate estimation or determination of these parameters is needed in order to accurately estimate the reliability at design conditions when we follow the testing design given by the optimum ALT plan.

To conduct the sensitivity analysis, we change the value of one of the model parameters $\hat{\gamma}_0, \hat{\gamma}_1, \hat{\beta}$, τ_2, MNF, and T, and keep the other values unchanged, then we solve the nonlinear optimization problem to obtain the corresponding optimum ALT plan. If a small change in any parameter results in a relatively large change in the optimum solution, then the ALT plan is sensitive to that parameter.

The result of sensitivity analysis is summarized in Tables 1.

Table 1. Sensitivity Analysis: Effect of Model Parameter Uncertainty on Stress Levels.

Parameter	Deviation	$Temp_1^*$	τ_1^*
$\hat{\gamma}_0$	−10%	145.2°C	262.7
	+10%	145°C	262.5
$\hat{\gamma}_1$	−10%	149°C	258.5
	+10%	145°C	262.5
$\hat{\beta}$	−10%	136°C	273.5
	+10%	153°C	252.5
T	−10%	145°C	262.5
	−10%	145°C	262.5

As shown in Table 1, the proposed optimum ALT plan derived in this example is robust to the deviations of the model parameters. Specially the ALT plan is robust to changes in the model parameters $\hat{\gamma}_0, \hat{\gamma}_1$, and strongly robust to the changes in the parameter T. But it is relatively more sensitive to the deviations of the stress coefficient $\hat{\beta}$ than other parameters. So the estimates of baseline values of model parameter $\hat{\beta}$ should be accurately estimated to ensure accurate optimum ALT plans.

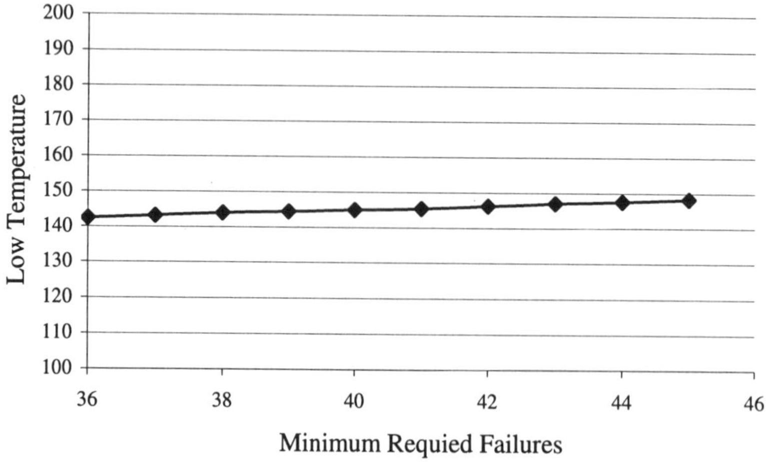

Figure 3. Low stress level vs. minimum required failures MNF .

Figure 3 describes the relationship between the optimum low stress level and the required minimum number of failures (MNF). The figure shows a minor increase in optimum stress level corresponding to the increase in the required minimum number of failures. Figures 4 shows the relationship between the optimum low stress level and the censoring time τ_2 . A decreasing trend of optimum stress level is displayed corresponding to the increases in censoring time. Although both trends exist, the increasing (decreasing) slopes are minor. The optimum ALT plan given in this example is still robust to the parameters MNF and τ_2 .

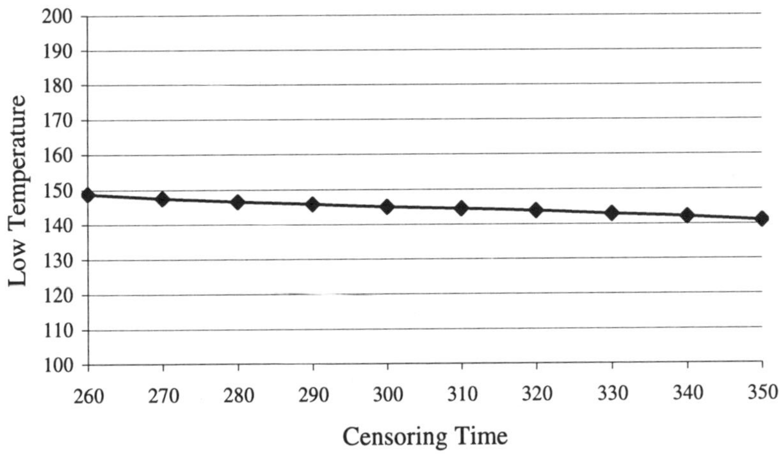

Figure 4. Low stress level vs. censoring time τ_2 .

4. Conclusions

In this paper, we present the PH-based optimum accelerated life testing plan with simple step-stress. The plan determines the optimum low stress level z_1^* and the optimum stress changing time τ_1^* such that the asymptotic variance of the reliability function estimate at design stress is minimized over a specified period of time. The constraints include the censoring time, total number of test units available, and the minimum number of failures at low stress level. The optimum ALT plans are based on the proportional hazards model. To relate the life distribution under step-stress to the distribution under a constant stress, we use cumulative exposure model in this paper. This optimization approach is verified by a numerical example, and the sensitivity analysis shows that the optimal solutions are robust to the deviations in the model parameters.

References

1. Alhadeed, A. A. and Yang, S. S. (2002), "Optimal simple step-stress plan for Khamis-Higgins model", *IEEE Trans. on Reliability*, 51, 212-215.
2. Bai, D. S. and Chun, Y. R. (1991), "Optimum simple step-stress accelerated life tests with competing causes of failure", *IEEE Transactions on Reliability* 40, 622-627.
3. Bai, D. S., Kim, M. S. and Lee, S. H. (1989), "Optimum simple step-stress accelerated life tests with censoring", *IEEE Transaction on Reliability* 38, 528-532.

4. Chernoff, H. (1962), "Optimal accelerated life design for estimation", *Technometrics*, 4, 381-408.
5. Cox, D. R. (1972), "Regression models and life tables (with discussion)", *Roy. Stat. Soc.* B34, 187-208.
6. Elsayed, E. A. (1996), *Reliability Engineering*, MA: Addison-Wesley Longman.
7. Elsayed, E. A. and Jiao L. (2002), "Optimal design of proportional hazards based accelerated life testing plans", *International J. of Materials & Product Technology*, 17 411-424.
8. Jiao. L. (2001), *Optimal Allocations of Stress Levels and Test Units in Accelerated Life Tests*, Ph.D. Dissertation, Dept. of Industrial and Systems Engineering, Rutgers University.
9. Kalbfleisch, J. D. and Prentice, R. L. (2002), *The Statistical Analysis of Failure Time Data*, New Jersey: John Wiley & sons, Inc.
10. Khamis, I. H. and Higgins, J. J. (1996), "Optimum 3-step step-stress tests", *IEEE Transactions on Reliability*, 45, 341-345.
11. Khamis, I. H. and Higgins, J. J. (1998), "New model for step-stress testing", *IEEE Transactions on Reliability*, 47, 131-134.
12. Meeker, W. Q. and Hahn, G. J. (1985). "How to plan an accelerated life test - some practical guidelines", *Statistical Techniques*, 10, ASQC Basic reference in QC.
13. Nelson, W. and Meeker, W. (1978), "Theory for optimum censored accelerated life tests for Weibull and extreme value distributions", *Technometrics*, 20, 171-177.
14. Nelson, W. (1990), *Accelerated Testing: Statistical Models, Test Plans, and Data Analyses*, New York: John Wiley & sons, Inc.
15. Powell, M.J.D., (1992) "A direct search optimization method that models the objective and constraint functions by linear interpolation", *DAMTP/NA5*, Cambridge, England.
16. Tang, L. C. (1999), "Planning for accelerated life tests", *International J. of Reliability, Quality, and Safety Engineering*, 6, 265-275.
17. Yang, G. B. (1994), "Optimum constant-stress accelerated life test plans", *IEEE Trans. on Reliability*, 43, 575-581.

A BDD-BASED ALGORITHM FOR COMPUTING THE K-TERMINAL NETWORK RELIABILITY

G. HARDY and C. LUCET*
LaRIA
Amiens, FRANCE
E-mail: {gary.hardy,corinne.lucet}@u-picardie.fr

N. LIMNIOS
LMAC, UTC
Compiegne, FRANCE
E-mail: nikolaos.limnios@utc.fr

This paper presents an exact method using Binary Decision Diagram for computing the K-terminal reliability of networks such as computer, communication or power networks. The K-terminal reliability is defined as the probability that a subset K of nodes in the network can communicate together, taking into account the random failures of network components. Some examples and experiments show the effectiveness of this approach.

1. Introduction

Nowadays, network reliability analysis receives considerable attention for the design, validation and maintenance of many real world systems, such as computer, communication or power networks. The reliability of these complex systems is an increasing concern as the failure of some of their components could lead to disastrous results. Our network model is an undirected stochastic graph $G = (V, E)$, where V is the vertex set and E is the edge set. Sites correspond to vertices and links to edges. Each edge can fail randomly and independently with known probability and we consider that vertices are perfectly reliable. The K-terminal reliability is the probability that a given subset K of vertices remain connected, *i.e.* there exists at least one path made of functioning edges linking each pair of K-nodes. The terminal nodes are essential to the system function and have to communicate with each other. The network is operational if and only if these K-nodes

*this research was supported by the conseil regional de picardie

are connected. This problem is well-known as NP-hard. Provan [8] showed that even for planar graphs this problem is still NP-hard. The problem of its evaluation has received considerable attention from the research community. We propose an algorithm based on Binary Decision Diagrams (BDD), a data structure to encode and manipulate boolean functions, for computing K-terminal reliability of large networks. This structure avoids huge storage and high computation time. In literature, two classes of methods are often used for computing the network reliability. The first class deals with the enumeration of all the minimum paths or cuts. A *path* is defined as a set of network components (edges and/or vertices) such that if these components are all fail-free, the system is up. A path is *minimal* if it has no proper subpaths. In the opposite, a *cut* is a set of network components such that if these components fail, the system is down. The *inclusion-exclusion* or *sum of disjoint products* (*SDP*) methods have to be applied since this enumeration provides non-disjoint events. The algorithms in the second class are factoring algorithms improved by reductions. It consists in reducing the size of the network while preserving its reliability. When no reduction is allowed, the factoring method is used. The idea is to choose a component and decompose the problem into two sub-problems: the first assumes the component has failed, the second assumes it is functioning. Satyanarayana and Chang [13] and Wood [12] have shown that the factoring algorithms with reductions are more efficient than the classical path or cut enumeration method for solving this problem. This was confirmed by the experimental works of Theologou and Carlier [14]. Our method can be seen as an extension to the computation of the all-terminal reliability measure of the method in [9].

This paper is organized as follows. First, we illustrate the preliminaries of BDD in Section 2. In Section 3, an algorithm for constructing the BDD of a K-terminal network is shown. This method avoids the redundant computations of isomorphic sub-problems during the computation process. In Section 4, experimental results on several networks are shown. Finally, we draw some conclusions and outline the direction of future works in Section 5.

2. Binary Decision Diagram (BDD)

Akers [1] first introduced BDD for representing boolean function. Bryant popularized the use of BDD by introducing a set of algorithms for efficient construction and manipulation of the BDD structure [2]. Nowadays, BDD are used in a wide range of area, including hardware synthesis and verification, model checking and protocol validation. Their use in the reliability analysis framework has been introduced by Madre and Coudert [5 4] and developped by Odeh [6] and Rauzy [3]. Sekine and Imai have introduced the BDD structure in network reliability [10 11]. The BDD structure provides compact representations of boolean expressions. A BDD is a directed acyclic graph (DAG) based on Shannon's decomposition. The Shannon's decomposition for a boolean function f is defined as follows:

$$f = xf_{x=1} + \bar{x}f_{x=0}$$

where x is one of decision variables and $f_{x=i}$ is the boolean function f evaluated at $x = i$.

The graph has two sink nodes labeled with 0 and 1 representing the two corresponding constant expressions. Each internal node is labeled with a boolean variable x and has two out-edges called 0-edge and 1-edge. The node linked by 1-edge represents the boolean expression when $x = 1$, *i.e.* $f_{x=1}$ while the node linked by 0-edge represents the boolean expression when $x = 0$, *i.e.* $f_{x=0}$. An ordered binary decision diagram (OBDD) is a BDD where variables are ordered according to a known total ordering and every path visits variables in an ascending order. Afterwards, BDDs will be considered as ordered. Leaves of the BDD give the value of f for the assignment corresponding to a path from the root to the leaf.

The size of a BDD structure (the number of nodes) depends critically on the chosen variable ordering. Figure 1 shows the effect of the variable ordering on the BDD size. If we consider the expression $(x_1 \Leftrightarrow x_3) \wedge (x_2 \Leftrightarrow x_4)$ the resulting BDD using the ordering $x_1 < x_2 < x_3 < x_4$ consists of 11 nodes (figure 1(a)) and not 8 nodes as for the ordering $x_1 < x_3 < x_2 < x_4$ (figure 1(b)). Finding an ordering that minimizes the size of BDD is also a NP-complete problem [7]. Several heuristics relying on different principles have been proposed in many domains. However, they both try to put close in the order the variables that are close in the formula as illustrated in figure 1.

3. K-Terminal Reliability Computation

3.1. *Definitions and notations*

The K-terminal reliability computation is the most general network reliability problem found in the literature. It consists in evaluating the probability that net-

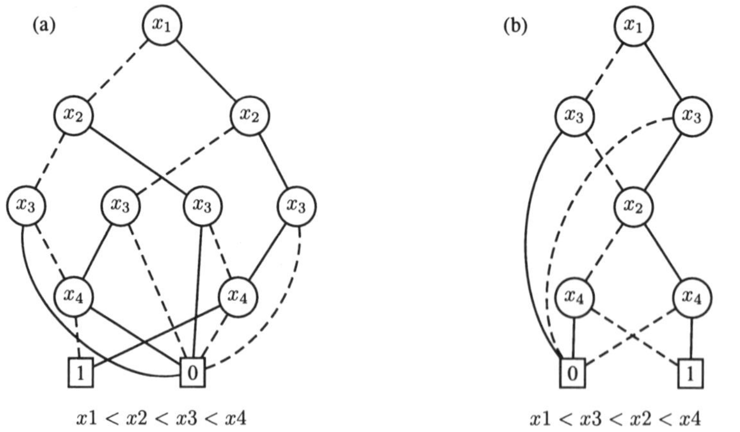

Figure 1. *function $f(x_1, x_2, x_3, x_4) = (x_1 \Leftrightarrow x_3) \wedge (x_2 \Leftrightarrow x_4)$ representing by BDD with two different orders: $x_1 < x_2 < x_3 < x_4$ (a) and $x_1 < x_3 < x_2 < x_4$ (b). A dashed (solid) line represents the value 0 (1).*

work components of a specified subset K remain connected when the components are subject to failure.

Our network model is an undirected stochastic graph $G = (V, E)$, with V its set of vertex (representing workstations, servers, routers ...) and $E \subseteq V \times V$ its set of edges (representing the links between these nodes). Each edge e_i of the stochastic graph is subject to failure with known probability q_i. We denote $p_i = 1 - q_i$ the probability that edge e_i functions, and assume that all the failure events are statistically independent. In the following, we consider the vertices as perfect, but the proposed algorithms are still functioning for such problem. In classical enumerative method, all the states of the graph are generated, evaluated as a fail state or a functioning state, and then probabilistic methods are used for computing the associated reliability. So, as there are two states for each edge, there are 2^m (with $m = |E|$) possible states for the graph. A state $\mathcal{G}$ of the stochastic graph G is denoted by $(x_1, x_2 \ldots, x_m)$ where x_i stands for the state of edge e_i, *i.e.* $x_i = 0$ when edge e_i fails and $x_i = 1$ when it functions. The associated probability of $\mathcal{G}$ is defined as:

$$Pr(\mathcal{G}) = \prod_{i=1}^{m} (x_i.p_i + (1 - x_i).q_i)$$

At each state $\mathcal{G}$ is associated a partial graph $G(\mathcal{G}) = (V, E')$ such that $e_i \in E'$ if and only if $e_i \in E$ and $x_i = 1$. A *path* is defined as a set of edges such that if these edges are all up, the system is up. A path is *minimal* if it has no proper subpaths. We define a subset of the nodes $K \subseteq V$ to be the "terminals" (with $2 \leq |K| \leq |V|$). If $|K| = 2$ this problem is well-known as the *2 -terminal reliability problem* and if $|K| = |V|$ it deals with the *all-terminal reliability problem*. The terminal nodes are essential to the system function and have to communicate with each other, *i.e.* the network is up if and only if there exists at least one path made of functioning edges linking nodes in K. The K-terminal reliability, denoted by $R_K(p; G)$ ($p = (p_1, \dots, p_m)$), is the probability that all vertices in K are connected and can be defined as follows:

$$R_K(p;G) = \sum_{\text{K-nodes are connected by working links in } G(\mathcal{G})} Pr(\mathcal{G})$$

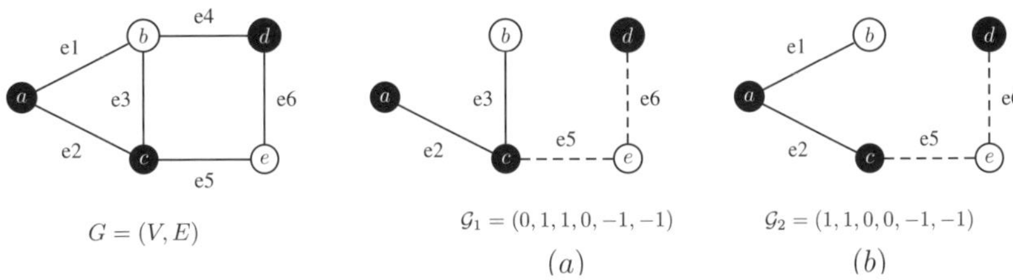

Figure 2. *Nodes in black represent terminal vertices ($K = \{a, c, d\}$). $G(\mathcal{G}_1)$ and $G(\mathcal{G}_2)$ represent sub-graphs in level 4 in the computation process illustrated in figure 3(a). $G(\mathcal{G}_1)$ and $G(\mathcal{G}_2)$ have the same corresponding partition, $[ac][d]$ during the computation. $e_i = -1$ means the state of e_i is not yet fixed.*

3.2. *Encoding and evaluating the network reliability by BDD*

The K-terminal network reliability function can be represented by a boolean function f defined as follows:

$$\begin{cases} f(x_1, x_2, \dots, x_m) = 1 \text{ if nodes in K are linked by edges } e_i \text{ with } x_i = 1 \\ f(x_1, x_2, \dots, x_m) = 0 \text{ otherwise} \end{cases}$$

where boolean variable x_i stands for the state of the link e_i ($1 \leq i \leq m$).
For instance, the boolean formula encoded by the BDD structure in figure 3 is:

$$x_1(\bar{x}_2(\bar{x}_3x_4x_5x_6 + x_3(\bar{x}_4x_5x_6 + x_4)) + x_2(\bar{x}_4x_5x_6 + x_4)) + \bar{x}_1x_2(x_3(\bar{x}_4x_5x_6 + x_4) + \bar{x}_3x_5x_6)$$

Our aim is to encode this reliability function by BDD. The algorithm is developed in Section 3.3. In figure 3(b), we explain the definition of BDD through an example of BDD representing the K-terminal reliability of network G (see figure

2). The BDD can represent the SDP implicitly avoiding huge storage for large number of SDP. A useful property of BDD is that all the paths from the root to the leaves are disjoint. If f represents the system reliability expression, based on this property, the K-terminal network reliability R_K of G can be recursively evaluated by:

$$\forall i \in \{1, \ldots, m\}:$$
$$R_K(p; G) = Pr(f = 1)$$
$$R_K(p; G) = Pr(x_i.f_{x_i=1} = 1) + Pr(\bar{x_i}.f_{x_i=0} = 1)$$
$$R_K(p; G) = p_i.Pr(f_{x_i=1} = 1) + q_i.Pr(f_{x_i=0} = 1)$$

with $p = (p_1, \ldots, p_m)$.
For instance, in figure 3(b), the K-terminal network reliability is then defined as follows:

$$R_K(p; G) = p_1(q_2(q_3p_4p_5p_6+p_3(q_4p_5p_6+p_4))+p_2(q_4p_5p_6+p_4))+q_1p_2(p_3(q_4p_5p_6+p_4)+q_3p_5p_6)$$

The next section presents our BDD-based algorithm for the K-terminal network reliability problem.

3.3. *Construction of the BDD representing the K-terminal reliability function*

We remind that the order of the variables is very important for BDD generation (see Section 2). Time and space complexity of BDD closely depend on variable ordering. This paper is not concerned with this kind of problem and we use a breadth-first-search (BFS) ordering.
In short, our algorithm follows three steps:

- 1 The edges are ordered by using a heuristic.
- 2 The BDD is generated to encode the network reliability. The following shows the construction of the BDD encoding the K-terminal network reliability.
- 3 From this BDD structure, we obtain the K-terminal network reliabilities (whatever p_i, $i \in [1 \ldots m]$) as shown in the previous section.

The top-down construction process can be represented as a binary tree such that the root corresponds to the original graph G and children correspond to graphs obtained by deletion /contraction of edges. Nodes in the binary tree correspond to subgraphs of G. At the root, we consider the edge e_1, construct the subgraph G_{-1}, that is G with e_1 deleted and the subgraph G_{*1} that is G with e_1 contracted. Then at the second step, from G_{-1}, we construct G_{-1-2} where e_2 is deleted and G_{-1*2} where e_2 is contracted and so on from each created subgraphs until the vertices of K are fully connected or at least one vertex of K is disconnected. There are 2^n possible states and isomorphic graphs appear in the computation process. For the graph G pictured in Fig. 2, its subgraphs G_{*1*2} and G_{-1*2*3} are isomorphic. Our aim is to provide an efficient method in order to avoid redundant computation due to the appearance of isomorphic subproblems during the process. We use the method introduced by Carlier and Lucet [15] for representing graph by partition which is an efficient way for solving this kind of problem. By identifying the isomorphic subgraphs an expansion tree is modified as a rooted acyclic graph which is a BDD (see figure 3(b)).

3.3.1. *Finding isomorphic graphs*

In this part, we present a method for efficiently recognize isomorphic graphs during the top-down construction of the reliability BDD. Consider that $E_k = \{e_1, \ldots, e_k\}$ and $\bar{E_k} = \{e_{k+1}, \ldots, e_m\}$. At level k in the BDD, each edge in E_k has fixed state (either deleted or contracted) and the state of edges in $\bar{E_k}$ is not yet fixed. In consequence, graphs in the k-th level of the BDD are subgraphs of G with the edge set $\bar{E_k}$. For each level k, we define the *boundary set* F_k as:

- a vertex set such that each non-terminal vertex of F_k is incident to at least one edge in E_k and one edge in $\bar{E_k}$ and each terminal vertex of F_k is incident to at least one edge in E_k.

The subgraphs will be represented by the partitions of F_k. For each subgraph, the corresponding partition is made by gathering vertices of F_k in blocks according to the following rules:

- two vertices x and y of F_k are in the same block if and only if there exists a path made of contracting (*i.e.* functioning) edges linking x to y.

For instance in figure 3(a), in the first level, the boundary set F_1 is equal to $\{a, b\}$. G_{*1} can be represented by partition $[ab]$ and G_{-1} by partition $[a][b]$. In level 3, the boundary set F_3 is equal to $\{a, b, c\}$ and the 3 possible network states are represented by partitions $[ab][c]$, $[abc]$ and $[ac][b]$. Now, we order partitions in the same level k in order to identify them in an efficient way. Moreover, we only keep the number representing the partition in order to reduce the space complexity since we can find again a partition from its associated number. We number the partition from 1 to $Bell(|F_k|)$ where $Bell(|F_k|)$ (known as the *Bell number*) is the theoretical maximum number of partitions of $|F_k|$ elements ($|F_k|$ represents the number of nodes in the boundary set of level k). This number grows exponentially with $|F_k|$, consequently the number of classes grows exponentially with the size of the boundary set. From now on, we only manipulate partitions instead of graphs during the K-terminal reliability computation. The order of the partitions is stemmed from the *Stirling numbers* of second kind, and more particulary from the following recursive formulae:

$$A_{i,j} = j.A_{i-1,j} + A_{i-1,j-1} \text{ if } 1 < j \leq i$$

with $A_{i,1} = 1$ and $A_{i,j} = 0$ if $0 < i < j$.

$A_{i,j}$ is the number of partitions of j blocks that can be made with i elements. For constructing and ordering partitions of i elements and j blocks, the order first follows the growing number of blocks, and then uses the recursive generation: we order partitions stemmed from partitions of $i-1$ elements and j blocks (by adding one element in each blocks), then we order partitions stemmed from partitions of $i-1$ elements and $j-1$ blocks (by adding a ith element in a new block) Thus, the *Bell number* for the boundary set F can be defined as:

$$Bell(|F|) = \sum_{j=1}^{|F|} A_{|F|,j}$$

$Bell(|F|)$ represents the theorical number of partitions of a boundary set F. Figure 4 gives the 15 possible partitions of 4 elements and their ordering according to the previous formulas. In this way, the BDD structure is constructed by converging isomorphic subproblems thanks to the representation of graphs by partitions and the representation of partitions by numbers. Figure 3(a) illustrates the K-terminal reliability computation process of graph G in figure 2.

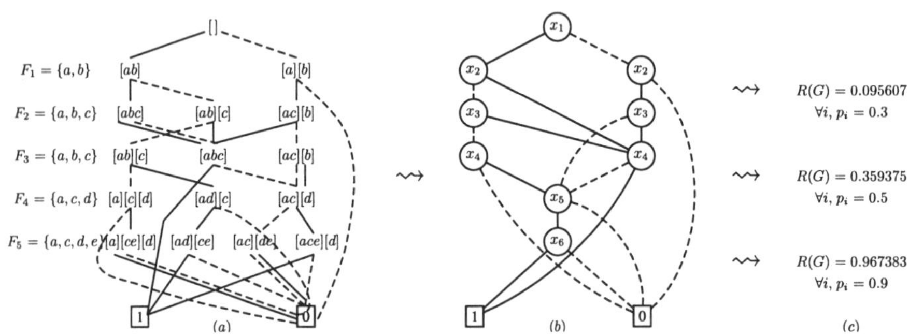

Figure 3. The network is shown figure 2. (a) illustrates the reliability computation process of the BDD. The terminal nodes are a, c and d ($K = \{a, c, d\}$). (b) shows the resulting BDD (after elimination of redundant nodes during the computation process). From this BDD structure, K-terminal network reliabilities can be obtain whatever the link functioning probabilities p_i (see (c))

1 - [1234]	9 - [14][2][3]
2 - [134][2]	10 - [1][24][3]
3 - [13][24]	11 - [1][2][34]
4 - [14][23]	12 - [13][2][4]
5 - [1][234]	13 - [1][23][4]
6 - [124][3]	14 - [12][3][4]
7 - [12][34]	15 - [1][2][3][4]
8 - [123][4]	

Figure 4. Partitions of 4 elements and their associated numbers.

4. Experimental Results

We considerer the benchmark of networks collected in literature by Ko, Lu and Yeh (see [16] for a complete description of these networks). For each network, the K-terminal reliability for $|K| = 2$, $|K| = |V|/2$ and $|K| = |V|$ is computed. Our algorithm has been tested on a Pentium 4 workstation with 512 MB memory. It has been written in C language. The experimental results for the K-terminal problem are shown in Table 1. The unit of time is in second. The running time includes the computation of BDD plus the K-terminal reliability computation. $|BDD|$ is the number of nodes of the constructed BDD.The heuristic used for ordering edges (and so variables in BDD) in the experiments is known as a breadth-first-search (BFS) ordering. The 2-terminal network reliability is always the highest among the three kinds of computed network reliablities. The computation speed heavily depends on the chosen edge ordering. According to the choosen ordering, the resulting BDD are of moderate size and the network reliability computations are immediate.

5. Conclusion

A method for evaluating the K terminal network reliability via BDD has been proposed in this paper. The difficult problem of efficiently identifying the isomorphic subgraphs met during the BDD computation process has been resolved. The algorithm was tested on literature instances and supplied good results. Based on this approach, our futur works will focus on computing other kinds of reliability and reusing the BDD structure in order to optimize design of network topology.

Table 1. Comparison results for K-terminal network reliabilities

Network			$\|K\| = 2$			$\|K\| = \|V\|/2$			$\|K\| = \|V\|$		
type	$\|V\|$	$\|E\|$	*time*	$\|BDD\|$	*rel.*	*time*	$\|BDD\|$	*rel.*	*time*	$\|BDD\|$	*rel.*
com.6	6	15	0.05	154	0.9999799	0.04	145	0.9999699	0.06	174	0.9999399
net2_6	20	30	0.40	65665	0.994395	1.61	424450	0.984685	0.25	45307	0.973502
net2_8	16	24	0.13	8452	0.9955528	0.6	31659	0.9889225	0.09	5961	0.9796585
net.18	14	23	0.07	1810	0.9864792	0.16	5170	0.9842904	0.1	2567	0.9692617
net.19	20	30	0.12	5370	0.9973206	0.40	74198	0.9890925	0.13	4813	0.9600691
st3x10	30	47	0.06	356	0.9644744	0.6	21454	0.9447255	0.09	236	0.9240546
st3x12	36	57	0.07	440	0.9617304	1.09	36586	0.9391320	0.07	377078	0.9173059
st5x5	25	40	0.13	2425	0.9755566	0.18	6858	0.9574274	0.09	1397	0.9398131
st2x20	40	58	0.10	152	0.7844822	0.08	989	0.7651158	0.06	1541	0.7452985
st2x100	200	298	0.03	792	0.3042932	9.83	15151	0.2765776	0.11	596	0.2510734

References

1. B. Akers, Binary Decision Diagrams, *IEEE Trans. On Computers*, vol. C-27, 509–516, (1978).
2. R. E. Bryant, Symbolic Boolean Manipulation with Ordered Binary-Decision Diagrams, *ACM Computing Surveys*, vol. 24 (3), 293–318, (1992).
3. A. Rauzy, New Algorithms for Fault Tolerant Trees Analysis, *Reliability Engineering and System Safety*, 203–211, (1993).
4. O. Coudert and J. C. Madre, Implicit and Incremental Computation of Primes and Essential Primes of Boolean functions, *Proceedings of the 29th ACM/IEEE Design Automation Conference (DAC'92)*, 36–39, IEEE Computer Society Press, (1992).
5. O. Coudert and J. C. Madre, A New Method to Compute Prime and Essential Prime Implicants of Boolean Functions, *Advanced Research in VLSI and Parallel Systems*, 113–128, (1992).
6. K. Odeh, Nouveaux algorithmes pour le traitement probabiliste et logique des arbres de défaillance, thesis, Université de Technologie de Compiègne, 1995.
7. S. J. Friedman and K. J. Supowit, Finding an optimal variable ordering for Binary Decision Diagrams, *IEEE Trans. On Computers*, vol. C-39, 710–713, (1990).
8. J.S. Provan, The complexity of reliability computations on planar and acyclic graphs, *SIAM J. Computing*, vol. 15 (3), 694–702, (1986).
9. G. Hardy, C. Lucet and N. Limnios, Computing all-terminal reliability of stochastic networks with Binary Decision Diagrams, *11th International Symposium on Applied Stochastic Models and Data Analysis*, 17-20 may 2005, Brest.
10. K. Sekine and H. Imai, Computation of the Network Reliability (Extended Abstract), *technical report*, Department of Information Science, University of Tokyo, (1998).
11. H. Imai, K. Sekine and K. Imai, Computational Investigations of All-Terminal Network Reliability via BDDs, *IEICE Trans. Fundamentals*, vol. E82-A, no. 5, 714–721.
12. R.K. Wood, A factoring algorithm using polygon-to-chain reductions for computing K-terminal network reliability, *Networks*, vol. 15, 173–190, (1985).
13. A. Satyanarayana and M.K. Chang, Network Reliability and the Factoring Theorem, *Networks*, vol. 13, 107–120, (1983).
14. O. Theologou and J. Carlier, Factoring and reductions for networks with imperfect vertices, *IEEE Transactions on Reliability*, vol. 40, 210–217, (1991).
15. J. Carlier and C. Lucet, A decomposition algorithm for network reliability evaluation, *Discrete Applied Mathematics*, vol. 65, 141–156, (1996).
16. F.-M. Yeh, S.-K. Lu and S.-Y. Kuo, OBDD-based evaluation of k-terminal network reliability, *IEEE Transactions on Reliability*, vol. 51, no. 4, (2002).

RELIABILITY EVALUATION OF A PACKET-LEVEL FEC BASED ON A CONVOLUTIONAL CODE CONSIDERING GENERATOR MATRIX DENSITY

T. HINO, M. ARAI, S. FUKUMOTO and K. IWASAKI

Department of Electrical Engineering
Graduate School of Engineering, Tokyo Metropolitan University
1–2 Minami-osawa, Hachioji, Tokyo 192–0397, Japan
E-mail: hino@info.eei.metro-u.ac.jp,
{arai, fukumoto, iwasaki}@eei.metro-u.ac.jp

As the Internet and its many applications become pervasive throughout the world, packet loss recovery is becoming an important technique for the reliable transmission of data. There are two types of packet loss recovery: the automatic repeat request (ARQ) and the forward error correction (FEC). In ARQ, receivers send acknowledgement messages and request packet retransmissions from senders. On the other hand, FEC employs proactive redundant packets without any retransmission. From the viewpoint of real-time applications, FEC is considered an extremely promising technique since there are no time-delays when retransmitting data . Several researchers conducted a study of Read-Solomon-code-based packet-level FEC's. We have shown that a convolutional-code-based packet-level FEC is more efficient under the low packet loss ratio. In this paper, we examine convolutional-code-based packet-level FEC's considering the density of the generator matrix. Stochastic analysis and simulations show the effect of our new FEC scheme.

1. Introduction

It is important to make transmissions over the Internet more reliable and many techniques for improving reliability have been proposed. Of these, packet-loss recovery is considered to be very important [1,2]. The means of recovery for the automatic repeat request (ARQ) mechanism is through the retransmission of lost packets. In ARQ, receivers use acknowledgement messages and request that the senders retransmit lost packets. Thus, from the viewpoint of end-to-end applications, packet losses cause long arrival delays, making it difficult to realize real-time transmissions [3,4,5]. Another promising technique for the recovery of lost packets is forward error correction (FEC). In FEC, the sender transmits redundant packets, and receivers use these redundant packets to recover lost packets. FEC operates without retransmission requests, and it is therefore considered suitable for real-time applications where as many packets as possible must be received or reconstructed without retransmission.

As with the FEC coding schemes, the Reed-Solomon (RS) code has been widely investigated [6,7]. The RS code is a kind of block code, which can recover as many lost packets as redundant packets generated for each block. Recently, there have also been studies of low-density parity-check (LDPC) codes over the erasure channel [8,9]. While LDPC codes do not always guarantee the same level of recovery as RS codes, LDPC shows smaller computational complexity because of its simple XOR-based calculations and low-density generator matrices.

We have proposed a scheme for packet-loss recovery based on convolutional codes [10]. Our scheme is similar to the LDPC codes except that redundant packets are generated as convolutions of the preceding code groups which correspond to blocks of the block codes. Our evaluations, using computer simulations and analysis [11] have shown that the proposed FEC has an improved ability for recovery in comparison with RS codes under given conditions of redundancy with given matrices. While the previous work assumed that the generator matrices and their densities were given, the effect of the density of generator matrices must be evaluated more precisely, especially in a case where a large number of redundant packets are transmitted. The density of matrices in the proposed scheme affects recovery and computation more strongly than in the RS or LDPC codes. Also, it becomes difficult or impossible to manually handle the matrices in a case where a larger amount of redundant packets are generated by larger matrices.

In this paper we discuss the effects of the density of generator matrices of convolutional-code-based FEC on recovery abilities. First, we theoretically analyze the post reconstruction receiving rate (PRRR), which is the probability that an information packet is received or recovered after the recovery process is performed, in terms of matrix density δ and packet loss probability p. We derive the PRRR's for all possible matrices over $(3, 2, 2)$ convolutional codes (where one redundant packet is generated for every two information packets), and calculate the expected PRRR assuming that the matrix is randomly chosen. Then, we use a computer simulation to assess the PRRR and computational complexity under the larger matrices.

This paper is organized as follows. Section 2 gives an overview of convolutional-code-based FEC. We theoretically analyze PRRR and show some numerical examples in Sec. 3, while in Sec. 4 we conduct a computer simulated evaluation. Section 5 offers a brief summary.

2. Convolutional-Code-Based Packet-Level FEC

Packet loss recovery using convolutional codes is one type of packet-level FEC, where redundant packets are generated based on convolutional codes. Here, we briefly explain the recovery scheme.

When a sender transmits k information packets, $(n-k)$ redundant packets are generated for each group. We call these n packets, which are a set of k information packets along with the $(n-k)$ redundant packets, a code group. The i-th group of the information packets is expressed as $\boldsymbol{u}_i = [u_{i,1} \cdots u_{i,k}]$, and the i-th code group is expressed as $\boldsymbol{v}_i = [v_{i,1} \cdots v_{i,n}]$, where each of $u_{i,1}$, ..., $u_{i,k}$, $v_{i,1}$, ..., and $v_{i,n}$ is a packet. The code group $\boldsymbol{v}_i$ is generated using the following equation:

$$\boldsymbol{v}_i = \boldsymbol{u}_i \cdot \boldsymbol{G}_i(D), \tag{1}$$

where $\boldsymbol{G}_i(D)$ is the following $k \times n$ generator matrix:

$$\begin{aligned} &\boldsymbol{G}_i(D) \\ &\quad = \left[\boldsymbol{I}_k \; \boldsymbol{g}(D) \right] \\ &\quad = \left[\boldsymbol{I}_k \; \boldsymbol{g}^{(1)}(D) \cdots \boldsymbol{g}^{(n-k)}(D) \right] \\ &\quad = \begin{bmatrix} & g_1^{(1)}(D) & \cdots & g_1^{(n-k)}(D) \\ \boldsymbol{I}_k & \vdots & \ddots & \vdots \\ & g_k^{(1)}(D) & \cdots & g_k^{(n-k)}(D) \end{bmatrix}. \end{aligned} \tag{2}$$

The elements in the matrix $\boldsymbol{g}(D)$, that is the right side of $\boldsymbol{G}_i(D)$, can be considered as polynomials of a degree that is at most m. For example, the polynomial $g_y^{(x)}(D)$ is shown below:

$$g_y^{(x)}(D) = g_{y,1}^{(x)} + g_{y,2}^{(x)} \cdot D + \cdots + g_{y,m+1}^{(x)} \cdot D^m. \tag{3}$$

Parameter m denotes the constraint length, which is the number of previous code groups that affect the redundant packets of a given code group. D is a delay operator, and the following equation holds that:

$$D \cdot v_{i,j} = v_{i-1,j}. \tag{4}$$

Therefore, the redundant packet $v_{i,k+j}$ is constructed from the following equation:

$$
\begin{aligned}
v_{i,k+j} &= g_{1,1}^{(j)} \cdot u_{i,1} + \cdots + g_{1,m+1}^{(j)} \cdot u_{i-m,1} \\
&+ g_{2,1}^{(j)} \cdot u_{i,2} + \cdots + g_{2,m+1}^{(j)} \cdot u_{i-m,2} \\
&+ \cdots + g_{k,m+1}^{(j)} \cdot u_{i-m,k}
\end{aligned} \tag{5}
$$

When each coefficient in the polynomials is defined as 0 or 1, that is the elements from $GF(2)$, the symbol " + " equals the bit-wise exclusive-OR calculation for each packet.

The generator matrix for redundant packet can also be expressed as a binary $k \times (n-k) \cdot (m+1)$ matrix as follows:

$$
\boldsymbol{g} = \left[\boldsymbol{g}^{(1)} \cdots \boldsymbol{g}^{(n-k)}\right], \tag{6}
$$

$$
\boldsymbol{g}^{(y)} = \begin{bmatrix} g_{1,1}^{(y)} & \cdots & g_{1,m+1}^{(y)} \\ \vdots & \ddots & \vdots \\ g_{k,1}^{(y)} & \cdots & g_{k,m+1}^{(y)} \end{bmatrix}. \tag{7}
$$

Implanted sequence numbers,information packets and redundant packets are transmitted to the receivers. From the gap between the sequence numbers of the received packets, receivers can locate the packet losses [2,7]. Therefore, Eq. (5) also holds for the receiver. Because the positions of the losses are known, packet losses in the equation are the unknown values. If Eq. (5) contains one unknown packet, it can be recovered directly from the equation.

Each redundant packet is generated from information packets in more than one code group, therefore the equation system holds for the continuous redundant packets. Even if more than one information packet is lost, all lost packets can be recovered under the condition that this equation system has a unique solution.

3. Analysis

In this section, we analyze the post reconstruction receiving rate of a convolutional-code-based packet-level FEC whose generator matrix is randomly decided. If the $(n, n-1, m)$ convolutional code is considered, the generator matrix can be described as follows:

$$g = \begin{pmatrix} g_1 & g_2 & \cdots & g_{m+1} \\ g_{m+2} & g_{m+2} & \cdots & g_{2(m+1)} \\ \vdots & \vdots & \ddots & \vdots \\ g_{n(m+1)+1} & g_{n(m+1)+2} & \cdots & g_{(n-1)(m+1)} \end{pmatrix}. \tag{8}$$

Each element in this matrix, g_x ($x = 1, 2, \ldots, (n-1)(m+1)$), is set to 1 with a probability δ, or a generator matrix density. We now introduce the expected post reconstruction receiving rate, $f_x(p)$, which is the probability that a data packet will be received or can be restored in the case of loss under the condition that the generator matrix has x elements and that each of transmitted packets are lost with the probability p. Thus, the expected post reconstruction receiving rate for a packet-level FEC with a randomly assigned generator matrix $\Phi_p(\delta)$, is derived as

$$\Phi_p(\delta) = \sum_{x=0}^{(n-1)(m+1)} \binom{(n-1)(m-1)}{x} \delta^x (1-\delta)^{(n-1)(m+1)-x} \cdot f_x(p). \tag{9}$$

3.1. *Analysis for (3, 2, 2) convolutional coding*

In the rest of this section, we concentrate on the analysis of a (3, 2, 2) convolutional code based packet level FEC. This enables us to explicitly yield the expected post reconstruction receiving rate, which consists of the conditional probabilities, $f_0(p), f_1(p), \cdots$, and $f_6(p)$. The basic idea of our analysis is to set up renewal equations that regard the packet transmission sequence as finite. As an example, we show the derivation of $f_2(p)$ below.

If the generator matrices of a (3, 2, 2) convolutional code have two elements of 1, we can classify these matrices into two groups depending on their recoverability, which are

$$\left\{ \begin{pmatrix} 1\,1\,0 \\ 0\,0\,0 \end{pmatrix}, \begin{pmatrix} 1\,0\,1 \\ 0\,0\,0 \end{pmatrix}, \begin{pmatrix} 0\,1\,1 \\ 0\,0\,0 \end{pmatrix}, \begin{pmatrix} 0\,0\,0 \\ 1\,1\,0 \end{pmatrix}, \begin{pmatrix} 0\,0\,0 \\ 1\,0\,1 \end{pmatrix}, \begin{pmatrix} 0\,0\,0 \\ 0\,1\,1 \end{pmatrix} \right\}$$

and

$$\left\{ \begin{pmatrix} 1\,0\,0 \\ 1\,0\,0 \end{pmatrix}, \begin{pmatrix} 1\,0\,0 \\ 0\,1\,0 \end{pmatrix}, \begin{pmatrix} 1\,0\,0 \\ 0\,0\,1 \end{pmatrix}, \begin{pmatrix} 0\,1\,0 \\ 1\,0\,0 \end{pmatrix}, \begin{pmatrix} 0\,1\,0 \\ 0\,1\,0 \end{pmatrix}, \begin{pmatrix} 0\,1\,0 \\ 0\,0\,1 \end{pmatrix}, \right.$$

$$\left. \begin{pmatrix} 0\,0\,1 \\ 1\,0\,0 \end{pmatrix}, \begin{pmatrix} 0\,0\,1 \\ 0\,1\,0 \end{pmatrix}, \begin{pmatrix} 0\,0\,1 \\ 0\,0\,1 \end{pmatrix} \right\}.$$

Any two matrices in a group have identical recoverability because of their symmetric properties. Let us call the first group shown above 'group a', and the second one 'group b'. We further define $f_{2a}(p)$ and $f_{2b}(p)$ as the post reconstruction receiving rates with matrices in the group a and b respectively. Weighting these rates by the group size factors, the conditional post reconstruction receiving rate $f_2(p)$ is obtained by

$$f_2(p) = \frac{6}{15} \cdot f_{2a}(p) + \frac{9}{15} \cdot f_{2b}(p). \tag{10}$$

In order to evaluate $f_{2a}(p)$, we use the first matrix of the group a, in which two types of redundant structures are provided; one is for the upper stand data packets in Figure 1 and the other is for the lower stand ones. Let $f_{2a}^{(u)}(p)$ and $f_{2a}^{(l)}(p)$ denote the post reconstruction receiving rate for the upper and lower stand data packets respectively. The conditional probability $f_{2a}(p)$ can be obtained as the average of these values, that is,

$$f_{2a}(p) = \frac{f_{2a}^{(u)}(p) + f_{2a}^{(l)}(p)}{2}. \tag{11}$$

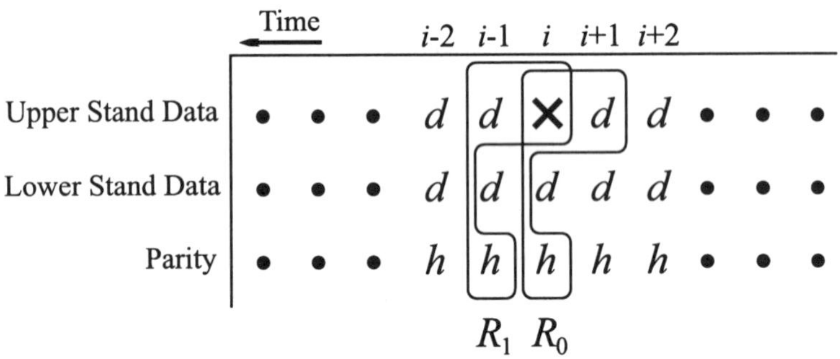

Figure 1. Code groups for packet loss recovery in a (3, 2, 2) convolutional code based packet level FEC with the generator matrix $\begin{pmatrix} 1 & 1 & 0 \\ 0 & 0 & 0 \end{pmatrix}$.

An upper stand data packet is contained in two code groups R_0 and R_1, in which a lost packet may be recovered by exclusive OR operations. Hence, considering the direction of time in Figure 1, let us make the following notations:

A_b: An event where a data packet is received or can be recovered in case of loss by the code group R_0 and its backward packets if necessary.

A_f: An event where a data packet is received or can be recovered in case of loss by the code group R_1 and its forward packets if necessary.

Assuming an infinite packet transmission sequence, we can easily obtain the renewal equations

$$Pr\{A_b\} = (1-p) + p(1-p)Pr\{A_b\} \tag{12}$$

and

$$Pr\{A_f\} = (1-p) + p(1-p)Pr\{A_f\}. \tag{13}$$

Substituting the solutions of Eqs. (12) and (13), $f_{2a}^{(u)}(p)$ is derived as

$$\begin{aligned} f_{2a}^{(u)}(p) &= Pr\{A_b \cup A_f\} \\ &= Pr\{A_b\} + Pr\{A_f\} - Pr\{A_b \cap A_f\} \\ &= Pr\{A_b\} + Pr\{A_f\} - \{(1-p) + p(1-p)^2 \cdot Pr\{A_b\} \cdot Pr\{A_f\}\} \\ &= \frac{p^4 - 3p^3 + 3p^2 - 2p + 1}{(p^2 - p + 1)^2}. \end{aligned} \tag{14}$$

On the other hand, the lower stand data packets are not contained in any of the code groups. This means that these data packets have no recovery scheme, i.e.

$$f_{2a}^{(l)}(p) = 1 - p. \tag{15}$$

Next, we use the first matrix of the group b to evaluate $f_{2b}(p)$. In this case, the upper and lower stand data packets are recovered using normal parity coding. Then, we have

$$f_{2b}^{(l)}(p) = f_{2b}^{(u)}(p) = f_{2b}(p) = (1-p) + p(1-p)^2. \tag{16}$$

The Eqs. (11), (14), (15), and (16) above explicitly yield the conditional post reconstruction receiving rate $f_2(p)$ as follows:

$$f_2(p) = \frac{3p^7 - 12p^6 + 20p^5 - 17p^4 + p^3 + 11p^2 - 11p + 5}{5\,(p^2 - p + 1)^2}. \tag{17}$$

We can similarly obtain the other conditional post reconstruction receiving rates. For instance, $f_3(p)$ is expressed as:

$$\begin{aligned} f_3(p) = {} & (20 - 81p + 155p^2 - 84p^3 - 363p^4 + 1438p^5 - 2980p^6 + 3470p^7 \\ & -263p^8 - 7382p^9 + 15540p^{10} - 18575p^{11} + 15024p^{12} - 8539p^{13} \\ & +3379p^{14} - 889p^{15} + 140p^{16} - 10p^{17}) \\ & /\{20(1 - p + 2p^2 - p^3)^2(1 - p - p^2 + 5p^3 - 4p^4 + p^5)^2\}. \end{aligned} \tag{18}$$

Although $f_4(p)$, $f_5(p)$, $f_6(p)$ are also obtained, we leave out the results since they are more complex expressions than that of $f_3(p)$. On the contrary, $f_0(p)$ and $f_1(p)$ are derived as quite simple formulae: $f_0(p) = 1 - p$ and $f_1(p) = (1-p)(2+p)/2$. The expected post reconstruction receiving rate for a (3, 2, 2) convolutional code based packet level FEC with a randomly assigned generator matrix is conclusively evaluated by

$$\Phi_p(\delta) = \sum_{x=0}^{6} \binom{6}{x} \delta^x (1-\delta)^{6-x} \cdot f_x(p). \tag{19}$$

3.2. *Numerical example*

Figure 2 shows the dependence of a matrix density δ on the expected post reconstruction receiving rate $\Phi_p(\delta)$ for the (3, 2, 2) convolutional coding obtained above, where the packet loss probability is $p = 0.1, 0.2, 0.3, 0.4,$

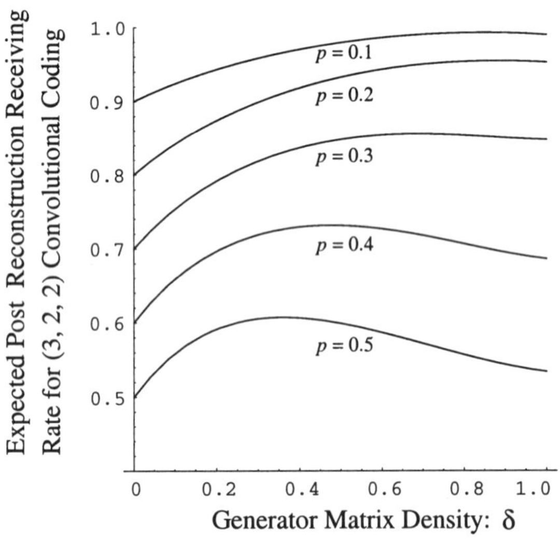

Figure 2. $\Phi_p(\delta)$ for a (3, 2, 2) convolutional code whose generator matrix is randomly assigned with the matrix density δ.

and 0.5. It is obvious that a small value of p yields large values of $\Phi_p(\delta)$ ranging over the whole gamut of δ. Moreover, we can see that there exists the critical value of δ, i.e., the optimal density for maximizing $\Phi_p(\delta)$. Let us denote the optimal density as δ^*. The more p decreases, the more δ^* increases.

4. Evaluation Using Simulation

Using Monte Carlo simulations, we evaluate the PRRR and computational complexity in terms of matrix density. The simulator runs in the following way. First, the sender generates $k \times L$ information packets and adds $(n-k)$ redundant packets for each group of k information packets. Each element in the generator matrix is randomly set to 0 or 1 by a given density probability δ. Next, the packets are transmitted to the receiver through a communication link. On the communication link, each packet is independently discarded according to the packet loss probability p. Then, the receiver recovers the lost information packets. We repeat the trial of transmission R times under different matrices, and calculate the average PRRR.

For computational complexity, we calculate the average number of operations per trial at the recovery process. We assume that the length of each packet is 1000 bytes, and the byte-wize XOR is regarded as one operation, as well as an operation for a control variable. In this paper, we set R = 1000 and L = 100 .

Figure 3 is the calculation result of the average PRRR for a (30, 20, 10) convolutional code where p is set to 1%, 5%, 10%, and 20% .

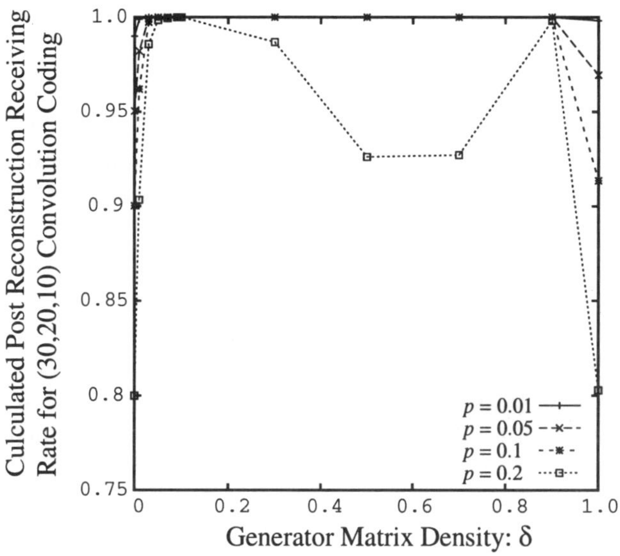

Figure 3. Calculation result of the average PRRR for (30, 20, 10) convolutional codes.

In Fig. 3, $\delta = 0$ means that all the elements in the generator matrix are 0, that is, any redundant packets have no contribution to the recovery. Thus, PRRR is equal to $(1-p)$. When δ is increased slightly from 0, the PRRR is drastically improved and comes close to reaching 1. The smaller p is, the smaller δ is at the point where the PRRR reaches 1. While the PRRR's for $p \leq 0.1$ continue to reach 1 in the range of $\delta \leq 0.9$, those for $p = 0.2$ fall quickly as δ increases and PRRR gains again. When the packet loss probability is low, a lost packet might be directly recovered using the equation shown in Eq. (5). Thus, a very sparse matrix is enough for efficient recovery if one information packet is included with at least one redundant packet. On the other hand, when the packet loss probability becomes higher, the loss of multiple packets in one or successive code groups occurs more frequently. In such cases, not only are the lost packets required to be included in the equations, but the equations must also be linearly independent. When δ is 1, the generator matrix becomes all-1 and thus the equations for all redundant packets in the same code group become linearly dependent. When this happens, the multiple packet losses in one code group are impossible to recover, which results in a decreased PRRR.

A (30, 20, 10) convolutional code has the same redundancy as the (3, 2, 2) codes analyzed in Sec. 2, but the associated PRRR shows quite different and complicated behavior . This is mainly because of an increased length constraint. As mentioned in [10], an increased length constraint improves the PRRR at a lower packet loss probability, instead of reducing it at a higher loss probability.

Figure 4 is the calculation result of the average number of operations at the recovery obtained using the same simulation results as Fig. 3.

The number of operations increases proportionally to δ and p in the range of $\delta \leq 0.9$, and increases significantly at $\delta = 1$. In the recovery process, Eq. (5) is derived for each of the redundant packets, and the known values, or received packets, are summed up if the equation contains at least one unknown value, or lost packet. The number of known values in Eq. (5) is proportional to δ, and the number of equations holding unknown values is proportional to p. As shown in Fig. 3, the case of $\delta = 1$ shows a reduced PRRR. This means that many equations may hold, but the equation system cannot be solved. When the equations are not solved, they are kept until the next code group arrives, because they might contribute to the recovery of the next code group. Therefore, those equations linearly dependent to each other drastically increase the computational complexity because they manipulate a large equation system.

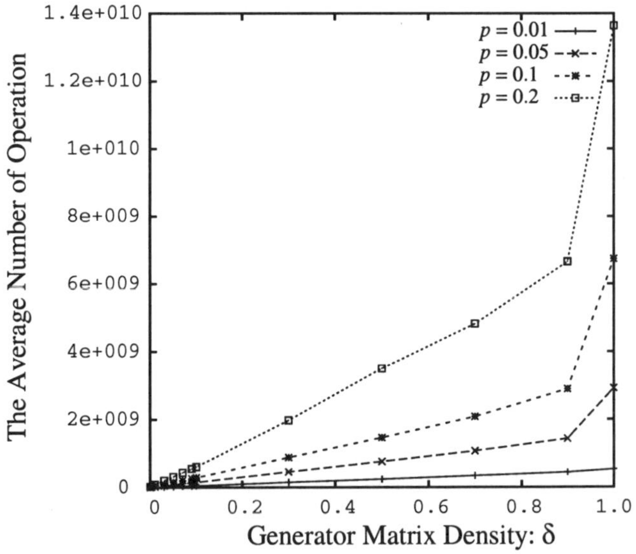

Figure 4. Calculation result of the average number of operations for (30, 20, 10) convolutional codes.

5. Conclusions

In this paper we discussed the effect of the density of the generator matrices of convolutional-code-based FEC on the abilities to recover lost packets. We theoretically analyzed the post reconstruction receiving rate $\Phi_p(\delta)$ for a (3, 2, 2) convolutional code in terms of matrix density δ and packet loss probability p. Numerical examples showed that an optimal density existed for a given p We then evaluated the PRRR and computational complexity for larger matrices using a computer simulation. At a given p, while δ had an optimal value which maximized the PRRR, the computational complexity increased in proportion to δ.

References

1. S. Tanenbaum, *Computer Networks, Third Edition*, Prentice Hall (1996).
2. H. Liu, H. Ma, M. E. Zarki, and S. Gupta, Error Control Schemes for Networks: An Overview, *ACM Mobile Networks & Applications*, **2**, (2), pp. 167–182 (1997).
3. T. Yokohira and T. Okamoto, A Delay Margin Assignment Method for EDD Connection Admission Control Scheme – An Assignment Proportional to Worstcase Link Delays –, *IEICE Trans. Commun.*, **J84-B**, (8), pp. 1484–

1493 (2001).
4. H. Obata, K. Ishida, J. Fnasaka, and K. Amano, Evaluation of TCP Performance on Asymmetric Networks Using Satellite and Terrestrial Links, *IEICE Trans. Commun.*, **E84-B**, (6), pp. 1480–1487 (2001).
5. K. Yasui, T. Nakagawa, and H. Sandoh, Modeling and Analysis for Data Communication System, *Communications of Operations Research Society of Japan*, **40**, (4), pp. 205–210 (1995)
6. L. Rizzo, Effective Erasure Codes for Reliable Computer Communication Protocols, *Computer Communication Review*, **27**, (2), pp 167–182 (1997).
7. J. Nonnenmacher, E. Biersack, and D. Towsley, Parity-Based Loss Recovery for Reliable Multicast Transmission, *IEEE Trans. Networking*, **6**, (4), pp. 349–361 (1998).
8. J. Byers, M. Luby, M. Mitzenmacher, and A. Rege, A Digital Fountain Approach to Reliable Distribution of Bulk Data, In *ACM SIGCOMM'98*, pp. 56–67 (1998).
9. J. S. Plank and M. G. Thomason, A Practical Analysis of Low-Density Parity-Check Erasure Codes for Wide-Area Storage Application, In *Dependable Systems and Networks*, pp. 115–124 (2004).
10. M. Arai, A. Yamaguchi, and K. Iwasaki, Method to Recover Packet Losses Using $(n, n-1, m)$ Convolutional Codes, In *Dependable Systems and Networks*, pp. 382–389 (2000).
11. A. Yamaguchi, M. Arai, S. Fukumoto, and K. Iwasaki, Fault-Tolerance Design for Multicast Using Convolutional-Code-Based FEC and Its Analytical Evaluation, *IEICE Trans. Info. & Sys.*, **E85-D**, (5), pp. 864–873 (2002).

A FRAMEWORK FOR DISCRETE SOFTWARE RELIABILITY MODELING WITH PROGRAM SIZE AND ITS APPLICATIONS

SHINJI INOUE and SHIGERU YAMADA

Tottori University,
4-101 Minami, Koyama-cho, Tottori-shi,
Tottori 680-8552, JAPAN
E-mail: ino@sse.tottori-u.ac.jp

We discuss a generalization framework for discrete software reliability growth models (SRGMs) with the effect of program size on software reliability growth process. Our framework enables us to develop a plausible discrete software reliability growth model with program size by applying a suitable software failure-occurrence times distribution to our framework. In this chapter generalized software reliability assessment measures are also derived. After that, we discuss a parameter estimation method of our modeling framework discussed in this chapter. Additionally, we discuss optimal software release problems based on a SRGM developed by using our framework as one of the application issues. Finally, we depict numerical illustrations for the software reliability assessment measures and for derived optimal release policies by using actual fault count data.

1. Introduction

Software reliability assessment in a testing-phase located in a final stage of a software development process is one of the important activities to develop a highly reliable software system. In the testing-phase, an implemented software system is tested to detect and correct software faults latent in the software system. The software development manager has to assess software reliability to ship a reliable software system to the user. A *software reliability growth model* (abbreviated as *SRGM*)[1-3] is known as one of the useful mathematical tools to assess software reliability quantitatively. As a role of software systems has been expanding rapidly, the size, complexity, and diversification of software systems have been growing drastically in recent years. Then, we need to develop more plausible SRGMs which enable us to assess software reliability more accurately. Recently, as one of the solutions, generalization or unified approaches for software reliability growth modeling have been proposed based on order-statistics[4,5], an infinite server queueing theory[6], Markov processes[1,7,8], and so on. These generalized and unified approaches provide us with frameworks for software reliability growth modeling, and treat existing SRGMs proposed so far theoretically and generally. Then, the software development manager can develop a plausible SRGM which has good performance for actual software reliability assessment by using these modeling framework. However, almost of the unified or generalization approaches have been discussed for continuous-time software reliability growth modeling because the SRGMs on the continuous-time domain is specifically applicable to the reliability analysis and better for mathematical manipulations.

On the other hand, for software reliability modeling or analysis, it is better to use discrete SRGMs which describe a software reliability growth process depending on the discrete-time domain, such as the number of executed test-cases and the calendar time, from the point of view of the model performance on software reliability assessment and the consistency in data collection activities in an actual testing-phase. On the discrete SRGMs, the discrete-time domain is regarded as the unit of software fault-detection period, and countable. Considering that there are discrete SRGMs to describe a software reliability growth process depending on discrete time-domain, we need to discuss a generalization or unified approach for plausible discrete software reliability growth modeling. Until now, a few generalization or unified approaches for discrete SRGMs have been discussed[5,9].

In this chapter we propose a new generalization approach for SRGMs with the effect of program size. The program size is one of the important metrics of software complexity which influences the software reliability growth process in the testing-phase. We then discuss parameter estimation for our modeling framework based on the method of maximum-likelihood. Further, as one of the interesting issues for practical applications of an SRGM, optimal software release problems under the simultaneous cost and reliability requirements are also discussed in this chapter. Finally, we depict numerical illustrations of software reliability assessment based on a discrete SRGM developed under our modeling framework and of derived optimal software release policies by using actual fault count data.

2. Generalized Modeling

We discuss a new generalization approach with the effect of program size. Our generalization approach is discussed under the basic assumptions for modeling on software failure-occurrence phenomenon in a testing-phase.

2.1. *Basic assumptions*

Our generalization approach is based on the following assumptions[4,5]:

(A1) Whenever a software failure is observed, the fault which caused it will be detected immediately, and no new faults are introduced in the fault-detection procedure.

(A2) Each software failure occurs at independently and identically distributed of random times with the discrete probability distribution $P(i) \equiv \Pr\{I \leq i\} = \sum_{k=0}^{i} p_I(k)$ $(i = 0, 1, 2, \cdots)$, where $p_I(k)$ and $\Pr\{A\}$ represent the probability mass function for I and the probability of event A, respectively.

(A3) The initial number of faults in the software system, $N_0(> 0)$, is a random variable, and is finite.

Now, let $\{N(i),\ i = 0, 1, \cdots\}$ denote a discrete stochastic process representing the number of faults detected up to i-th testing-period. Then, the conditional probability that m faults are detected up to i-th testing-period given that the initial fault content, N_0, equals n is derived as:

$$\Pr\{N(i) = m \mid N_0 = n\} = \binom{n}{m} \{P(i)\}^m \{1 - P(i)\}^{n-m}. \tag{1}$$

From Eq. (1), we can derive the probability mass function that m faults are detected up to i-th testing-period as

$$\Pr\{N(i) = m\} = \sum_{n} \binom{n}{m} \{P(i)\}^m \{1 - P(i)\}^{n-m} \Pr\{N_0 = n\}$$
$$(m = 0, 1, 2, \cdots). \tag{2}$$

The stochastic behavior of software fault-detection or failure-occurrence phenomenon in the testing-phase can be characterized by giving a suitable probability mass function of the initial fault content N_0. Okamura et al.[5] have discussed a generalization framework for SRGMs following nonhomogeneous Poisson processes (abbreviated as NHPP) for software reliability assessment by assuming that the initial fault content, N_0, obeys a Poisson distribution, and proposed a unified parameter estimation method based on the EM algorithm.

2.2. *Modeling framework with effect of program size*

We propose a generalization framework for software reliability growth modeling with the effect of program size by considering the case that the probability distribution of the initial fault content, N_0, follows a binomial distribution with parameters (K, λ) given as

$$\Pr\{N_0 = n\} = \binom{K}{n} \lambda^n (1-\lambda)^{K-n}$$
$$(0 < \lambda < 1\,;\, n = 0, 1, \cdots, K). \qquad (3)$$

Eq. (3) has the following physical assumptions:

(a) The software system consists of K lines of code (LOC) at the beginning of the testing-phase.
(b) Each code has a fault with a constant probability λ.
(c) Each software failure caused by a fault remaining in the software system occurs independently and randomly.

These assumptions are useful to apply a binomial distribution as a probability mass function of the initial fault content in the software system to software reliability growth modeling, and to incorporate the effect of the program size into the proposed model[7,10].

Substituting Eq. (3) into Eq. (2), we can derive the probability mass function of the number of faults detected up to i-th testing-period as

$$\Pr\{N_B(i) = m\} = \binom{K}{m} \{\lambda P(i)\}^m \{1 - \lambda P(i)\}^{K-m}$$
$$(m = 0, 1, 2, \cdots K). \qquad (4)$$

From Eq. (4), And discrete SRGM with the effect of program size can be developed easily by giving a suitable probability distribution function to the software failure-occurrence times distribution.

3. Software Reliability Assessment Measures

Software reliability assessment measures are well-known as useful metrics of quantitative measures for software reliability assessment. We derive several generalized software reliability assessment measures based on the basic assumptions of discrete software reliability growth modeling discussed in **2.1**.

3.1. Expectation and variance of the number of detected faults

Information on the current number of detected faults is one of the important metrics to estimate the degree of testing-progress. Therefore, the expectation and variance of the number of detected faults are useful measures because the number of faults detected up to i-th testing-period, $N(i)$, is treated as a random variable.

The expectation of the number of detected faults, $\mathrm{E}[N(i)]$, is derived as

$$\begin{aligned}\mathrm{E}[N(i)] &= \sum_{z=0}^{n} z \sum_{n} \binom{n}{z} \{P(i)\}^{z}\{1-P(i)\}^{n-z}\Pr\{N_0=n\} \\ &= \mathrm{E}[N_0]P(i). \end{aligned} \tag{5}$$

And its variance, $\mathrm{Var}[N(i)]$, is also derived as

$$\begin{aligned}\mathrm{Var}[N(i)] &= \mathrm{E}[N(i)^2] - (\mathrm{E}[N(i)])^2 \\ &= \mathrm{Var}[N_0]\{P(i)\}^2 + \mathrm{E}[N_0]P(i)\{1-P(i)\}. \end{aligned} \tag{6}$$

Therefore, if N_0 follows the binomial distribution in Eq. (3), they are given as

$$\mathrm{E}[N_B(i)] = K\lambda P(i), \tag{7}$$

$$\mathrm{Var}[N_B(i)] = K\lambda P(i)\{1-\lambda P(i)\}, \tag{8}$$

respectively. We can see that $K\lambda$ in Eq. (7) represents the expected initial fault content when N_0 follows the binomial distribution.

3.2. Software reliability function

A software reliability function is one of the well-known software reliability assessment measures. Given that the testing or the operation has been going up to i-th testing-period, the discrete software reliability function is defined as the probability that a software failure does not occur in the time-interval $(i, i+h](i, h = 0, 1, \cdots)$[12,11]. Then, we can formulate a generalized discrete software reliability function $R(i,h)$ as

$$\begin{aligned}R(i,h) &= \sum_{k} \Pr\{N(i+h)=k \mid N(i)=k\}\Pr\{N(i)=k\} \\ &= \sum_{k} \Big[\{P(i)\}^{k}\{1-P(i+h)\}^{-k} \\ &\qquad \cdot \sum_{n} \binom{n}{k} \{1-P(i+h)\}^{n} \cdot \Pr\{N_0=n\}\Big], \end{aligned} \tag{9}$$

by using Eq. (2). Then, if N_0 obeys the binomial distribution in Eq. (3), the discrete software reliability function can be derived as

$$R_B(i,h) = [1 - \lambda\{P(i+h) - P(i)\}]^K , \tag{10}$$

by using Eq. (9).

3.3. *Instantaneous and cumulative MTBFs*

We also derive instantaneous and cumulative mean time between software failures (abbreviated as MTBFs) which are substitutions for an ordinary MTBF. Under the basic assumptions discussed in **2.1**, the ordinary MTBF can not be derived because the basic assumptions have the following properties:

$$F(i,0) = 1 - R(i,0) = 0, \tag{11}$$

$$\begin{aligned} F(i,\infty) &= 1 - R(i,\infty) \\ &= 1 - \sum_n \{P(i)\}^n \cdot \Pr\{N_0 = n\}, \end{aligned} \tag{12}$$

where $F(i,h)$ represents a probability that a software failure occurs in the time-interval $(i, i+h]$. These equations above implies that the probability distribution function, $F(i,h)$, does not satisfy the property of ordinary probability distribution functions. Accordingly, we need to utilize discrete instantaneous and cumulative MTBFs as substitutions for the ordinary MTBF. Using Eq. (5), we can formulate the discrete instantaneous MTBF as

$$MTBF_I(i) = \frac{1}{\mathrm{E}[N(i+1)] - \mathrm{E}[N(i)]}. \tag{13}$$

And the discrete cumulative MTBF can also given as

$$MTBF_C(i) = \frac{i}{\mathrm{E}[N(i)]}. \tag{14}$$

4. Parameter Estimation

We discuss parameter estimation for discrete SRGM developed by using our modeling framework in Eq. (4) based on the method of maximum-likelihood. Suppose that we have observed K data pairs $(t_i,\ y_i)(i = 0, 1, 2, \cdots, K)$ with respect to the cumulative number of faults, y_i, detected during a constant time-interval $(0,\ t_i](0 < t_1 < t_2 < \cdots < t_K)$. The likelihood function l for the number of detected faults, $N_B(i)$, can be derived as

$$l = \Pr\{N_B(t_1) = y_1, N_B(t_2) = y_2 \cdots, N_B(t_N) = y_N\}$$
$$= \prod_{i=2}^{N} \Pr\{N_B(t_i) = y_i \mid N_B(t_{i-1}) = y_{i-1}\} \cdot \Pr\{N_B(t_1) = y_1\}, \qquad (15)$$

by using the Bayes' formula and the Markov property [13,14]. The conditional probability in Eq. (15), $\Pr\{N_B(t_i) = y_i \mid N_B(t_{i-1}) = y_{i-1}\}$, can be shown as

$$\Pr\{N_B(t_i) = y_i \mid N_B(t_{i-1}) = y_{i-1}\}$$
$$= \binom{K - y_{i-1}}{y_i - y_{i-1}} \{z(t_{i-1}, t_i)\}^{y_i - y_{i-1}} \{1 - z(t_{i-1}, t_i)\}^{K - y_i}, \quad (16)$$

by considering that we can regard t_{i-1} as the initial time and that the distribution range of $N_B(i)$ is $0 \leq N_B(i) \leq K - y_{i-1}$. In above equation, setting

$$z(t_{i-1}, t_i) \equiv \frac{\lambda\{P(t_i) - P(t_{i-1})\}}{1 - \lambda P(t_{i-1})}, \qquad (17)$$

we can rewrite Eq. (15) as

$$l = \prod_{i=1}^{N} \binom{K - y_{i-1}}{y_i - y_{i-1}} \{z(t_{i-1}, t_i)\}^{y_i - y_{i-1}} \{1 - z(t_{i-1}, t_i)\}^{K - y_i}, \qquad (18)$$

by using Eq. (16), where $t_0 = 0$, $y_0 = 0$, and $P(0) = 0$. Accordingly, the logarithmic likelihood function can be derived as

$$L \equiv \log l$$
$$= \log K! - \log\{(K - y_N)!\} - \sum_{i=1}^{N} \log\{(y_i - y_{i-1})!\} + y_N \log \lambda$$
$$+ \sum_{i=1}^{N} (y_i - y_{i-1}) \log\{P(t_i) - P(t_{i-1})\} + (K - y_N) \log\{1 - \lambda P(t_N)\}, \qquad (19)$$

by taking the natural logarithm of Eq. (18). Then, when we apply the geometric distribution given as

$$P(i) = 1 - (1 - p)^i \qquad (i = 0, 1, 2, \cdots;\ 0 < p < 1), \qquad (20)$$

to the software failure-occurrence times distribution, the logarithmic likelihood function can be derived as

$$L = \log K! - \log\{(K - y_N)!\} + y_N \log \lambda - \sum_{i=1}^{N}\{(y_i - y_{i-1})!\}$$
$$+ \sum_{i=1}^{N}(y_i - y_{i-1}) \log\{(1-p)^{t_{i-1}} - (1-p)^{t_i}\}$$
$$+ (K - y_N) \log\left[1 - \lambda\{1 - (1-p)^{t_N}\}\right], \tag{21}$$

by using Eq. (19). In this case we have to estimate the parameters λ and p if we can know the program size K. The simultaneous likelihood equations with respect to the parameters λ and p can be derived as

$$\frac{\partial L}{\partial \lambda} = \frac{y_N}{\lambda} + \cdot \frac{(K - y_N) \cdot \{(1-p)^{t_N} - 1\}}{1 - \lambda\{1 - (1-p)^{t_N}\}} = 0, \tag{22}$$

$$\frac{\partial L}{\partial p} = \sum_{i=1}^{N}(y_i - y_{i-1}) \frac{\{t_i(1-p)^{t_i - 1} - t_{i-1}(1-p)^{t_{i-1}-1}\}}{\{(1-p)^{t_{i-1}} - (1-p)^{t_i}\}}$$
$$- \frac{(K - y_N)\{t_N \lambda (1-p)^{t_N - 1}\}}{1 - \lambda\{1 - (1-p)^{t_N}\}} = 0, \tag{23}$$

respectively. By solving Eq. (22) with respect to λ, we can obtain

$$\lambda = \frac{y_N}{K\{1 - (1-p)^{t_N}\}}. \tag{24}$$

Substituting Eq. (24) into Eq. (23), we can obtain the following equation:

$$\frac{t_N y_N (1-p)^{t_N - 1}}{1 - (1-p)^{t_N}}$$
$$= \sum_{i=1}^{N}(y_i - y_{i-1}) \frac{\{t_i(1-p)^{t_i - 1} - t_{i-1}(1-p)^{t_{i-1}-1}\}}{\{(1-p)^{t_{i-1}} - (1-p)^{t_i}\}}. \tag{25}$$

Accordingly, we can obtain the maximum-likelihood estimates $\widehat{\lambda}$ and $\widehat{p}$ of the parameters λ and p, respectively, by solving the simultaneous likelihood functions in Eqs. (24) and (25) numerically.

5. Optimal Software Release Problems

Software development managers have a great interest in how to develop a reliable software product economically or when to release the software to the customers[15]. We discuss discrete cost-optimal software release policies based on a discrete SRGM developed under our modeling framework in Eq. (4). And then, we also discuss discrete optimal software release policies with the simultaneous cost and reliability requirements in consideration of the software quality control. In the discussion on optimal software release problems in this chapter, we assume that the software failure-occurrence times follows the geometric distribution in Eq. (20).

5.1. *Cost-optimal software release policies*

We discuss cost-optimal software release policies based on an SRGM developed by using our modeling framework. First of all, we define the following notations:

c_1 : debugging cost per one fault in the testing phase.

c_2 : debugging cost per one fualt in the operational phase, where $0 < c_1 < c_2$.

c_3 : testing cost per constant period.

Let Z be the software release period. Then, the expected total software cost, $C(Z)$, which indicates the expected total cost during the testing and operational phases is formulated as

$$C(Z) = c_1 \mathrm{E}[N_B(Z)] + c_2(K\lambda - \mathrm{E}[N_B(Z)]) + c_3 Z. \tag{26}$$

The cost-optimal software release period is the test-termination period minimizing the expected total software cost $C(Z)$ in Eq.(26). From Eq. (26), we can derive the following equation by taking the forward difference in terms of Z:

$$C(Z+1) - C(Z) = (c_2 - c_1)\left[\frac{c_3}{c_2 - c_1} - W(Z)\right], \tag{27}$$

where $W(Z)$ represents the expected number of detected faults during an Z-th period. And, we need to define the following notation to discuss the discrete optimal software release policies:

$$< n >= \begin{cases} [n] & (\text{if } C([n]) \leq C([n]+1) \\ [n]+1 & (\text{otherwise}), \end{cases} \tag{28}$$

where $[\,n\,]$ represents the Gaussian symbol for any real number n.

In the case that the software failure-occurrence times distribution follows the geometric distribution, we can say that $W(Z)$ has the following properties:

$$\left.\begin{array}{l} W(Z+1) < W(Z) \\ W(0) = K\lambda p \\ W(\infty) = 0 \end{array}\right\}, \tag{29}$$

for any nonnegative interger $Z(\geq 0)$ since $0 < p < 1$. That is, we can see that $W(Z)$ is a monotonically decreasing function in terms of the testing-period $Z(\geq 0)$. Therefore, we can obtain the cost-optimal software release policies as follows:

[**Cost-Optimal Software Release Policy**]

Suppose that $c_2 > c_1 > 0$ and $c_3 > 0$.

(1) If $W(0) \leq c_3/(c_2 - c_1)$, then the cost-optimal software release period is $Z^* = 0$.

(2) If $c_3/(c_2 - c_1) < W(0)$, then we have the following only solution $Z = Z_0$ minimizing Eq.(26):

$$Z_0 = \frac{\log\left[c_3/\{(c_2 - c_1)K\lambda p\}\right]}{\log(1-p)}. \tag{30}$$

Thus, the optimal software release period $Z^* =< Z_0 >$.

5.2. *Cost-reliability-optimal software release policies*

We also discuss an optimal software release problem which takes both total software cost and reliability criteria into consideration simultaneously. In an actual software development, the software development manager has to spend and control the testing resources minimizing the total software cost and satisfying the software reliability requirement rather than only minimizing the cost in **5.1**.

Now, let R_0 $(0 < R_0 \leq 1)$ be the software reliability objective. By using the discrete software reliability function in Eq. (10), we can discuss the optimal software release policies which minimize the total expected software cost in Eq. (26) with satisfying the software reliability objective R_0. That is, the cost-reliability-optimal software release problem can be formulated as follows:

$$\left.\begin{array}{l} \text{minimize } C(Z) \\ \text{subject to } R(Z,h) \geq R_0,\ Z \geq 0 \end{array}\right\}. \tag{31}$$

Supposing h is a constant nonnegative integer, we can see that the discrete software reliability function, $R(Z, h)$, is a monotonically increasing function in terms of the testing-period Z when the software failure-occurrence times

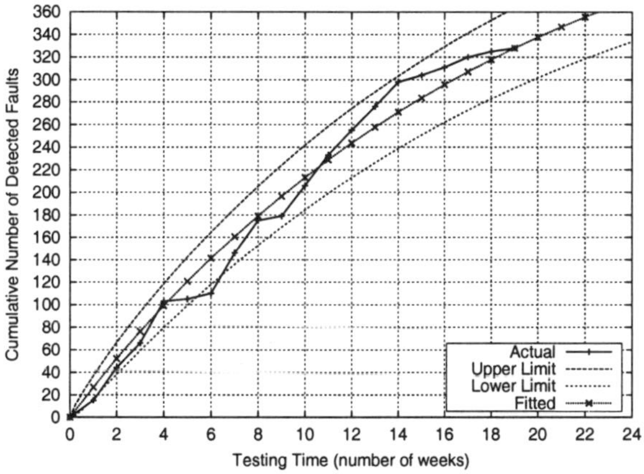

Figure 1. The estimated expected number of detected faults, $\widehat{\mathrm{E}}[N_B(i)]$, and its 95 % confidence limits.

follow the geometric distribution. Accordingly, if $R(0,h) < R_0$, then we have an only finite solution Z_1 satisfying $R(Z-1,h) < R_0$ and $R(Z,h) \geq R_0$. Furthermore, if $R(0,h) \geq R_0$, then $R(Z,h) \geq R_0$ for any nonnegative integer Z. In this case, we only have to discuss optimal software release policies based on only the cost criterion.

From the above discussion, the cost-reliability-optimal software release policies can be obtained as follows:

[Cost-Reliability-Optimal Software Release Policy]

Suppose that $c_2 > c_1 > 0$, $c_3 > 0$, $0 < R_0 < 1$, and $h \geq 0$.

(1) If $W(0) \leq \frac{c_3}{c_2-c_1}$ and $R_B(0,h) \geq R_0$, then the cost-reliability-optimal software release period $Z^* = 0$.

(2) If $W(0) \leq \frac{c_3}{c_2-c_1}$ and $R_B(0,h) < R_0$, then the cost-reliability-optimal software release period $Z^* = Z_1$.

(3) If $W(0) > \frac{c_3}{c_2-c_1}$ and $R_B(0,h) \geq R_0$, then the cost-reliability-optimal software release period $Z^* = < Z_0 >$.

(4) If $W(0) > \frac{c_3}{c_2-c_1}$ and $R_B(0,h) < R_0$, then the optimal software release period $Z^* = \max\{< Z_0 >, Z_1\}$.

6. Numerical Examples

We show numerical examples for a discrete SRGM with the effect of program size which is developed under our modeling framework in Eq.(4) by

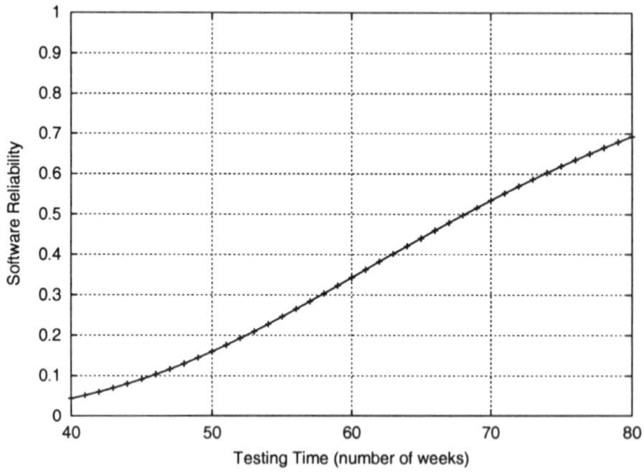

Figure 2. The estimated software reliability function, $\widehat{R_B}(i, 1)$.

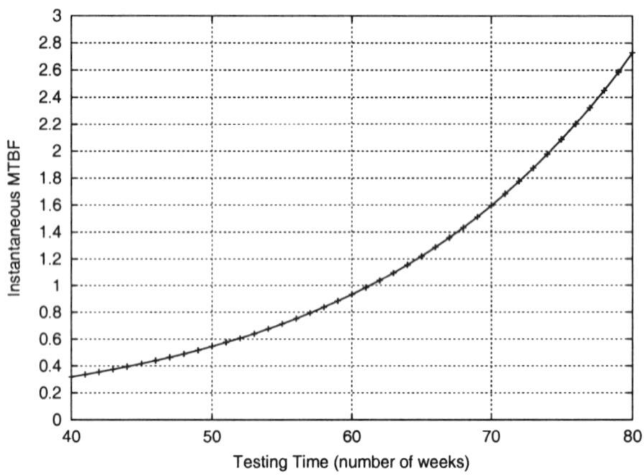

Figure 3. The estimated instantaneous MTBF, $\widehat{MTBF_B}(i)$.

using actual fault count data cited by Ohba [16]. The data consists of 19 data pairs $(t_i,\ y_i)(i = 0, 1, 2, \cdots, 19;\, t_{19} = 19 \text{ (weeks)}, y_{19} = 328)$. And the program size K of this software system is 1.317×10^6 (LOC). In this numerical examples, we assume that the software failure-occurrence times distribution follows the geometric distribution in Eq. (20). Applying the geometric distribution to the software failure-occurrence times distribution means that a probability that a software failure occurs at any testing-period decreases geometrically, which represents the case that the internal program

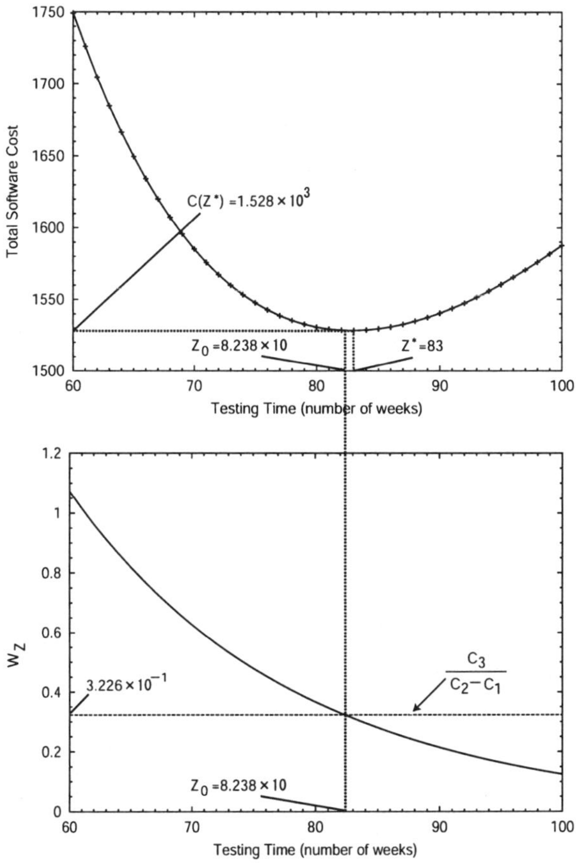

Figure 4. The optimal software release policy based on the cost criterion under $c_1 = 1$, $c_2 = 32$, and $c_3 = 10$.

structure is simple or the testing-skill of test-case designers is high[17].

Figure 1 depicts the estimated expected number of detected faults, $\widehat{\mathrm{E}}[N_B(i)]$, and its 95% confidence limits. As to Figure 1, the parameter estimates of λ and p have been obtained that $\widehat{\lambda} = 0 \cdot 340 \times 10^{-3}$ and $\widehat{p} = 0 \cdot 052$ by using the method of maximum-likelihood discussed in **4**. The $100\gamma\%$ confidence limits for $\widehat{\mathrm{E}}[N_B(i)]$ are derived as

$$\widehat{\mathrm{E}}[N_B(i)] \pm K_\gamma \sqrt{\widehat{\mathrm{Var}}[N_B(i)]}, \tag{32}$$

where K_γ indicates the $100(1+\gamma)/2$ percent point of the standard normal distribution[18]. By using the estimates, the expected initial fault content

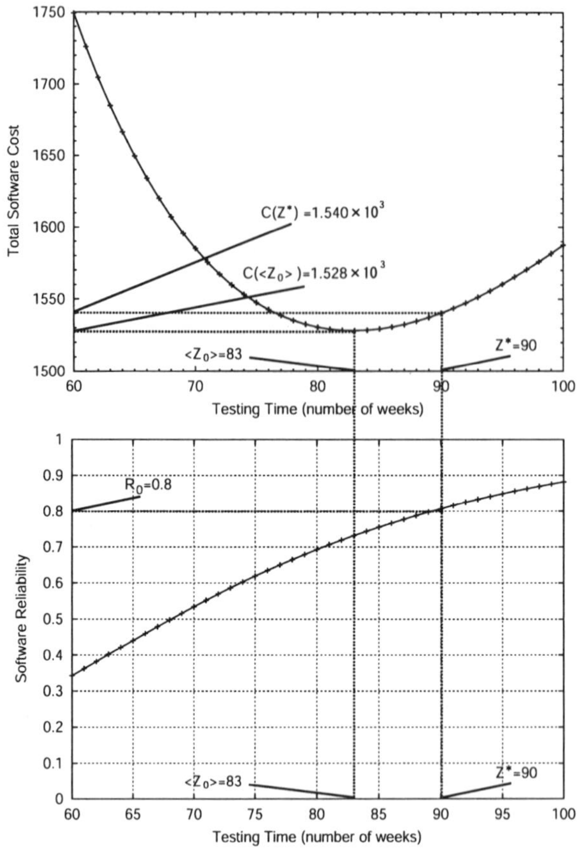

Figure 5. The optimal software release policy based on the cost and reliability criteria under $c_1 = 1$, $c_2 = 32$, and $c_3 = 10$.

can be estimated as $\widehat{K} \cdot \widehat{\lambda} \approx 513$. Figure 2 shows the estimated software reliability function, $\widehat{R_B}(i, 1)$, by using the parameter estimates. From Figure 2, we can estimate the software reliability at the 60-th testing-period to be about $0{\cdot}342$. And, Figure 3 also shows the estimated instantaneous MTBF, $\widehat{MTBF_B}(i)$. From Figure 3, we can estimate the instantaneous MTBF at the 60-th testing-period to be about $0 \cdot 933$ (weeks) or to be about 157 (hours).

Figure 4 depicts the cost-optimal software release policy for $c_1 = 1$, $c_2 = 32$, and $c_3 = 10$. In this case, **Cost-Optimal Software Release Policy** (2) is applied. And we can estimate that the cost-optimal software release period $Z^* = 83$ (weeks). Additionally, we show numerical examples for the cost-reliability-optimal software release policy. For the specific operational period $h = 1$ and the reliability objective $R_0 = 0 \cdot 8$, the cost-reliability-optimal software release problem can be discussed in the followings. Suppose that the cost-optimal software release policy have been discussed in the case of $c_1 = 1$, $c_2 = 32$, and $c_3 = 10$. We can estimate $Z_1 = 90$ because $R(89, 1) = 0.798 < R_0$ and $R(90, 1) = 0.807 > R_0$. Since $W(Z) > c_3/(c_2 - c_1)$ and $R(0, 1) = 2 \cdot 280 \times 10^{-12} < R_0$, Z^* is estimated as $Z^* = \max\{< Z_0 >, Z_1\} = \max\{83, 90\} = 90$ by using the **Cost-Reliability-Optimal Software Release Policy** (4) (see Figure 5). In Figure 5, we can show that the software development managers should estimate the optimal software release period by considering not only minimizing the total expected software cost but also satisfying the reliability objective simultaneously.

7. Concluding Remarks

We have discussed a modeling framework for discrete SRGMs with effect of program size. After that, we have derived generalized software reliability assessment measures, such as expectation and variance of the number of detected faults, a discrete software reliability function, and instantaneous and cumulative MTBFs. Then, we have also discussed a parameter estimation method for a discrete SRGMs developed by using our modeling framework. Additionally, as one of the interesting issue on project management of software development, we have discussed optimal software release policies under the criteria on simultaneous cost and reliability objective.

Our modeling framework enables us to obtain a suitable discrete SRGM easily by analyzing the software failure-occurrence times distribution in the actual testing-phase and applying its suitable probability distribution function to the modeling framework. In this chapter, though we have applied the geometric distribution as the software failure-occurrence times distribution, we plan to develop a plausible software failure-occurrence times distribution which enables us to describe the distribution of software failure-occurrence times flexibly. And then, we have to discuss the validity of our modeling framework for actual software reliability assessment in the future studies.

Acknowledgements

This work was supported in part by the Grant-in-Aid for Scientific Research (C), Grant No. 18510124, from the Ministry of Education, Sports, Science, and Technology of Japan.

References

1. J.D. Musa, D. Iannio, and K. Okumoto, *Software Reliability: Measurement, Prediction, Application*, (McGraw-Hill, New York, 1987).
2. S. Yamada, Software reliability models, in *Stochastic Models in Reliability and Maintenance*, S. Osaki ed. (Springer-Verlag, Berlin, 2002), 253–280.
3. H. Pham, *Software Reliability*, (Springer-Verlag, Singapore, 2000).
4. N. Langberg and N.D. Singpurwalla, A unification of some software reliability models, *SIAM Journal on Scientific Computing*, **6**(3) (1985) 781–790.
5. H. Okamura, A. Murayama, and T. Dohi, EM algorithm for discrete software reliability models: a unified parameter estimation method, *Proc. 8th IEEE International Symposium on High Assurance Systems Engineering*, 219–228 (2004).
6. T. Dohi, T. Matsuoka, and S. Osaki, An infinite server queueing model for assessment of the software reliability, *Electronics and Communications in Japan (Part 3)*, **85**(3), (2002) 43–51.
7. M. Kimura, S. Yamada, H. Tanaka, and S. Osaki, Software reliability measurement with prior-information on initial fault content, *Transactions of Information Processing Society Japan*, **34**(7), (1993) 1601–1609.
8. J.G. Shanthikumar, A general software reliability model for performance prediction, *Microelectronics and Reliability*, **21**(5), (1981) 671–682.
9. C.Y. Huang, M.R. Lyu, and S.Y. Kuo, A unified scheme of some nonhomogeneous Poisson process models for software reliability estimation, *IEEE Transactions on Software Engineering*, **29**(3), (2003) 261–269.
10. S. Inoue and S. Yamada, Generalized discrete software reliability modeling with effect of program size, *IEEE Transactions on Systems, Man, and Cybernetics, Part A*, to be published in 2006.
11. S. Yamada and S. Osaki, Discrete software reliability growth models, *Journal of Applied Stochastic Models and Data Analysis*, **1**(1), (1985) 65–77.
12. T. Kitaoka, S. Yamada, and S. Osaki, A discrete non-homogeneous error detection rate model for software reliability, *Transactions of IECE of JAPAN*, **E-69**(8), (1986) 859–865.
13. S. Osaki, *Applied Stochastic System Modeling* (Springer-Verlag, Berlin, Heidelberg, 1992).
14. K.S. Trivedi, *Probability and Statistics with Reliability, Queueing and Computer Science* (Second Ed.), (John Wiley & Sons, New York, 2002).
15. S. Yamada and S. Osaki, Cost-reliability optimal release policies for software systems, *IEEE Transactions on Reliability*, **R-34**(5), (1985) 422–424.
16. M. Ohba, Software reliability analysis models, *IBM Journal of Research and Development*, **R-34**, (1985) 422–424.
17. S. Inoue and S. Yamada, Testing-coverage dependent software reliability growth modeling, *International Journal of Reliability, Quality and Safety Engineering*, **11**(4), (2004) 303–312.
18. S. Yamada and S. Osaki, Software reliability growth modeling: Models and applications, *IEEE Transactions on Software Engineering*, **SE-11**(12), (1985) 1431–1437.

PART B

Maintenance

DISCRETE-TIME OPPORTUNISTIC REPLACEMENT POLICIES AND THEIR APPLICATION *

T. DOHI†, N. KAIO‡ and S. OSAKI††

†*Department of Information Engineering, Graduate School of Engineering*
Hiroshima University, Higashi-Hiroshima-shi 739–8527, Japan
E-mail: dohi@rel.hiroshima-u.ac.jp
‡*Department of Economic Informatics, Faculty of Economic Sciences*
Hiroshima Shudo University, Hiroshima-shi 731-3195, Japan
E-mail: kaio@shudo-u.ac.jp
††*Department of Information and Telecommunication Engineering*
Faculty of Mathematical Science and Information Engineering
Nanzan University, Seto-shi 489-0863, Japan
E-mail: shunji@nanzan-u.ac.jp

In this article, we consider discrete-time opportunistic age replacement models with application to a scheduled maintenance problem for a section switch to distribute the electric power. It is shown that a replacement model with three maintenance options can be classified into six models by taking account of priority of maintenance options. Further, we develop new stochastic models with probabilistic priority to unify six models with deterministic priority, and derive the optimal opportunistic age replacement policies minimizing the expected costs per unit time in the steady state. A numerical example with real failure data of section switches is presented.

1. Introduction

In this article, we consider discrete-time opportunistic age replacement models with application to a scheduled maintenance problem for a section switch which distributes the electric power to other places. The section switches equipped with telegraph poles have to be replaced preventively before they fail and the electric current is off over an extensive area. On

*The present research was partially supported by the Ministry of Education, Science, Sports and Culture, Grant-in-Aid for Scientific Research (B); Grant No. 16310116 (2004–2006), the Research Program 2005 under the Institute for Advanced Studies of the Hiroshima Shudo University and the Nanzan University Pache Research Subsidy I-A-2 for 2005.

the other hand, the section switch can be replaced if the telegraph pole is removed for any construction before its age has elapsed a threshold level. This problem is reduced to a simple opportunity-based age replacement model. In the earlier literature, many authors analyzed several opportunistic replacement models. Radner and Jorgenson [1] was the seminal work on the opportunistic replacement model for a single unit. Berg [2], Pullen and Thomas [3] and Zheng [4] discussed opportunity-triggered replacement policies for multiple-unit systems. Further, Dekker and Smeithink [5,6], Dekker and Dijkstra [7], and Zheng and Fard [8] extended the original models from a variety of standpoints. Recently, simple but somewhat different opportunity based age replacement models were considered by Iskandar and Sandoh [9,10]. In fact, their model [10] is essentially same as ours in this paper except that it is considered in a discrete-time setting.

In general, the discrete-time models are considered as trivial analogies of the continuous-time ones. First, Nakagawa and Osaki [11] formulated a discrete-time model for the classical age replacement problem. Kaio and Osaki [12,13,14] derived some discrete maintenance policies along the same line of Nakagawa and Osaki [11]. Nakagawa [15,16,17,18] summarized and generalized the discrete-time maintenance models by taking account of the significant concept of minimal repair. For the details of discrete models, see Kaio and Osaki [19]. The main reasons to consider the discrete-time model for the scheduled maintenance problem for a section switch are as follows. (i) In the Japanese electric power company under investigation, the failure time data of section switches are recorded as group data (the number of failures per year). (ii) It is not easy to carry out the preventive replacement schedule of section switches at the unit of week or month, since the service team is engaged in other works, too. From our questionnaire, it would be helpful for practitioners that the preventive replacement schedule should be determined roughly at the unit of year. These would motivate our discrete-time opportunistic age replacement model. In addition, we show in this paper that a replacement model with more than two maintenance options can be classified into some kinds of model by taking account of the priority of maintenance options. This implies that the discrete-time model has more delicate aspects for analysis than the continuous one.

The rest part of this article is organized as follows. In Section 2, the discrete-time opportunistic age replacement models are described with notation and assumptions. According to the priority of maintenance options, we introduce six models. In Section 3, the optimal age replacement times which minimize the expected costs per unit time in the steady state are

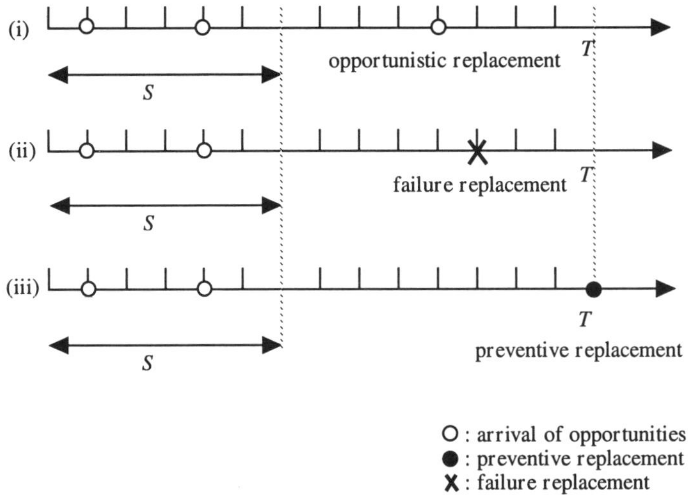

Figure 1. Configuration of the opportunity-based age replacement model.

derived for respective models. Section 4 develops the stochastic models with probabilistic priority to unify six models with deterministic priority. A numerical example with real data is presented in Section 5, where the optimal age replacement time measured by year is estimated. Finally, we conclude our discussion in Section 6.

2. Model Description

First, we reformulate a continuous-time model considered by Iskandar and Sandoh [10] to a discrete-time one. Let us consider the single-unit system with a non-repairable item in a discrete-time setting. Suppose that the time interval between opportunities for replacements, X, obeys the geometric distribution $\Pr\{X = x\} = g_X(x) = p(1-p)^{x-1}$ $(x = 1, 2, \cdots; 0 < p < 1)$ with survivor function $\Pr\{X \geq x\} = (1-p)^{x-1} = \overline{G}_X(x-1)$, mean $\mathrm{E}[X] = 1/p$ and variance $\mathrm{Var}[X] = (1-p)/p^2$, where in general $\overline{\phi}(\cdot) = 1 - \phi(\cdot)$. Then, the unit may be replaced at the first opportunity after elapsed time S, which is a non-negative integer, even if it does not fail. The failure time (lifetime), Y, follows the common probability mass function $\Pr\{Y = y\} = f_Y(y)$ $(y = 1, 2, \cdots)$ with survivor function $\Pr\{Y \geq y\} = \overline{F}_Y(y-1)$ and failure rate $r_Y(y) = f_Y(y)/\overline{F}_Y(y-1)$. Without any loss of generality, we assume that $f_Y(0) = g_X(0) = 0$. If the failure occurs before a pre-specified preventive replacement time T $(= 1, 2, \cdots)$, then the corrective replacement may be executed. On the other hand, if the unit does not fail up to the time T, then the preventive replacement may take place at time T. The configuration of the opportunistic age replacement model is depicted in Fig. 1.

The cost components under consideration are the following:

c_1 (> 0): corrective replacement cost per failure,
c_2 (> 0): cost for each preventive replacement,
c_3 (> 0): cost for each opportunistic replacement.

From the above notation, we make the following two assumptions:

Assumption (A-1): $c_1 > c_3 > c_2$

Assumption (A-2): $c_1 > c_2 > c_3$

It would be valid to assume that the corrective replacement cost is most expensive. The relationship between the preventive replacement cost and the opportunistic replacement one has to be ordered taking account of the economic justification.

Note that the discrete-time model above has to be treated carefully. At an arbitrary discrete point of time, the decision maker has to select one decision among three options; failure (corrective) replacement $\mathrm{F_a}$, preventive replacement $\mathrm{S_c}$ and opportunistic replacement $\mathrm{O_p}$. We introduce the following symbol for the priority relationship:

Definition 2.1: The option P has a priority to the option Q if $\mathrm{P} \succ \mathrm{Q}$.

From Definition 2.1, if two options occur at the same time point, the option with higher priority will be selected. In our model setting, it is possible to consider totally six different models as follows:

(1) Model 1: $\mathrm{S_c} \succ \mathrm{F_a} \succ \mathrm{O_p}$,
(2) Model 2: $\mathrm{F_a} \succ \mathrm{S_c} \succ \mathrm{O_p}$,
(3) Model 3: $\mathrm{S_c} \succ \mathrm{O_p} \succ \mathrm{F_a}$,
(4) Model 4: $\mathrm{O_p} \succ \mathrm{S_c} \succ \mathrm{F_a}$,
(5) Model 5: $\mathrm{F_a} \succ \mathrm{O_p} \succ \mathrm{S_c}$,
(6) Model 6: $\mathrm{O_p} \succ \mathrm{F_a} \succ \mathrm{S_c}$.

For Model 1, Model 2 and Model 5, the probability that the system is replaced at time n $(= 0, 1, 2, \cdots)$ is given by

$$h_1(n) = h_2(n) = h_5(n) = \begin{cases} f_Y(n) & (0 \le n \le S) \\ f_Y(n)\overline{G}_X(n-1-S) + \overline{F}_Y(n)g_X(n-S) & \\ & (S+1 \le n \le T-1) \\ \overline{F}_Y(T-1)\overline{G}_X(T-1-S) & (n = T) \\ 0 & (n \ge T+1). \end{cases} \quad (1)$$

In a fashion similar to Eq.(1), the probability that the system is replaced at time n $(= 0, 1, 2, \cdots)$ for the other models is obtained as

$$h_3(n) = h_4(n) = h_6(n) = \begin{cases} f_Y(n) & (0 \le n \le S) \\ f_Y(n)\overline{G}_X(n-S) + g_X(n-S)\overline{F}_Y(n-1) & \\ & (S+1 \le n \le T-1) \\ \overline{F}_Y(T-1)\overline{G}_X(T-1-S) & (n = T) \\ 0 & (n \ge T+1), \end{cases} \quad (2)$$

where $\sum_{n=0}^{\infty} h_j(n) = 1$ $(j = 1, \cdots, 6)$.

From Eqs. (1) and (2), the mean time length of one cycle, $A_j(T)$, for Model j $(= 1, \cdots, 6)$ are all same, that is, $A_1(T) = A_2(T) = A_3(T) = A_4(T) = A_5(T) = A_6(T)$, where

$$\begin{aligned} A_1(T) &= \sum_{n=0}^{S} n f_Y(n) + \sum_{n=S+1}^{T-1} n\Big\{ f_Y(n)\overline{G}_X(n-1-S) + \overline{F}_Y(n)g_X(n-S)\Big\} \\ &\quad + T\overline{F}_Y(T-1)\overline{G}_X(T-1-S) \\ &= \sum_{k=1}^{S} \overline{F}_Y(k-1) + \sum_{k=S+1}^{T} \overline{F}_Y(k-1)\overline{G}_X(k-S-1) \end{aligned} \quad (3)$$

are statistically independent of priorities.

On the other hand, the expected total costs during one cycle, $B_j(T)$, for Model j $(= 1, \cdots, 6)$ are given by

$$\begin{aligned} B_1(T) &= c_1 \sum_{n=0}^{S} f_Y(n) + c_1 \sum_{n=S+1}^{T-1} f_Y(n)\overline{G}_X(n-1-S) \\ &\quad + c_2\overline{F}_Y(T-1)\overline{G}_X(T-1-S) + c_3 \sum_{n=S+1}^{T-1} \overline{F}_Y(n)g_X(n-S), \end{aligned} \quad (4)$$

$$B_2(T) = c_1 \sum_{n=0}^{S} f_Y(n) + c_1 \sum_{n=S+1}^{T} f_Y(n)\overline{G}_X(n-1-S)$$

$$+c_2\overline{F}_Y(T)\overline{G}_X(T-1-S)+c_3\sum_{n=S+1}^{T-1}\overline{F}_Y(n)g_X(n-S), \tag{5}$$

$$B_3(T)=c_1\sum_{n=0}^{S}f_Y(n)+c_1\sum_{n=S+1}^{T-1}f_Y(n)\overline{G}_X(n-S)$$
$$+c_2\overline{F}_Y(T-1)\overline{G}_X(T-1-S)+c_3\sum_{n=S+1}^{T-1}\overline{F}_Y(n-1)g_X(n-S), \tag{6}$$

$$B_4(T)=c_1\sum_{n=0}^{S}f_Y(n)+c_1\sum_{n=S+1}^{T-1}f_Y(n)\overline{G}_X(n-S)$$
$$+c_2\overline{F}_Y(T-1)\overline{G}_X(T-S)+c_3\sum_{n=S+1}^{T}\overline{F}_Y(n-1)g_X(n-S), \tag{7}$$

$$B_5(T)=c_1\sum_{n=0}^{S}f_Y(n)+c_1\sum_{n=S+1}^{T}f_Y(n)\overline{G}_X(n-1-S)$$
$$+c_2\overline{F}_Y(T)\overline{G}_X(T-S)+c_3\sum_{n=S+1}^{T}\overline{F}_Y(n)g_X(n-S), \tag{8}$$

$$B_6(T)=c_1\sum_{n=0}^{S}f_Y(n)+c_1\sum_{n=S+1}^{T}f_Y(n)\overline{G}_X(n-S)+c_2\overline{F}_Y(T)\overline{G}_X(T-S)$$
$$+c_3\sum_{n=S+1}^{T}\overline{F}_Y(n-1)g_X(n-S), \tag{9}$$

respectively.

Then the expected costs per unit time in the steady state, $C_j(T)$, for Model j $(=1,2,\cdots,6)$ are, from the familiar renewal reward argument [20],

$$C_j(T)=\lim_{n\to\infty}\frac{\mathrm{E}[\text{total cost on }(0,n]]}{n}=\frac{B_j(T)}{A_j(T)}, \tag{10}$$

and the problem is to determine the optimal preventive replacement time T^* which minimizes the expected cost $C_j(T)$ for a fixed S.

When the scheduled maintenance problem for a section switch is considered, it is meaningful to assume that the variable S is determined in advance. Because the threshold age to start the opportunistic age replacement should be estimated from the efficiency and price of the section switch. Hence, throughout this article, we suppose that the variable S is fixed from any physical or economical reason.

3. Optimal Opportunistic Replacement Policies

In this section, we consider six models, Model 1 $\sim$ Model 6, and derive the respective optimal opportunistic age replacement policies which minimize the expected costs per unit time in the steady state. Define the non-linear functions:

$$q_1(T) = \frac{1}{1-p}\Big\{(c_1 - c_2)R_Y(T) + p(c_3 - c_2)\Big\}A_1(T) - B_1(T), \tag{11}$$

$$q_2(T) = \Big\{(c_1 - c_2)r_Y(T+1) + \frac{p(c_3 - c_2)}{1-p}\Big\}A_2(T) - B_2(T), \tag{12}$$

$$q_3(T) = \Big\{\Big[(c_1 - c_2) + \frac{p}{1-p}(c_3 - c_2)\Big]R_Y(T) + \frac{p}{1-p}(c_3 - c_2)\Big\}A_3(T) \\ -B_3(T), \tag{13}$$

$$q_4(T) = \Big\{(c_1 - c_2)R_Y(T) + p(c_3 - c_2)\Big\}A_4(T) - B_4(T), \tag{14}$$

$$q_5(T) = \Big\{\Big[(c_1 - c_2) + p(c_2 - c_3)\Big]r_Y(T+1) + p(c_3 - c_2)\Big\}A_5(T) - B_5(T) \tag{15}$$

$$q_6(T) = \Big\{(1-p)(c_1 - c_2)r_Y(T+1) + p(c_3 - c_2)\Big\}A_6(T) - B_6(T), \tag{16}$$

where

$$R_Y(T) = f_Y(T)/\overline{F}_Y(T). \tag{17}$$

Lemma 3.1. *The function $R_Y(T)$ is strictly increasing [decreasing] if the failure time distribution is strictly IFR (Increasing Failure Rate) [DFR (Deceasing Failure Rate)].*

Proof. Note first that $R_Y(T)$ is different from $r_T(T)$. From the definition, it turns out that

$$R_Y(T) = \frac{f_Y(T)}{\overline{F}_Y(T)} = \frac{\overline{F}_Y(T-1)}{\overline{F}_Y(T)} \cdot \frac{f_Y(T)}{\overline{F}_Y(T-1)} \\ = \Big\{\frac{\overline{F}_Y(T-1) - f_Y(T)}{\overline{F}_Y(T-1)}\Big\}^{-1} r_Y(T) = \{1 - r_Y(T)\}^{-1} r_Y(T). \tag{18}$$

A simple algebraic manipulation yields

$$R_Y(T+1) - R_Y(T) = \frac{r_Y(T+1) - r_Y(T)}{(1 - r_Y(T+1))(1 - r_Y(T))} \tag{19}$$

and the proof is completed. □

Theorem 3.1. *(i) For Model 1, Model 2 and Model 3, suppose that the failure time distribution is strictly IFR and the assumption (A-1) holds.*

(1) *If* $q_j(S+1) < 0$ *and* $q_j(\infty) > 0$ $(j = 1, 2, 3)$*, then there exists at least one (at most two) optimal preventive replacement time* T^* $(S+1 < T^* < \infty)$ *which satisfies* $q_j(T^* - 1) < 0$ *and* $q_j(T^*) \geq 0$.

(2) *If* $q_j(\infty) \leq 0$ $(j = 1, 2, 3)$*, then the optimal preventive replacement time is* $T^* \to \infty$ *and it is optimal to carry out either the failure replacement or the opportunistic one.*

(3) *If* $q_j(S+1) \geq 0$ $(j = 1, 2, 3)$*, then the optimal preventive replacement time is* $T^* = S + 1$ *and it is optimal to carry out either the failure replacement or the preventive one.*

(ii) For Model 1, Model 2 and Model 3, suppose that the failure time distribution is DFR and the assumption (A-1) holds. Then the optimal preventive replacement time is $T^* \to \infty$ *or* $T^* = S + 1$.

Theorem 3.2. *(i) For Models 4, 5 and 6, suppose that the failure time distribution is strictly IFR and the assumption (A-2) holds.*

(1) *If* $q_j(S+1) < 0$ *and* $q_j(\infty) > 0$ $(j = 4, 5, 6)$*, then there exists at least one (at most two) optimal preventive replacement time* T^* $(S+1 < T^* < \infty)$ *which satisfies* $q_j(T^* - 1) < 0$ *and* $q_j(T^*) \geq 0$.

(2) *If* $q_j(\infty) \leq 0$ $(j = 4, 5, 6)$*, then the optimal preventive replacement time is* $T^* \to \infty$.

(3) *If* $q_j(S+1) \geq 0$ $(j = 4, 5, 6)$*, then the optimal preventive replacement time is* $T^* = S + 1$.

(ii) For Models 4, 5 and 6, suppose that the failure time distribution is DFR and the assumption (A-2) holds. Then the optimal preventive replacement time is $T^* \to \infty$ *or* $T^* = S + 1$.

Proof. Here, we give the proof for Model 1. Taking the difference of $C_1(T)$, we have

$$C_1(T+1) - C_1(T) = \frac{\overline{F}_Y(T)\overline{G}_X(T-S)}{A_1(T+1)A_1(T)} q_1(T), \tag{20}$$

where

$$q_1(T) = \Delta B_1(T)A_1(T) - B_1(T)\Delta A_1(T), \tag{21}$$

$$\Delta A_1(T) = \{A_1(T+1) - A_1(T)\}/\{\overline{F}_Y(T)\overline{G}_X(T-S)\} = 1 \tag{22}$$

and

$$\Delta B_1(T) = \{B_1(T+1) - B_1(T)\}/\{\overline{F}_Y(T)\overline{G}_X(T-S)\}$$

$$= \frac{c_1 - c_2}{1-p} R_Y(T) + \frac{p(c_3 - c_2)}{1-p}. \tag{23}$$

Further taking the difference leads to

$$q_1(T+1) - q_1(T) = \frac{c_1 - c_2}{1-p}\Big\{R_Y(T+1) - R_Y(T)\Big\}A_1(T+1). \tag{24}$$

If the failure time distribution is strictly IFR and the assumption (A-1) holds, $q_1(T+1) - q_1(T) > 0$. Further, if $q_1(S+1) < 0$ and $q_1(\infty) > 0$, then the function $C_1(T)$ is strictly convex in T and there exists at least one (at most two) optimal preventive replacement time T^* $(S+1 < T^* < \infty)$ which satisfies $q_1(T^*-1) < 0$ and $q_1(T^*) \geq 0$. On the other hand, if $q_1(\infty) \leq 0$ and $q_1(S+1) \geq 0$, then the function $C_1(T)$ is monotonically decreasing and increasing, respectively, and the optimal preventive replacement times are $T^* \to \infty$ and $T^* = S+1$, where

$$C_1(S+1) = B_1(S+1)/A_1(S+1) = \Big(c_1 F_Y(S) + c_2 \overline{F}_Y(S)\Big) \Big/ \sum_{j=1}^{S+1} \overline{F}_Y(j-1), \tag{25}$$

$$C_1(\infty) = B_1(\infty)/A_1(\infty), \tag{26}$$

$$A_1(\infty) = \sum_{j=1}^{S} \overline{F}_Y(j-1) + \sum_{j=S+1}^{\infty} \overline{F}_Y(j-1)\overline{G}_X(j-S-1), \tag{27}$$

$$B_1(\infty) = c_1\Big\{F_Y(S) + \sum_{n=S+1}^{\infty} f_Y(n)\overline{G}_X(n-1-S)\Big\} + c_3 \sum_{n=S+1}^{\infty} \overline{F}_Y(n) g_X(n-S). \tag{28}$$

If the failure time distribution is DFR, then the function $C_1(T)$ is concave in T. Thus, if $C_1(S+1) < C_1(\infty)$, then $T^* = S+1$, otherwise, $T^* \to \infty$. The other proofs for Model 2 $\sim$ Model 6 are similar to the above. □

In Table 1, the relationship between six models and the corresponding necessary conditions of optimality are summarized. From this table, it is found that the optimal preventive replacement schedule for each model should be characterized under different cost assumptions.

Table 1. Classification of discrete-time opportunistic replacement models.

Model	Priority	Necessary conditions
Model 1	$\mathrm{S_c} \succ \mathrm{F_a} \succ \mathrm{O_p}$	$c_1 > c_2$
Model 2	$\mathrm{F_a} \succ \mathrm{S_c} \succ \mathrm{O_p}$	$c_1 > c_2$
Model 3	$\mathrm{S_c} \succ \mathrm{O_p} \succ \mathrm{F_a}$	$c_1 > c_2$ $c_3 > c_2$
Model 4	$\mathrm{O_p} \succ \mathrm{S_c} \succ \mathrm{F_a}$	$c_1 > c_2$
Model 5	$\mathrm{F_a} \succ \mathrm{O_p} \succ \mathrm{S_c}$	$c_1 > c_2$ $c_2 > c_3$
Model 6	$\mathrm{O_p} \succ \mathrm{F_a} \succ \mathrm{S_c}$	$c_1 > c_2$

4. Unified Models with Probabilistic Priority

In this section, we unify six replacement models considered in Section 2. Now suppose that one of the multiple maintenance options at any time may be selected with random priority. Under the assumption (A-1), define the probabilities p_a, p_b and p_c to select the priorities $\mathrm{S_c} \succ \mathrm{F_a} \succ \mathrm{O_p}$, $\mathrm{F_a} \succ \mathrm{S_c} \succ \mathrm{O_p}$ and $\mathrm{S_c} \succ \mathrm{O_p} \succ \mathrm{F_a}$, respectively, where $0 \le p_a \le 1$, $0 \le p_b \le 1$, $0 \le p_c \le 1$ and $p_a + p_b + p_c = 1$. Also, under the assumption (A-2), we define p_d, p_e and p_f to select the priorities $\mathrm{O_p} \succ \mathrm{S_c} \succ \mathrm{F_a}$, $\mathrm{F_a} \succ \mathrm{O_p} \succ \mathrm{S_c}$ and $\mathrm{O_p} \succ \mathrm{F_a} \succ \mathrm{S_c}$, respectively, where $0 \le p_d \le 1$, $0 \le p_e \le 1$, $0 \le p_f \le 1$ and $p_d + p_e + p_f = 1$. We call these two models with triplets (p_a, p_b, p_c) and (p_d, p_e, p_f) Model 7 and Model 8, respectively.

In Model 7 and Model 8, the probabilities that the system is replaced at time n $(= 0, 1, 2, \cdots)$ are given by

$$h_7(n) = \begin{cases} f_Y(n) & (0 \le n \le S) \\ (p_a + p_b)\{f_Y(n)\overline{G}_X(n-S-1) + g_X(n-S)\overline{F}_Y(n)\} & \\ +p_c\{f_Y(n)\overline{G}_X(n-S) + g_X(n-S)\overline{F}_Y(n-1)\} & \\ & (S+1 \le n \le T-1) \\ \overline{F}_Y(T-1)\overline{G}_X(T-1-S) \; (n = T) & \\ 0 & (n \ge T+1) \end{cases} \tag{29}$$

and

$$h_8(n) = \begin{cases} f_Y(n) & (0 \le n \le S) \\ (p_d + p_f)\{f_Y(n)\overline{G}_X(n-S) + g_X(n-S)\overline{F}_Y(n-1)\} & \\ +p_e\{f_Y(n)\overline{G}_X(n-1-S) + g_X(n-S)\overline{F}_Y(n)\} & \\ & (S+1 \le n \le T-1) \\ \overline{F}_Y(T-1)\overline{G}_X(T-1-S) \; (n = T) & \\ 0 & (n \ge T+1), \end{cases} \tag{30}$$

respectively, where $\sum_{n=0}^{\infty} h_7(n) = p_a + p_b + p_c = 1$ and $\sum_{n=0}^{\infty} h_8(n) = p_d + p_e + p_f = 1$.

The mean time lengths of one cycle and the expected total costs during one cycle for Model 7 and Model 8 are given by

$$A_7(T) = A_8(T) = \sum_{j=1}^{S} \overline{F}_Y(j-1) + \sum_{j=S+1}^{T} \overline{F}_Y(j-1)\overline{G}_X(j-S-1), \tag{31}$$

$$\begin{aligned} B_7(T) = c_1 \sum_{n=0}^{S} f_Y(n) + p_a\Big\{c_1 \sum_{n=S+1}^{T-1} f_Y(n)\overline{G}_X(n-S-1) \\ +c_2\overline{F}_Y(T-1)\overline{G}_X(T-S-1) + c_3 \sum_{n=S+1}^{T-1} \overline{F}_Y(n)g_X(n-S)\Big\} \\ +p_b\Big\{c_1 \sum_{n=S+1}^{T} f_Y(n)\overline{G}_X(n-S-1) + c_2\overline{F}_Y(T)\overline{G}_X(T-S-1) \\ +c_3 \sum_{n=S+1}^{T-1} \overline{F}_Y(n)g_X(n-S)\Big\} + p_c\Big\{c_1 \sum_{n=S+1}^{T-1} f_Y(n)\overline{G}_X(n-S) \\ +c_2\overline{F}_Y(T-1)\overline{G}_X(T-S-1) + c_3 \sum_{n=S+1}^{T-1} \overline{F}_Y(n-1)g_X(n-S)\Big\}, \end{aligned} \tag{32}$$

$$\begin{aligned} B_8(T) = c_1 \sum_{n=0}^{S} f_Y(n) + p_d\Big\{c_1 \sum_{n=S+1}^{T-1} f_Y(n)\overline{G}_X(n-S) \\ +c_2\overline{F}_Y(T-1)\overline{G}_X(T-S) + c_3 \sum_{n=S+1}^{T} \overline{F}_Y(n-1)g_X(n-S)\Big\} \\ +p_e\Big\{c_1 \sum_{n=S+1}^{T} f_Y(n)\overline{G}_X(n-S-1) + c_2\overline{F}_Y(T)\overline{G}_X(T-S) \\ +c_3 \sum_{n=S+1}^{T} \overline{F}_Y(n)g_X(n-S)\Big\} + p_f\Big\{c_1 \sum_{n=S+1}^{T} f_Y(n)\overline{G}_X(n-S) \\ +c_2\overline{F}_Y(T)\overline{G}_X(T-S) + c_3 \sum_{n=S+1}^{T} \overline{F}_Y(n-1)g_X(n-S)\Big\}, \end{aligned} \tag{33}$$

respectively.

Then the problem is to determine the optimal preventive replacement time T^* which minimizes the expected cost $TC_j(T)(j = 7, 8)$ for a fixed S, where

$$TC_j(T) = \frac{B_j(T)}{A_j(T)}. \tag{34}$$

Define the following non-linear functions:

$$q_7(T) = \Big\{\Big[p_a(c_1-c_2)/(1-p) + p_c\{(c_1-c_2)+p(c_3-c_2)/(1-p)\}\Big]R_Y(T)$$
$$+p_b(c_1-c_2)r_Y(T+1)+p(c_3-c_2)/(1-p)\Big\}A_7(T) - B_7(T), \tag{35}$$

$$q_8(T) = \Big\{p_d(c_1-c_2)R_Y(T) + \Big[p_e\{(c_1-c_2)+p(c_2-c_3)\}$$
$$+p_f(1-p)(c_1-c_2)\Big]r_Y(T+1)+p(c_3-c_2)\Big\}A_8(T) - B_8(T). \tag{36}$$

Theorem 4.1. *(i) For Models 7 and 8, suppose that the failure time distribution is strictly IFR and both the assumptions (A-1) and (A-2) hold.*

(1) If $q_j(S+1) < 0$ and $q_j(\infty) > 0$ $(j = 7, 8)$, then there exists at least one (at most two) optimal preventive replacement time T^* $(S+1 < T^* < \infty)$ which satisfies $q_j(T^*-1) < 0$ and $q_j(T^*) \geq 0$.

(2) If $q_j(\infty) \leq 0$ $(j = 7, 8)$, then the optimal preventive replacement time is $T^* \to \infty$.

(3) If $q_j(S+1) \geq 0$ $(j = 7, 8)$, then the optimal preventive replacement time is $T^* = S+1$.

(ii) For Models 7 and 8, suppose that the failure time distribution is DFR and both the assumptions (A-1) and (A-2) hold. Then the optimal preventive replacement time is $T^* \to \infty$ or $T^* = S+1$.

The proof is omitted for brevity.

From Theorem 4.1, the age replacement models with probabilistic priority involve the deterministic priority models as special cases. For instance, it is seen that Model 7 is reduced to Model 1 if $(p_a, p_b, p_c) = (1, 0, 0)$. Although the earlier models in the literature[11–18] assumed the priority unconsciously in accordance with the order of costs, the rigorous treatment for modeling will be needed if the priority is uncertain.

5. A Numerical Illustration

An empirical study for the preventive replacement of electric devices in the continuous-time setting was reported by Holland and McLean[21]. Here, we calculate the discrete optimal preventive replacement time T^* for section switches equipped with telegraph poles for a fixed S. The failure data

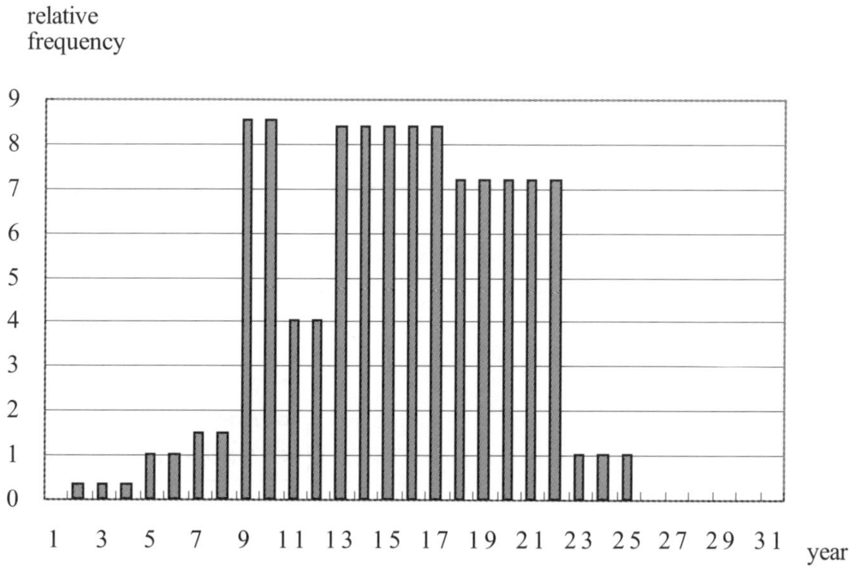

Figure 2. Failure time data of section switches.

used are recorded in Hiroshima City, Japan, during past twenty five years. Figure 2 illustrates the relative frequency of the failure data. Suppose that the (discrete) failure time obeys the following discrete Weibull distribution:

$$\Pr\{Y = y\} = (q)^{(y-1)^{\beta}} - (q)^{(y)^{\beta}}, \tag{37}$$

where $0 < q < 1$, $\beta > 0$ and $y = 1, 2, \cdots$. From the definition above, the survivor function and its failure rate are given by

$$\Pr\{Y \geq y\} = (q)^{(y-1)^{\beta}} \tag{38}$$

and

$$r_Y(y) = 1 - (q)^{(y)^{\beta} - (y-1)^{\beta}}, \tag{39}$$

respectively. This interesting discrete distribution was introduced first by Nakagawa and Osaki[22]. Later, Stein and Dattero[23] defined a somewhat different discrete Weibull distribution. Ali Khan, Khalique and

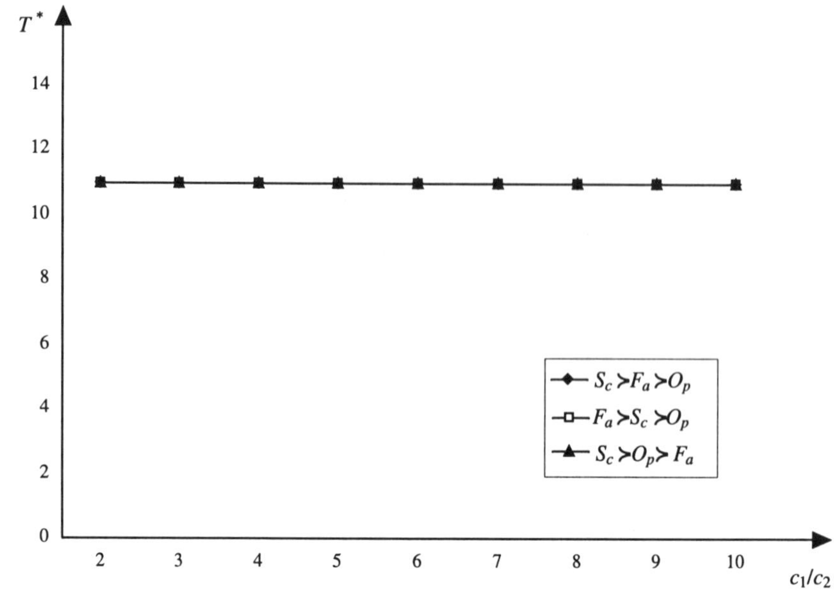

Figure 3. Dependence of the optimal preventive replacement time for varying cost ratio c_1/c_2: case of Models 1, 2 and 3 ($c_2 = 1$, $c_3 = 1.5$, $\hat{q} = 0.9995$, $\hat{\beta} = 2.8547$, $S = 10$, $p = 0.05$).

Abousammoh[24] developed an intuitive but simple parameter estimation method as well as the moment method and the maximum likelihood method for the original discrete Weibull distribution[22]. It is obvious that the discrete Weibull distribution is reduced to the geometric distribution when $\beta = 1$, and is much attractive to characterize the discrete failure mechanism.

Using the 112 failure data, we estimate two parameters, q and β, by the classical moment method[24]. From $\mathrm{E}[Y] = 13.40$ [year] and $\mathrm{Var}[Y] = 24.36$, we have

$$\hat{q} = 0.9995, \qquad \hat{\beta} = 2.8547.$$

Tables 2 and 3 present the optimal preventive replacement time and its associated minimum expected cost for varying S in the deterministic priority models. From thiese results, it is observed that the optimal replacement times for respective models tend to take same values in most cases. This is because the model under consideration is the discrete-time model, and the difference between model structures is not so remarkable. But, it can

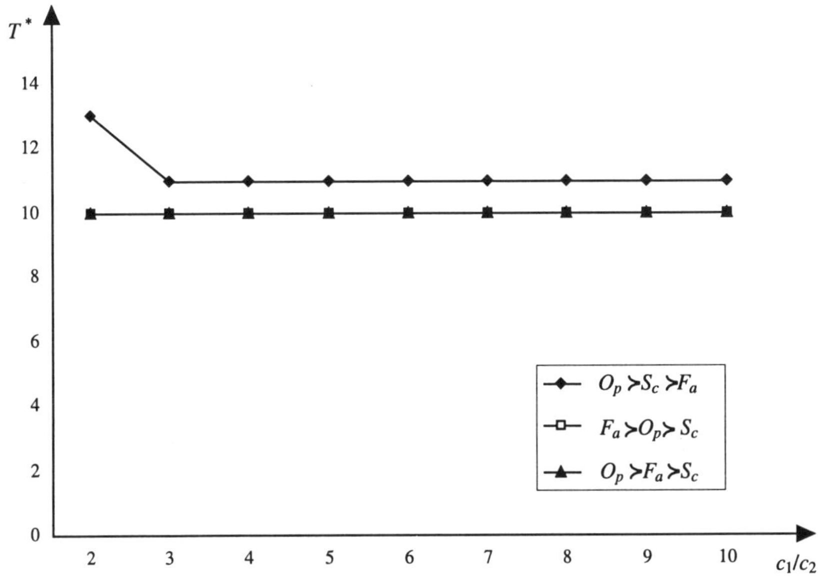

Figure 4. Dependence of the optimal preventive replacement time for varying cost ratio c_1/c_2: case of Models 4, 5 and 6 ($c_2 = 1$, $c_3 = 0.5$, $\hat{q} = 0.9995$, $\hat{\beta} = 2.8547$, $S = 10$, $p = 0.05$).

be seen that the corresponding expected cost values are rather different. In order to clarify the model performance, we investigate the dependence of the optimal preventive replacement time and its associated minimum expected cost for varying cost ratio c_1/c_2 in Figs. 3-6. From Fig. 3, the optimal replacement times for Models 1 $\sim$ 3 take the similar values, but Model 4 shows the different behavior from Models 5 and 6 when the cost ratio is relatively small in Fig. 4. Also, it is found in Fig. 5 that Model 2 overestimates the expected cost comparing with the other models. Thus, if one determines the preventive maintenance plan without taking account of the priority of maintenance options, the resulting decision making may not be suitable from the economical point of view.

Of our next concern is the investigation for the probabilistic priority models. Table 4 presents the optimal preventive replacement time and its associated minimum expected cost for probabilistic priority models. Comparing with Model 7 and Model 8, one sees that the remarkable difference between two preventive replacement times is not found. However, it is observed that the corresponding expected costs are slightly difference.

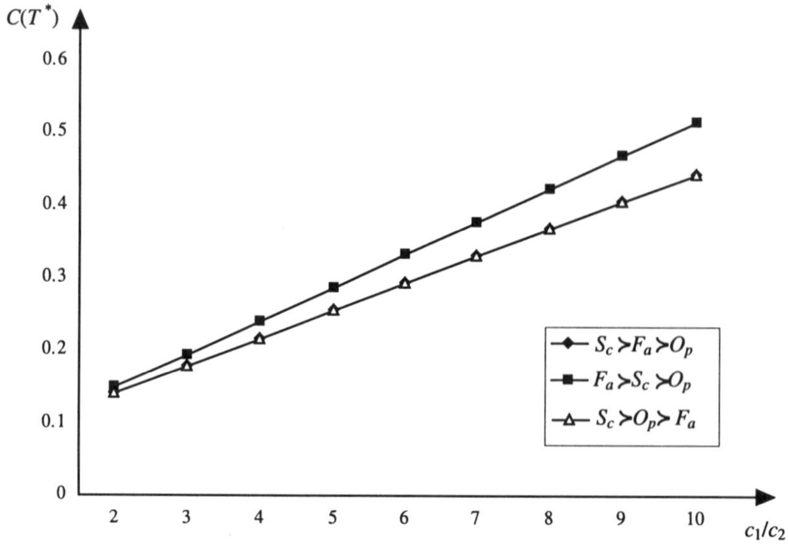

Figure 5. Dependence of the minimum expected cost for varying cost ratio c_1/c_2: case of Models 1, 2 and 3 ($c_2 = 1$, $c_3 = 1.5$, $\hat{q} = 0.9995$, $\hat{\beta} = 2.8547$, $S = 10$, $p = 0.05$).

Although we did not prove the convex property of the expected cost function in the threshold age S analytically, the results show that $TC_7(T)$ and $TC_8(T)$ may take their minimum values at $S = 6$. Therefore, if two dimensional optimization problem $\min_{T,S} TC_j(T, S)$ ($j = 7, 8$) has to be solved, any computation algorithm will be needed for calculation.

6. Concluding Remarks

In this paper, we have developed discrete-time opportunistic replacement models, taking account of the priority of multiple maintenance options. We have classified the underlying problem into six models with deterministic priority and have characterized the optimal preventive replacement policies under two different cost assumptions. Further, by introducing the probability that each maintenance option may be selected, the unified replacement models with probabilistic priority have been developed. In a numerical example, we have applied the results to a scheduled maintenance problem for section switches and have calculated the optimal preventive maintenance times based on the real failure time data.

In the future, the earlier discrete-time models have to be reformulated

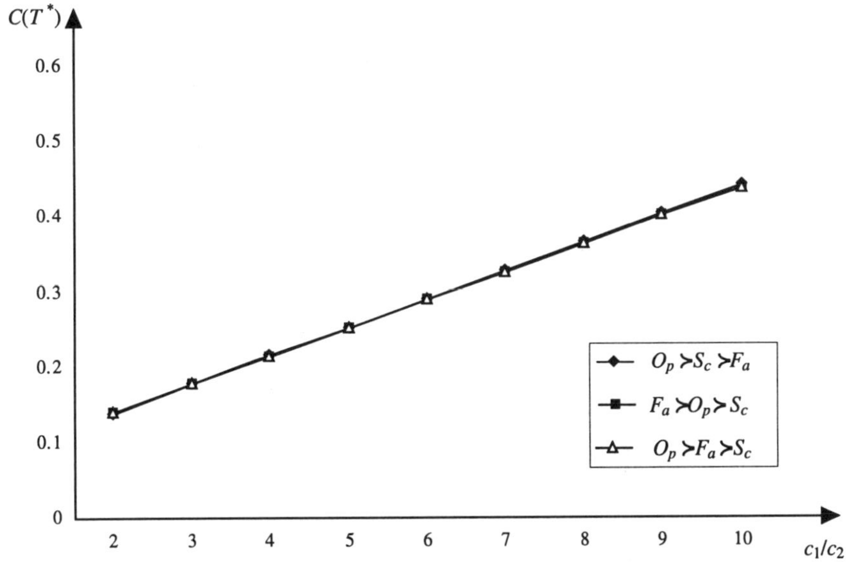

Figure 6. Dependence of the minimum expected cost for varying cost ratio c_1/c_2: case of Models 4, 5 and 6 ($c_2 = 1$, $c_3 = 0.5$, $\hat{q} = 0.9995$, $\hat{\beta} = 2.8547$, $S = 10$, $p = 0.05$).

Table 2. The optimal preventive replacement time and its associated minimum expected cost for deterministic priority models (Model 1, Model 2 and Model 3: c_1= 5, c_2=1, c_3=1.5, p=0.05, $\hat{q}$= 0.9995, $\hat{\beta}$= 2.8547).

	$S_c \succ F_a \succ O_p$		$F_a \succ S_c \succ O_p$		$S_c \succ O_p \succ F_a$	
S	T^*	$C(T^*)$	T^*	$C(T^*)$	T^*	$C(T^*)$
5	7	0.223806	6	0.252167	7	0.222749
6	7	0.219053	7	0.248751	7	0.219053
7	8	0.221330	8	0.252530	8	0.221330
8	9	0.229066	9	0.261118	9	0.229066
9	10	0.240683	10	0.272888	10	0.240683
10	11	0.254986	11	0.286630	11	0.254986
11	12	0.270996	12	0.301384	12	0.270996
12	13	0.287870	13	0.316363	13	0.287870
13	14	0.304864	14	0.330921	14	0.304864
14	15	0.321334	15	0.344542	15	0.321334
15	16	0.336741	16	0.356843	16	0.336741

Table 3. The optimal preventive replacement time and its associated minimum expected cost for deterministic priority models (Model 4, Model 5 and Model 6: $c_1 = 5$, $c_2=1$, $c_3=0.5$, $p=0.05$, $\hat{q}= 0.9995$, $\hat{\beta}= 2.8547$).

	$O_p \succ S_c \succ F_a$		$F_a \succ O_p \succ S_c$		$O_p \succ F_a \succ S_c$	
S	T^*	$C(T^*)$	T^*	$C(T^*)$	T^*	$C(T^*)$
5	7	0.212699	5	0.221487	5	0.221487
6	8	0.215658	6	0.216195	6	0.216195
7	8	0.218574	7	0.218542	7	0.218542
8	9	0.226743	8	0.226274	8	0.226274
9	10	0.238726	9	0.237854	9	0.237854
10	11	0.253346	10	0.252119	10	0.252119
11	12	0.269633	11	0.268116	11	0.268116
12	13	0.286749	12	0.285020	12	0.285020
13	14	0.303955	13	0.302103	13	0.302103
14	15	0.320610	14	0.318726	14	0.318726
15	16	0.336175	15	0.334346	15	0.334346

Table 4. The optimal preventive replacement time and its associated minimum expected cost for probabilistic priority models (p=0.05, $\hat{q}$=0.9995, $\hat{\beta}$=2.8547, $p_a = p_b = p_d = p_e = 0.3$, $p_c = p_f = 0.4$).

	Model 7 $(c_1, c_2, c_3) = (5, 1, 3)$		Model 8 $(c_1, c_2, c_3) = (5, 1, 0.5)$	
S	T^*	$TC_7(T^*)$	T^*	$TC_8(T^*)$
5	6	0.232840	5	0.223045
6	6	0.227376	6	0.217970
7	7	0.229077	7	0.220499
8	8	0.236255	8	0.228371
9	9	0.247302	9	0.240043
10	10	0.261014	10	0.254350
11	11	0.276409	11	0.270336
12	12	0.292651	12	0.287175
13	13	0.309011	13	0.304144
14	14	0.324861	14	0.320608
15	15	0.339678	15	0.336036

according to the concept of priority introduced in this paper. Then, the formulation should provide the practical meaning with applications. Especially, the estimation method for probabilities on priority should be developed in a consistent way.

References

1. R. Radner and D. W. Jorgenson, Opportunistic replacement of a single part in the presence of several monitored parts, *Management Science*, **10**, 70–84 (1963).
2. M. Berg, General trigger-off replacement procedures for two-unit systems, *Naval Research Logistics*, **25**, 15–29 (1978).
3. K. Pullen and M. Thomas, Evaluation of an opportunistic replacement policy for a 2-unit system, *IEEE Transactions on Reliability*, **R-53** 320–323, (1986).
4. X. Zheng, All opportunity-triggered replacement policy for multiple-unit systems, *IEEE Transactions on Reliability*, **R-44**, 648–652 (1995).
5. R. Dekker and E. Smeithink, Opportunity-based block replacement, *European Journal of Operational Research*, **53**, 46–63 (1991).
6. R. Dekker and E. Smeithink, Preventive maintenance at opportunities of restricted duration, *Naval Research Logistics*, **41**, 335–353 (1994).
7. R. Dekker and M. C. Dijkstra, Opportunity based age replacement: exponentially distributed times between opportunities, *Naval Research Logistics*, **39**, 175–190 (1992)
8. X. Zheng and N. Fard, A maintenance policy for repairable systems based on opportunistic failure-rate tolerance, *IEEE Transactions on Reliability*, **R-40**, 237–244 (1991).
9. B. P. Iskandar and H. Sandoh, An opportunity-based age replacement policy considering warranty, *International Journal of Reliability, Quality and Safety Engineering*, **6**, 229–236 (1999).
10. B. P. Iskandar and H. Sandoh, An extended opportunity-based age replacement policy, *Revue Francaise d'Automatique, Informatique et Recherche Operationnelle* (*Recherche operationnelle/Operations Research*), **34**, 145–154 (2000).
11. T. Nakagawa and S. Osaki, Discrete time age replacement policies, *Operational Research Quarterly*, **28**, 881–885 (1977).
12. N. Kaio and S. Osaki, Discrete-time ordering policies, *IEEE Transactions on Reliability*, **R-28**, 405–406 (1979).
13. N. Kaio and S. Osaki, Discrete time ordering policies with minimal repair, *Revue Francaise d'Automatique, Informatique et Recherche Operationnelle* (*Recherche operationnelle/Operations Research*), **14**, 257-263 (1980).
14. N. Kaio and S. Osaki, A discrete repair limit policy, *Advances in Management Science*, **1**, 157–160 (1982).
15. T. Nakagawa, A summary of discrete replacement policies, *European Journal of Operational Research*, **17**, 382–392 (1984).
16. T. Nakagawa, Optimal policy of continuous and discrete replacement with minimal repair at failure, *Naval Research Logistics Quarterly*, **31**, 543–550 (1984).
17. T. Nakagawa, Continuous and discrete age replacement policies, *Journal of Operational Research Society*, **36**, 147–154 (1985).
18. T. Nakagawa, Modified discrete preventive maintenance policies, *Naval Research Logistics Quarterly*, **33**, 703–715 (1986).

19. N. Kaio and S. Osaki, Review of discrete and continuous distributions in replacement models, *International Journal of Systems Science*, **19**, 171–177 (1988).
20. S. M. Ross, *Applied Probability Models with Optimization Applications*, Holden–Day, San Francisco (1970).
21. C. W. Holland and R. A. McLean, Applications of replacement theory, *AIIE Transactions*, **7**, 42–47 (1975).
22. T. Nakagawa and S. Osaki, The discrete Weibull distribution, *IEEE Transactions on Reliability*, **R-24**, 300–301 (1975).
23. W. E. Stein and R. Dattero, A new discrete Weibull distribution,*IEEE Transactions on Reliability*, **R-33**, 196–197 (1984).
24. M. S. Ali Khan, A. Khalique and A. M. Abousammoh, On estimating parameters in a discrete Weibull distribution, *IEEE Transactions on Reliability*, **R-38**, 348-350 (1989).

RELIABILITY CONSIDERATION OF WINDOW FLOW CONTROL SCHEME FOR A COMMUNICATION SYSTEM WITH EXPLICIT CONGESTION NOTIFICATION

MITSUTAKA KIMURA

Department of International Cultural Studies,
Gifu City Women's College
7–1 Hitoichiba Kita-matchi Gifu City 501–0192, Japan
E-mail: kimura@gifu-cwc.ac.jp

MITSUHIRO IMAIZUMI

College of Business Administration,
Aichi Gakusen University
1 Shiotori, Ohike-cho, Toyota City, Aichi 471–8532, Japan
E-mail: imaizumi@gakusen.ac.jp

KAZUMI YASUI

Faculty of Management and Information Science,
Aichi Institute of Technology,
1247 Yachigusa, Yagusa-cho, Toyota City, Aichi 470–0392, Japan
E-mail: yasuilab@aitech.ac.jp

In the packet delivery process, packets are sometimes dropped by network congestion (packet loss). Several authors have studied some protocols for dissolving packet loss. For example, a window flow control scheme which sets the window to half of the first window size when a sender detects congestion, has already been considered. Recently, a window flow control scheme with Explicit Congestion Notification, by which a sender detects congestion during connection, has been proposed and several authors have shown that ECN mechanisms is effective by simulation. This paper considers a stochastic model of a window flow control scheme for a communication system with Explicit Congestion Notification. The mean time until the data transmission succeeds is analytically derived and an optimal policy which maximizes the throughput is discussed. Finally, numerical examples are given.

1. Introduction

As the Internet has been widely used, its network scheme has been urgently needed for a high reliable communication. For example, some packets may be discarded at a receiver due to buffer overflow. The window flow control mechanism to defuse this situation has been implemented [12]. That is, a receiver can throttle a sender by specifying a limit on the amount of data that it can transmit. The limit is determined by a window size at a receiver. On the other hand, a problem of packet loss is sometimes caused by network congestion. In order to defuse this case, some protocols, such that when a sender detects congestion, it sets to half of the first window size, have been already proposed [23].

Congestion is detected by packet drops in current TCP/IP networks, and dropped packets are detected either from the receipt of three duplicate acknowledgements or after time out of a retransmit timer [1~3]. For such detection of congestion, unnecessary packet drops can result in unnecessary delays for the receiver [4]. ECN (Explicit Congestion Notification) mechanisms prevent unnecessary packet drops. That is, routers set the ECN bit in packet headers when the average queue size exceeds a certain threshold and the sender detects incipient congestion during connection [4]. In Floyd et al., they have shown that some advantages of ECN mechanism is in avoiding unnecessary packet drops by simulation [4 ~ 6].

This paper considers a stochastic model of a communication system using a window flow control scheme with ECN. Further, the mean time until the data transmission succeeds is analytically derived and an optimal policy which maximizes the throughput is discussed: When the server receives the request for data packets from a client, the server makes a connection. Then, the server and the client confirm whether the ECN bit is set. If it is not set, the number of packets, which corresponds to a window size, is successively transmitted to a client by a web server. If it is set, the server notices that congestion has happened in the network and the number of packets, which corresponds to half of the first window size, are transmitted. When the client has accepted all packets correctly, the transmission succeeds. The mean time until packet transmissions succeed is obtained. Further, an optimal policy which maximizes the amount of packets per unit of time until the transmission succeeds is analytically discussed. Finally, numerical examples are given.

2. Model and Analysis

We consider a communication system which consists of several clients and a web server, and formulate the stochastic model as follows:

(1) Congestion in a network system occurs intermittently and is disappear. Congestion happens in the network according to an exponential distribution $(1 - e^{-\lambda t})(0 < \lambda < \infty)$ and continues according to an exponential distribution $(1 - e^{-\beta t})(0 < \beta < \infty)$. We define the following states of a network system:

State 0: No congestion occurs and the network system is in a normal condition.

State 1: Congestion occurs.

The network system states defined above form a two-state Markov process [7], where a state transition diagram of a communication system is shown in Figure 1.

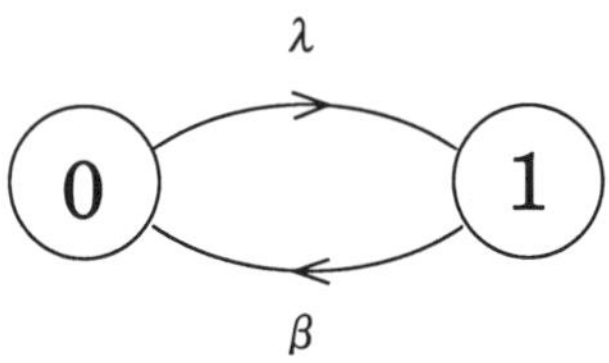

Figure 1. A state transition diagram of a communication system.

Thus, we have the following probabilities under the initial condition that $P_{00}(0) = P_{11}(0) = 1, P_{01}(0) = P_{10}(0) = 0$:

$$P_{00}(t) \equiv \frac{\beta}{\lambda+\beta} + \frac{\lambda}{\lambda+\beta} e^{-(\lambda+\beta)t},$$
$$P_{11}(t) \equiv \frac{\lambda}{\lambda+\beta} + \frac{\beta}{\lambda+\beta} e^{-(\lambda+\beta)t},$$
$$P_{01}(t) = 1 - P_{00}(t),$$
$$P_{10}(t) = 1 - P_{11}(t),$$

where, $P_{i,j}(t)$ are probabilities that the network system is in state $i(i = 0, 1)$ at time 0 and state $j(j = 0, 1)$ at time $t(> 0)$.

(2) A client transmits the request for data packets to the server. The request information is included in a window size, and the request

requires a time according to the general distribution $A(t)$ with mean a.

(3) The server establishes connection when a client requests data packets. Then, the server and the client confirm whether the ECN bit is set. That is, when congestion happens in the network, routers have set the ECN bit in the packet header. We assume the probability that ECN bit is set, is α. The server transmits the notification for connection completion to the client, and the notification requires the time according to an exponential distribution $A_1(t)$ with mean a_1. The client transmits the acknowledgement for the notification to the server, and the acknowledgement requires the time according to an exponential distribution $A_2(t)$ with mean a_2.

(i) The data transmission is implemented by the Selective-Repeat Protocol which is the usual retransmission control between the server and the client [8].

(ii) If the ECN bit is not set, the number of packets n_1, which corresponds to a window size, is successively transmitted to the client from the server. Then, if no congestion occurs, the probability that a packet loss occurs is p_0 $(0 < p_0 < 1)$. If congestion occurs, the probability that a packet loss occurs is p_1 $(> p_0)$.

(a) When the client has received n_1 packets correctly, it returns ACK. When the server has received NAK, the retransmission for only loss packets is made. The time the server takes to transmit the last packet to receive ACK or NAK has a general distribution $D(t)$ with mean d.

(b) If the retransmission has failed at k times again, the server interrupts, and the transmission are made again from the beginning of its initial state after a constant time μ where $G(t) \equiv 0$ for $t < \mu$ and 1 for $t \geq \mu$, because the ECN bit is needed to be checked intermittently.

(iii) If ECN bit is set, $n_2(< n_1)$ packets, which corresponds to half of the first window size, are transmitted. Then, the probability that a packet loss occurs is p_0.

(c) If the server has received ACK for the first time, the remaining packets n_2 are transmitted again. If the server has received ACK for all packets n_1, the transmission succeeds.

(iv) The process of editing and transmitting the data requires the time according to a general distribution $B(t)$ with mean b.

Under the above assumptions, we define the following states of the system:

State 2: System begins to operate.
State 3: Connection establishment from the client begins.
State 4: n_1 packet transmission begins (no congestion occurs and the network system is in a normal condition).
State 5: n_1 packet transmission begins (congestion occurs and no ECN bit has been set).
State 6: n_2 packet transmission begins (congestion occurs and ECN bit has been set).
State F: Retransmission fails k times and interrupted.
State S_2: n_2 packet transmission of first time succeeds and second time n_2 packet transmission begins.
State S_1: n_1 packet transmission succeeds.

The system states defined above form a Markov renewal process [78], where S_1 is an absorbing state. A transition diagram between system states is shown in Figure 1.

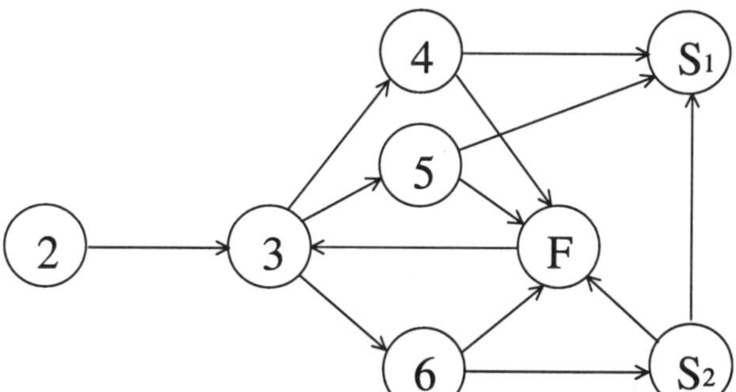

Figure 2. Transition diagram between system states.

Transition probabilities $Q_{i,j}(t)$ from state $i(i = 3, 4, 5, 6, S_2)$ to state $j(j = 4, 5, 6, F, S_1, S_2)$ are given by the following equations:

$$Q_{3,4}(t) = \left[\int_0^t P_{00}(x)dA_1(x)\right] * \left[\int_0^t P_{00}(x)dA_2(x)\right], \tag{1}$$

$$\begin{aligned} Q_{3,5}(t) = &\left[(1-\alpha)\int_0^t P_{01}(x)dA_1(x)\right] * \left[\int_0^t P_{10}(x)dA_2(x)\right] \\ &+ \left[\int_0^t P_{00}(x)dA_1(x)\right] * \left[(1-\alpha)\int_0^t P_{01}(x)dA_2(x)\right] \\ &+ \left[(1-\alpha)\int_0^t P_{01}(x)dA_1(x)\right] * \left[(1-\alpha)\int_0^t P_{11}(x)dA_2(x)\right], \end{aligned} \tag{2}$$

$$\begin{aligned} Q_{3,6}(t) = &\left[\alpha\int_0^t P_{01}(x)dA_1(x)\right] * \left[\int_0^t P_{10}(x)dA_2(x)\right] + \left[\int_0^t P_{00}(x)dA_1(x)\right] \\ &* \left[\alpha\int_0^t P_{01}(x)dA_2(x)\right] + \left[\alpha\int_0^t P_{01}(x)dA_1(x)\right] * \left[(1-\alpha)\int_0^t P_{11}(x)dA_2(x)\right] \\ &+ \left[(1-\alpha)\int_0^t P_{01}(x)dA_1(x)\right] * \left[\alpha\int_0^t P_{11}(x)dA_2(x)\right] \\ &+ \left[\alpha\int_0^t P_{01}(x)dA_1(x)\right] * \left[\alpha\int_0^t P_{11}(x)dA_2(x)\right], \end{aligned} \tag{3}$$

$$Q_{4,S_1}(t) \equiv Q_S(t|n_1, p_0, k), \tag{4}$$

$$Q_{5,S_1}(t) \equiv Q_S(t|n_1, p_1, k), \tag{5}$$

$$Q_{6,S_2}(t) = Q_{S_2,S_1}(t) \equiv Q_S(t|n_2, p_0, k), \tag{6}$$

$$Q_{4,F}(t) \equiv Q_F(t|n_1, p_0, k), \tag{7}$$

$$Q_{5,F}(t) \equiv Q_F(t|n_1, p_1, k), \tag{8}$$

$$Q_{6,F}(t) = Q_{S_2,F}(t) \equiv Q_F(t|n_2, p_0, k), \tag{9}$$

where,

$$
\begin{aligned}
Q_S(t|n,p,k) \equiv{}& (1-p)^n B^{(n)}(t) * D(t) \\
&+ \sum_{m_1=1}^{n} \binom{n}{m_1} p^{m_1}(1-p)^{n-m_1} B^{(n)}(t) * D(t) * (1-p)^{m_1} B^{(m_1)}(t) * D(t) \\
&+ \sum_{m_1=1}^{n} \binom{n}{m_1} p^{m_1}(1-p)^{n-m_1} B^{(n)}(t) * D(t) \\
&\quad * \sum_{m_2=1}^{m_1} \binom{m_1}{m_2} p^{m_2}(1-p)^{m_1-m_2} B^{(m_1)}(t) * D(t) * (1-p)^{m_2} B^{(m_2)}(t) * D(t) \\
&+ \\
&\vdots \\
&+ \\
&+ \sum_{m_1=1}^{n} \binom{n}{m_1} p^{m_1}(1-p)^{n-m_1} B^{(n)}(t) * D(t) \\
&\quad * \sum_{m_2=1}^{m_1} \binom{m_1}{m_2} p^{m_2}(1-p)^{m_1-m_2} B^{(m_1)}(t) * D(t) * \cdots \\
&\quad * \sum_{m_{k-1}=1}^{m_{k-2}} \binom{m_{k-2}}{m_{k-1}} p^{m_{k-1}}(1-p)^{m_{k-2}-m_{k-1}} B^{(m_{k-2})}(t) * D(t) \\
&\quad * (1-p)^{m_{k-1}} B^{(m_{k-1})}(t) * D(t),
\end{aligned} \tag{10}
$$

$$
\begin{aligned}
Q_F(t|n,p,k) \equiv{}& \sum_{m_1=1}^{n} \binom{n}{m_1} p^{m_1}(1-p)^{n-m_1} B^{(n)}(t) * D(t) * \cdots \\
& * \sum_{m_{k-1}=1}^{m_{k-2}} \binom{m_{k-2}}{m_{k-1}} p^{m_{k-1}}(1-p)^{m_{k-2}-m_{k-1}} B^{(m_{k-2})}(t) * D(t) \\
& * [1-(1-p)^{m_{k-1}}] B^{(m_{k-1})}(t) * D(t),
\end{aligned} \tag{11}
$$

$\Phi^{(i)}(t)$ is the i-fold convolution of $\Phi(t)$ and $\Phi^{(i)}(t) \equiv \Phi^{(i-1)}(t) * \Phi(t)$, $\Phi_1(t) * \Phi_2(t) \equiv \int_0^t \Phi_2(t-u)d\Phi_1(u)$, $\Phi^{(0)}(t) \equiv 1$. The asterisk mark denotes the Stieltjes convolusion. First, we derive the mean time ℓ_{2,S_1} until n_1 packets transmission succeeds. Let $H_{2,S_1}(t)$ be the time distribution from state 2 to state S_1 . Then we have

$$H_{2,S_1}(t) = A(t) * \sum_{i=1}^{\infty} \Bigg\{ \Big[Q_{3,4}(t) * Q_{4,F}(t) + Q_{3,5}(t) * Q_{5,F}(t) + Q_{3,6}(t) * \Big[Q_{6,F}(t) + Q_{6,S_2}(t) * Q_{S_2,F}(t) \Big] \Big] * G(t) \Bigg\}^{(i-1)} * \Big[Q_{3,4}(t) * Q_{4,S_1}(t) + Q_{3,5}(t) * Q_{5,S_1}(t) + Q_{3,6}(t) * Q_{6,S_2}(t) * Q_{S_2,S_1}(t) \Big]. \tag{12}$$

Thus, Laplace-Stieltjes (LS) transform $h_{2,S_1}(s)$ of $H_{2,S_1}(t)$ in (12) is

$$h_{2,S_1}(s) = \frac{a(s)\Big[q_{3,4}(s) q_{4,S_1}(s) + q_{3,5}(s) q_{5,S_1}(s) + q_{3,6}(s) q_{6,S_2}(s) q_{S_2,S_1}(s) \Big]}{1 - \Big\{ q_{3,4}(s) q_{4,F}(s) + q_{3,5}(s) q_{5,F}(s) + q_{3,6}(s) \Big[q_{6,F}(s) + q_{6,S_2}(s) q_{S_2,F}(s) \Big] \Big\} g(s)}$$

$$= \frac{a(s)\Big[q_{3,4}(s) q_S(s|n_1,p_0,k) + q_{3,5}(s) q_S(s|n_1,p_1,k) + q_{3,6}(s) \Big[q_S(s|n_2,p_0,k) \Big]^2 \Big]}{1 - \left\{ \begin{array}{l} q_{3,4}(s) q_F(s|n_1,p_0,k) + q_{3,5}(s) q_F(s|n_1,p_1,k) \\ + q_{3,6}(s) q_F(s|n_2,p_0,k) \Big[1 + q_S(s|n_2,p_0,k) \Big] \end{array} \right\} g(s)}, \tag{13}$$

where $\phi(s) \equiv \int_0^\infty e^{-st} d\Phi(t)$.

By a method similar to [8], Laplace-Stieltjes (LS) transforms $q_{i,j}(s)(i = 3, j = 4, 5, 6)$,
$q_S(s|n,p,k)$ and $q_F(s|n,p,k)$ of transition probabilities $Q_{i,j}(t)(i = 3, j = 4, 5, 6)$, $Q_S(s|n,p,k)$ and $Q_F(s|n,p,k)$ are given by the following equations :

$$q_{3,4}(s) = a_1 a_2 \left[\widetilde{P}_{0,0}(s+a_1)\widetilde{P}_{0,0}(s+a_2) \right], \tag{14}$$

$$q_{3,5}(s) = a_1 a_2 \begin{bmatrix} (1-\alpha)\widetilde{P}_{0,1}(s+a_1)\widetilde{P}_{1,0}(s+a_2) + (1-\alpha)\widetilde{P}_{0,0}(s+a_1)\widetilde{P}_{0,1}(s+a_2) \\ + (1-\alpha)^2\widetilde{P}_{0,1}(s+a_1)\widetilde{P}_{1,1}(s+a_2) \end{bmatrix}, \tag{15}$$

$$q_{3,6}(s) = a_1 a_2 \begin{bmatrix} \alpha\widetilde{P}_{0,1}(s+a_1)\widetilde{P}_{1,0}(s+a_2) + \alpha\widetilde{P}_{0,0}(s+a_1)\widetilde{P}_{0,1}(s+a_2) \\ + [2\alpha(1-\alpha)+\alpha^2]\widetilde{P}_{0,1}(s+a_1)\widetilde{P}_{1,1}(s+a_2) \end{bmatrix}, \tag{16}$$

$$\begin{aligned} q_S(s|n,p,k) &= [(1-p)b(s)]^n \sum_{j=1}^{k} [d(s)]^j \left[\left(\sum_{i=0}^{j-1} [pb(s)]^i \right)^n - \left(\sum_{i=0}^{j-2} [pb(s)]^i \right)^n \right] \\ &\quad (k = 1, 2, \cdots), \end{aligned} \tag{17}$$

$$\begin{aligned} q_F(s|n,p,k) &= [b(s)]^n [d(s)]^k \left[\left((1-p)\sum_{i=0}^{k-2} [pb(s)]^i + [pb(s)]^{k-1} \right)^n - \left((1-p)\sum_{i=0}^{k-1} [pb(s)]^i \right)^n \right], \\ &\quad (k = 1, 2, \cdots), \end{aligned} \tag{18}$$

where,

$$\widetilde{P}_{i,j}(s) \equiv \int_0^\infty e^{-st} P_{i,j}(t)dt, (i,j = 0,1)$$

Hence, the mean time ℓ_{2,S_1} is given by

$$\ell_{2,S_1} = \int_0^\infty t dH_{2,S_1}(t) \equiv \lim_{s\to 0} \frac{-dh_{2,S_1}(s)}{ds}$$

$$= (a-\mu) + \frac{\begin{bmatrix} \mu + \frac{1}{a_1} + \frac{1}{a_2} + b \begin{bmatrix} n_1 \left\{ q_{3,4}(0) \sum_{i=0}^{k-1} p_0{}^i + q_{3,5}(0) \sum_{i=0}^{k-1} p_1{}^i \right\} \\ + n_2 q_{3,6}(0)[1 + (1-p_0{}^k)^{n_2}] \sum_{i=0}^{k-1} p_0{}^i \end{bmatrix} \\ + d \begin{bmatrix} q_{3,4}(0)[k - \sum_{i=0}^{k-1}(1-p_0{}^i)^{n_1}] + q_{3,5}(0)[k - \sum_{i=0}^{k-1}(1-p_1{}^i)^{n_1}] \\ + q_{3,6}(0)[1 + (1-p_0{}^k)^{n_2}][k - \sum_{i=0}^{k-1}(1-p_0{}^i)^{n_2}] \end{bmatrix} \end{bmatrix}}{\begin{bmatrix} 1 - q_{3,4}(0)[1 - (1-p_0{}^k)^{n_1}] \\ -q_{3,5}(0)[1 - (1-p_1{}^k)^{n_1}] - q_{3,6}(0)[1 - (1-p_0{}^k)^{2n_2}] \end{bmatrix}}. \tag{19}$$

where,

$$q_{3,4}(0) \equiv \frac{(a_1+\beta)(a_2+\beta)}{(a_1+\lambda+\beta)(a_2+\lambda+\beta)},$$
$$q_{3,5}(0) \equiv \frac{(1-\alpha)\lambda[a_1+2\beta+(1-\alpha)(a_2+\lambda)]}{(a_1+\lambda+\beta)(a_2+\lambda+\beta)},$$
$$q_{3,6}(0) \equiv \frac{\alpha\lambda[a_1+2\beta+(2-\alpha)(a_2+\lambda)]}{(a_1+\lambda+\beta)(a_2+\lambda+\beta)}.$$

3. Optimal Policy

We discuss an optimal winodw size which maximizes amount of packets per unit of time until the transmission succeeds when $n_1 \equiv 2n_2$ because it is a tradeoff between the window size n_1 and the mean transmission times $\ell_{2,S_1}(n_1)$. We define the throughput $E(n_2)$, which represents the rate of n_1 packets to their mean transmission times, as the following equation:

$$E(n_2) \equiv \frac{2n_2}{\ell_{2,S_1}(n_2)} = \frac{2n_2}{X(n_2)+(a-\mu)}, \tag{20}$$

where

$$X(n_2) \equiv \frac{\left[\begin{array}{l} \mu+\frac{1}{a_1}+\frac{1}{a_2}+bn_2\left[\begin{array}{r}\left\{2q_{3,4}(0)+q_{3,6}(0)[1+(1-{p_0}^k)^{n_2}]\right\}\sum_{i=0}^{k-1}{p_0}^i \\ +\,2q_{3,5}(0)\sum_{i=0}^{k-1}{p_1}^i\end{array}\right] \\ +\,d\left[\begin{array}{r} q_{3,4}(0)[k-\sum_{i=0}^{k-1}(1-{p_0}^i)^{2n_2}]+q_{3,5}(0)[k-\sum_{i=0}^{k-1}(1-{p_1}^i)^{2n_2}] \\ +\,q_{3,6}(0)[1+(1-{p_0}^k)^{n_2}][k-\sum_{i=0}^{k-1}(1-{p_0}^i)^{n_2}]\end{array}\right]\end{array}\right]}{1-q_{3,5}(0)[1-(1-{p_1}^k)^{2n_2}]-(q_{3,4}(0)+q_{3,6}(0))[1-(1-{p_0}^k)^{2n_2}]}.$$

We seek an optimal window size ${n_2}^*$ which maximizes $E(n_2)$. From the inequality $1/E(n_2+1)-1/E(n_2) \geq 0$, we have

$$n_2X(n_2+1)-(n_2+1)X(n_2)-(a-\mu) \geq 0, \tag{21}$$

Denoting the left side of (21) by $L(n_2)$, we have

$$L(n_2+1) - L(n_2) = (n_2+1)Y(n_2),$$

where

$$Y(n_2) \equiv [X(n_2+2) - X(n_2+1)] - [X(n_2+1) - X(n_2)].$$

Hence, when $Y(n_2) > 0$, $L(n_2)$ is strictly increasing in n_2 from $L(1)$ to ∞.

Therefore, we have the following optimal policy:

(i) If $Y(n_2) > 0$ and $L(1) < 0$ then there exists a finite and unique ${n_2}^*(> 1)$ which satisfies (21).
(ii) If $Y(n_2) > 0$ and $L(1) \geq 0$ then ${n_2}^* = 1$.

4. Numerical Examples and Remarks

We compute numerically the optimal window size ${n_2}^*$. Suppose that the mean time b until editing the data and transmitting one packet is a unit time. It is assumed that the mean time required for data packets is $a/b = 10$, the mean generation interval of network congestion is $(1/\lambda)/b = 60, 600$, the mean time until the congestion clears up is $(1/\beta)/b = 10, 100$, the mean time required for the notification of connection completion is $(1/a_1)/b = 5$, the mean time required for the acknowledgement of connection completion is $(1/a_2)/b = 5$, the probability that the ECN bit is set, is $\alpha = 0 \sim 1.0$, the mean time for the server to transmit all packets to receive ACK or NAK is $d/b = 2 \sim 16$, the mean time from editing the data to n_2 transmit again is $w/b = 10$, the mean time for the server to interrupt n_2 retransmission to restart again is $\mu/b = 30$ and the probability that loss packets occur is $p_0 = 0.04, 0.05$ and $p_1 = 0.1 \sim 0.4$.

Table 1 gives the optimal window size n_2^* which maximizes the throughput, the mean time $\ell_{2,S_1}({n_2}^*)$ and the throughput $E({n_2}^*)$. This indicates that n_2^* increases with d/b and decreases with p_0. Under the same value p_0, n_2^* shows little dependence with p_1. Moreover, $E(n_2^*)$ increases with $(1/\lambda)/b$. Further, Figure 2 gives the throughput $E(n_2^*)$ for α and $(1/\lambda)/b$ when $a_1 = 5, a_2 = 5, k = 2, d/b = 2, p_0 = 0.05$ and $p_1 = 0.2$. This indicates $E(n_2^*)$ increases with α. But when $(1/\lambda)/b$ is large, $E(n_2^*)$ shows little dependence with α.

Table 1: Optimal window size $n_2{}^*$ to maximize $E(n_2)$ when $k = 2$ and $\alpha = 0.9$.

			$(1/\lambda)/b = 60, (1/\beta)/b = 10$			$(1/\lambda)/b = 600, (1/\beta)/b = 100$		
p_0	p_1	d/b	$n_2{}^*$	$\ell_{2S_1}(n_2{}^*)$	$E(n_2{}^*)$	$n_2{}^*$	$\ell_{2S_1}(n_2{}^*)$	$E(n_2{}^*)$
0.04	0.1	2	54	166.5	0.6487	54	165.9	0.6511
		4	58	184.1	0.6300	57	180.0	0.6333
		6	62	202.2	0.6133	61	197.6	0.6173
		8	65	217.4	0.5981	64	212.4	0.6028
		16	78	284.5	0.5484	76	274.0	0.5548
	0.2	2	54	167.1	0.6464	54	166.0	0.6508
		4	59	187.9	0.6279	58	183.2	0.6330
		6	62	202.8	0.6113	61	197.7	0.6171
		8	66	221.4	0.5963	64	212.4	0.6025
		16	78	285.2	0.5470	76	274.1	0.5546
0.05	0.1	2	42	143.5	0.5852	41	139.8	0.5867
		4	45	159.3	0.5650	44	155.1	0.5675
		6	48	175.5	0.5471	47	170.8	0.5504
		8	50	188.4	0.5308	50	187.0	0.5349
		16	59	246.6	0.4785	58	239.5	0.4843
	0.2	2	42	144.2	0.5826	41	139.8	0.5864
		4	45	160.0	0.5626	44	155.1	0.5672
		6	48	176.2	0.5448	47	170.9	0.5501
		8	51	192.9	0.5288	50	187.1	0.5346
		16	60	251.6	0.4769	58	239.6	0.4841

5. Conclusions

We have considered a stochastic model that when the server receives requests for data packets from a client, the server makes a connection. Then, the server and the client confirm whether the ECN bit is set. If it is not set, n_1 packets, which correspond to a window size, is successively transmitted to a client by a web server. If it is set, the server notices that congestion has happened in the network, n_2 packets, which correspond to half of the first window size, are transmitted. We have derived the mean time until packet transmission succeeds. Further, we have analytically derived the optimal policy which maximizes the throughput.

From numerical examples, we have shown that the optimal window size decreases with the probability that a loss packet occurs. Further, the optimal throughput increases with the probability that the ECN bit is set. In this way, it is shown that the ECN mechanism is effective.

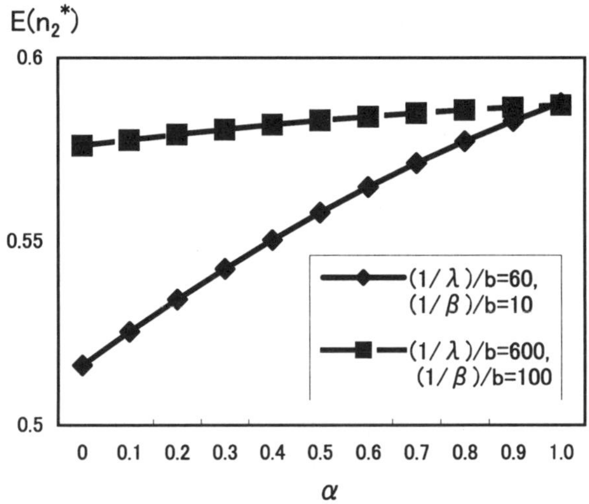

Figure 3. Throughput $E(n_2^*)$ when $a_1 = 5, a_2 = 5, p_0 = 0.05, p_1 = 0.2, d/b = 2, k = 2$,$(1/\lambda)/b = 60, (1/\beta)/b = 10, (1/\lambda)/b = 600, (1/\beta)/b = 100$

References

1. W. Stevens,, TCP Slow Start, Congestion Avoidance, Fast Retransmit and Fast Recovery Algorithms, *RFC2001*, (1997).
2. V. Jacobson, Congestion Avoidance and Control, *Computer Communication Review*, **vol.18**, (4), pp.314-329, (1988).
3. M. Mathis, J. Mahdavi, S. Floyd, A. Romanow, TCP Selective Acknowledgements Options, *RFC2018*, (1996).
4. S. Floyd, Tcp and explicit cogestion notification, *ACM Comput. Commun. Rev.*, **vol. 24**, (5), pp. 10-23, (1994).
5. S. Floyd and K. Fall, Promotimg the Use of End-to-End Cogestion Control in the Internet, *IEEE/ACM Transaction on Newtworking*, **vol.7**, (4), pp. 458-472, (1999).
6. K. Ramakrishnan and S. Floyd, A Proposal to add Explicit Cogestion Notification (ECN) to IP, *RFC 2481*, (1999).
7. S. Osaki,*Applied Stochastic System Modeling*, Springer-Verlag, Berlin, (1992).
8. K. Yasui, T. Nakagawa and H. Sandoh, Reliability models in data communication systems, *Stochastic Models in Reliability and Maintenance* (edited by S.Osaki), pp. 281-301, Springer-Verlag, Berlin (2002).

OPTIMAL AVAILABILITY MODELS OF A PHASED ARRAY RADAR

T. NAKAGAWA

Department of Marketing and Information Systems,
Aichi Institute of Technology,
1247 Yachigusa, Yagusa–cho, Toyota 470–0392, Japan
E-mail: toshi-nakagawa@aitech.ac.jp

K. ITO

Technology Training Center, Technical Headquarters,
Mitsubishi Heavy Industries, LTD.,
1–ban–50, Daikouminami 1–chome, higashi-ku, Nagoya 461–0047, Japan
E-mail: kodo_ito@mhi.co.jp

A phased array radar (PAR) antenna embodies a large number of small element antennas which radiate electromagnetic wave, and it directs electromagnetic wave direction by shifting individual wave phases of these elements. Failed elements have to be detected, diagnosed, localized and replaced at appropriate times to maintain a required radar system performance. However, the maintenance of an antenna should not be made so frequently, because it suspends the radar operation and degrades the radar system availability.

This paper considers two typical maintenances (cyclic and delayed maintenances) and two modified maintenances of a PAR where a certain number of survival elements is needed to retain a required performance. When failures of a PAR element antennas occur at a Poisson process, the availability is obtained and optimal policies for these maintenances which maximize them are analytically discussed. Several numerical examples are presented.

1. Introduction

A phased array radar (PAR) is the radar which steers the electromagnetic wave direction electrically. Comparing with conventional radars which steer their electromagnetic wave direction by moving their antennas mechanically, a PAR has no mechanical portion to steer its wave direction, and hence, it can steer very quickly. Most anti-aircraft missile systems and early warning systems have presently adopted PARs because they can acquire and track multiple targets simultaneously.

A PAR antenna consists of a large number of small and homogeneous element antennas which are arranged flatly and regularly, and steers its electromagnetic wave direction by shifting signal phases of waves which are radiated from these individual elements [10,11].

The increase in the number of failed elements degrades the radar performance, and at last, this may cause an undesirable situation such as the omission of targets [1]. The detection, diagnosis, localization and replacement of failed elements of a PAR antenna are indispensable to hold a certain required level of radar performance. A digital computer system controls a whole PAR system, and it detects, diagnoses and localizes failed elements. However, such maintenance actions intermit the radar operation and decrease its availability. So that, the maintenance should not be made so frequently. From the above reasons, it would be important to decide an optimal maintenance policy for a PAR antenna, by comparing the downtime loss caused by its maintenance with the degradational loss caused by its performance downgrade.

Recently, a new method of failure detection for PAR antenna elements has been proposed by measuring the electromagnetic wave pattern [3]. This method could detect some failed elements even when a radar system is operating, *i.e.*, it could be applied to the detection of confined failure modes such as power on-off failures. However, it would be generally necessary to stop the PAR operation for the detection of all failed elements.

Keithley [7] showed by Monte Carlo simulation that the maintenance time of PAR with 1024 elements had a strong influence on its availability. Hevesh [5] discussed the following three maintenances of PAR in which all failed elements could be detected immediately, and calculated the average times to failures of its equipments, and its availability in immediate maintenance :

1) *Immediate maintenance*: Failed elements are detected, localized and replaced immediately.
2) *Cyclic maintenance*: Failed elements are detected, localized and replaced periodically.
3) *Delayed maintenance*: Failed elements are detected and localized periodically, and replaced when their number has exceeded a predesignated one.

Further, Hesse [4] analyzed the field maintenance data of U.S. Army prototype PAR, and clarified that the repair times have a log-normal distribution. In the actual maintenance, the immediate maintenance is rarely adopted because frequent maintenances degrade a radar system availability. Either cyclic or delayed maintenances is commonly adopted.

We have already studied the comparison of cyclic and delayed maintenances of PAR considering the financial optimum [6]. In the study, we derived the expected costs per unit of time and discussed the optimal policies which minimize them analytically in these two maintenances, and concluded that the delayed maintenance is better than the cyclic one in suitable conditions by comparing these two costs numerically. Although the financial optimum takes priority for non-military systems and military systems in the non-combat condition, the operational availability should take more priority than economy for military systems in the combat condition. Therefore, maintenance policies which maximize the availability should be considered.

In this paper, we perform the periodic detection of failed elements of a PAR where it is consisted of N_0 elements and failures are detected at scheduled time interval : If the number of failed elements has exceeded a specified number N $(0 < N \leq N_0)$, a PAR cannot hold a required level of radar performance, and it causes the operational loss such as the target oversight to a PAR. We assume that failed elements occur at a Poisson process, and consider cyclic, delayed and two modified maintenances. Applying the method[8] to such maintenances, the availabilities are obtained, and optimal policies which maximize them are analytically discussed in cyclic and delayed maintenances. In a numerical example, we decide which maintenance is better, by comparing the availabilities.

2. Cyclic Maintenance

2.1. *Problem formulation*

We consider the following cyclic maintenance of a PAR :

1) A PAR is consisted of N_0 elements which are independent and homogeneous on all plains of PAR, and have an identical constant hazard rate λ_0. The number of failed elements at time t has a binomial distribution with mean $N_0[1 - exp(-\lambda_0 t)]$. Since N_0 is large and λ_0 is very small, it might be assumed that failures of elements occur approximately at a Poisson process with mean $\lambda \equiv N_0\lambda_0$. That is, the probability that j failures occur during $(0, t]$ is

$$p_j(t) \equiv \frac{(\lambda t)^j e^{-\lambda t}}{j!} \qquad (j = 0, 1, 2, \cdots).$$

2) When the number of failed elements has exceeded a specified number N, a PAR cannot hold a required level of radar performance such as maximum detection range and resolution.
3) Failed elements cannot be detected during operation and can be ascertained only according to the diagnosis software executed by a PAR system computer. Failed elements are usually detected at periodic diagnosis. The diagnosis is performed at time interval T and a single diagnosis spends time T_0.
4) All failed elements are replaced by new ones at the M-th diagnosis or at the time when the number of failed elements has exceeded N, whichever occurs first. The replacement spends time T_1.

When the replacement time of failed element antennas is assumed to be the regeneration point, the availability of system is denoted as[9]

$$A = \frac{Effective\ time\ between\ regeneration\ points}{Total\ time\ between\ regeneration\ points}. \tag{1}$$

When the number of failed elements is below N at the M-th diagnosis, the expected effective time until replacement is

$$MT \sum_{j=0}^{N-1} p_j(MT). \tag{2}$$

When the number of failed elements exceeds N at the i $(i = 1, 2, \cdots M)$-th diagnosis, the expected effective time until replacement is (Appendix A)

$$\sum_{i=1}^{M} \sum_{j=0}^{N-1} p_j[(i-1)T] \sum_{k=N-j}^{\infty} \int_{(i-1)T}^{iT} t dp_k[t-(i-1)T]. \tag{3}$$

Thus, from (2) and (3), the total expected effective time until replacement is (Appendix B)

$$T \sum_{i=0}^{M-1} \sum_{j=0}^{N-1} p_j(iT) - \sum_{i=0}^{M-1} \sum_{j=0}^{N-1} p_j(iT) \sum_{k=N-j+1}^{\infty} \frac{k-N+j}{\lambda} p_k(T). \tag{4}$$

Next, when the number of failed elements is below N at the M-th diagnosis, the expected time between two adjacent regeneration points is

$$\sum_{j=0}^{N-1} p_j(MT)[M(T+T_0)+T_1]. \tag{5}$$

When the number of failed elements exceeds N at the $i\,(i = 1, 2, \cdots M)$-th diagnosis, the expected time between two adjacent regeneration points is

$$\sum_{i=1}^{M}\sum_{j=0}^{N-1} p_j[(i-1)T] \sum_{k=N-j}^{\infty} [i(T+T_0)+T_1]p_k(T)\,. \tag{6}$$

Thus, from (5) and (6), the total expected time between two adjacent regeneration points is (Appendix C)

$$T_1 + (T+T_0)\sum_{i=0}^{M-1}\sum_{j=0}^{N-1} p_j(iT)\,. \tag{7}$$

Therefore, by dividing (4) by (7), the availability of cyclic maintenance $A_1(M)$ is

$$\begin{aligned} &A_1(M)\\ &= \frac{T\sum_{i=0}^{M-1}\sum_{j=0}^{N-1} p_j(iT) - \sum_{i=0}^{M-1}\sum_{j=0}^{N-1} p_j(iT)\sum_{k=N-j+1}^{\infty}\frac{k-N+j}{\lambda}p_k(T)}{T_1+(T+T_0)\sum_{i=0}^{M-1}\sum_{j=0}^{N-1} p_j(iT)}\\ &= \frac{T}{T+T_0}\left[1-\frac{\frac{T_1}{T+T_0}+\frac{1}{T}\sum_{i=0}^{M-1}\sum_{j=0}^{N-1} p_j(iT)\sum_{k=N-j+1}^{\infty}\frac{k-N+j}{\lambda}p_k(T)}{\frac{T_1}{T+T_0}+\sum_{i=0}^{M-1}\sum_{j=0}^{N-1} p_j(iT)}\right]\\ &\equiv \frac{T}{T+T_0}[1-\overline{A_1}(M)] \qquad (M=1,2,\cdots)\,. \end{aligned} \tag{8}$$

2.2. *Optimal policy*

Because maximizing the availability $A_1(M)$ is equal to minimizing the unavailability $\overline{A_1}(M)$ from (8), we consider M^* which minimizes $\overline{A_1}(M)$.

Forming the inequality $\overline{A_1}(M+1) - \overline{A_1}(M) \geq 0$, we have

$$\begin{aligned} &L_1(M)\left[\frac{T_1}{T+T_0}+\sum_{i=0}^{M-1}\sum_{j=0}^{N-1} p_j(iT)\right]\\ &\qquad -\sum_{i=0}^{M-1}\sum_{j=0}^{N-1} p_j(iT)\sum_{k=N-j+1}^{\infty}\frac{k-N+j}{\lambda}p_k(T) \geq \frac{T_1T}{T+T_0}\,, \end{aligned} \tag{9}$$

where

$$L_1(M) \equiv \frac{\sum_{j=0}^{N-1} p_j(MT)\sum_{k=N-j+1}^{\infty}\frac{k-N+j}{\lambda}p_k(T)}{\sum_{j=0}^{N-1} p_j(MT)}\,. \tag{10}$$

Letting $Q_1(M)$ denote the left-hand side of (9), we have

$$Q_1(M+1) - Q_1(M)$$
$$= [L_1(M+1) - L_1(M)]\left[\frac{T_1}{T+T_0} + \sum_{i=0}^{M}\sum_{j=0}^{N-1} p_j(iT)\right]. \quad (11)$$

Thus, if $L_1(M)$ is strictly increasing in M, then $Q_1(M)$ is strictly increasing in M.

Therefore, we have the following optimal policy :

Theorem 1.

(i) *If $L_1(M)$ is strictly increasing in M and $Q_1(\infty) > T_1T/(T+T_0)$ then there exists a finite and unique M^* which satisfies (9).*

(ii) *If $L_1(M)$ is strictly increasing in M and $Q_1(\infty) \leq T_1T/(T+T_0)$ then $M^* = \infty$.*

(iii) *If $L_1(M)$ is decreasing in M then $M^* = 1$ or $M^* = \infty$.*

3. Delayed Maintenance

3.1. *Problem formulation*

We consider the delayed maintenance of a PAR :

4)' All failed elements are replaced by new ones only when failed elements have exceeded a managerial number $N_c (< N)$ at diagnosis. The replacement spends time T_1.

Assumptions 1)-3) in Section 2.1, remain the same and Assumption 4) is replaced by 4)'.

When the number of failed elements is between N_c and N, the expected effective time until replacement is

$$\sum_{i=1}^{\infty}\sum_{j=0}^{N_c-1} p_j[(i-1)T] \sum_{k=N_c-j}^{N-j-1} iTp_k(T). \quad (12)$$

When the number of failed elements exceeds N, the expected effective time until replacement is

$$\sum_{i=1}^{\infty}\sum_{j=0}^{N_c-1} p_j[(i-1)T] \sum_{k=N-j}^{\infty} \int_{(i-1)T}^{iT} t dp_k[t-(i-1)T]. \quad (13)$$

Thus, the total expected effective time until replacement is, from (12) and (13) (Appendix D),

$$T\sum_{i=0}^{\infty}\sum_{j=0}^{N_c-1} p_j(iT) - \sum_{i=0}^{\infty}\sum_{j=0}^{N_c-1} p_j(iT) \sum_{k=N-j+1}^{\infty} \frac{k-N+j}{\lambda} p_k(T)\,. \qquad (14)$$

Similarly, when the number of failed elements is between N_c and N, the expected time between two adjacent regeneration points is

$$\sum_{i=1}^{\infty}\sum_{j=0}^{N_c-1} p_j[(i-1)T] \sum_{k=N_c-j}^{N-j-1} [i(T+T_0)+T_1]p_k(T)\,. \qquad (15)$$

When the number of failed elements exceeds N, the expected time between two adjacent regeneration points is

$$\sum_{i=1}^{\infty}\sum_{j=0}^{N_c-1} p_j[(i-1)T] \sum_{k=N-j}^{\infty} [i(T+T_0)+T_1]p_k(T)\,. \qquad (16)$$

Thus, the total expected time between two adjacent regeneration points is, from (15) and (16) (Appendix D),

$$(T+T_0)\sum_{i=0}^{\infty}\sum_{j=0}^{N_c-1} p_j(iT) + T_1\,. \qquad (17)$$

Therefore, the availability of delayed maintenance $A_2(N_c)$ is, by dividing (14) by (17),

$$\begin{aligned} &A_2(N_c) \\ &= \frac{T}{T+T_0}\left[1 - \frac{\frac{T_1}{T+T_0} + \sum_{i=0}^{\infty}\sum_{j=0}^{N_c-1} p_j(iT)\sum_{k=N-j}^{\infty}\left(1-\frac{N-j}{k+1}\right)p_k(T)}{\frac{T_1}{T+T_0} + \sum_{i=0}^{\infty}\sum_{j=0}^{N_c-1} p_j(iT)}\right] \\ &\equiv \frac{T}{T+T_0}[1-\overline{A_2}(N_c)] \qquad (N_c = 0,1,2,\cdots,N)\,. \qquad (18) \end{aligned}$$

3.2. *Optimal policy*

Forming the inequality $\overline{A_2}(N_c+1) - \overline{A_2}(N_c) \geq 0$, we have

$$\frac{E_{N_c}}{D_{N_c}-E_{N_c}} \sum_{j=0}^{N_c-1}(D_j - E_j) - \sum_{j=0}^{N_c-1} E_j \geq \frac{T_1}{T+T_0}\,, \qquad (19)$$

where $D_j \equiv \sum_{i=0}^{\infty} p_j(iT)$ and $E_j \equiv \sum_{i=0}^{\infty} p_j(iT)\sum_{k=N-j}^{\infty}[1-(N-j)/(k+1)]p_k(T)$.

Let $Q_2(N_c)$ denote the left-hand side of (19) and $L_2(N_c) \equiv E_{N_c}/(D_{N_c} - E_{N_c})$. Then,

$$Q_2(N_c+1) - Q_2(N_c) = [L_2(N_c+1) - L_2(N_c)] \sum_{j=0}^{N_c} (D_j - E_j) . \quad (20)$$

Because $D_j - E_j > 0$, the sign of $Q_2(N_c+1) - Q_2(N_c)$ depends on $L_2(N_c+1) - L_2(N_c)$.

Therefore, we have the following optimal policy :

Theorem 2.

(i) *If $L_2(N_c)$ is strictly increasing in N_c and $Q_2(N) > T_1/(T+T_0)$ then there exists a finite and unique N_c^* $(1 \le N_c^* < N)$ which satisfies (19).*

(ii) *If $L_2(N_c)$ is strictly increasing in N_c and $Q_2(N) \le T_1/(T+T_0)$ then $N_c^* = N$, i.e., the planned maintenance should not be done.*

4. Other Maintenances

In this section, we consider following two modified maintenance models and derive the availabilities of each model.

4.1. *Model 1*

We consider the combined maintenance of cyclic and delayed ones :

4)" All failed elements are replaced by new ones when failed elements have exceeded a managerial number $N_c(< N)$ at diagnosis. The replacement spends time T_1. Furthermore, when failed elements have exceeded a number N_c by the M_c-th diagnosis, the replacement spends time $T_2(> T_1)$.

Assumptions 1)-3) in Section 2.1, remain the same and Assumption 4) is replaced by 4)".

When the number of failed elements exceeds N_c and is below N, the expected effective time until replacement is

$$\sum_{i=1}^{\infty} \sum_{j=0}^{N_c-1} p_j[(i-1)T] \sum_{k=N_c-j}^{N-j-1} iTP_k(T) . \quad (21)$$

When the number of failed elements exceeds N, the expected effective time until replacement is

$$\sum_{i=1}^{\infty}\sum_{j=0}^{N_c-1} p_j[(i-1)T] \sum_{k=N-j}^{\infty} \int_{(i-1)T}^{iT} t dP_k[t-(i-1)T] . \tag{22}$$

Thus, the total expected effective time until replacement is, from (21) and (22),

$$T\sum_{i=0}^{\infty}\sum_{j=0}^{N_c-1} p_j(iT) - \sum_{i=0}^{\infty}\sum_{j=0}^{N_c-1}\sum_{k=N-j+1}^{\infty} \frac{k-N+j}{\lambda} P_k(T) . \tag{23}$$

Next, when the number of failed elements exceeds N_c and is below N by the M_c-th diagnosis, the expected time between two adjacent regeneration points is

$$\sum_{i=1}^{M_c}\sum_{j=0}^{N_c-1} p_j[(i-1)T] \sum_{k=N_c-j}^{N-j-1} [i(T+T_0)+T_2] P_k(T) . \tag{24}$$

When the number of failed elements exceeds N by the M_c-th diagnosis, the expected time between two adjacent regeneration points is

$$\sum_{i=1}^{M_c}\sum_{j=0}^{N_c-1} p_j[(i-1)T] \sum_{k=N-j}^{\infty} [i(T+T_0)+T_2] P_k(T) . \tag{25}$$

When the number of failed elements exceeds N_c and is below N afterward the M_c-th diagnosis, the expected time between two regeneration points is

$$\sum_{i=M_c+1}^{\infty}\sum_{j=0}^{N_c-1} p_j[(i-1)T] \sum_{k=N_c-j}^{N-j-1} [i(T+T_0)+T_1] P_k(T) . \tag{26}$$

When the number of failed elements exceeds N afterward the M_c-th diagnosis, the expected time between two regeneration points is

$$\sum_{i=M_c+1}^{\infty}\sum_{j=0}^{N_c-1} p_j[(i-1)T] \sum_{k=N-j}^{\infty} [i(T+T_0)+T_1] P_k(T) . \tag{27}$$

Thus, the total expected time between two adjacent regeneration points is, from (24), (25), (26) and (27),

$$(T+T_0)\sum_{i=0}^{\infty}\sum_{j=0}^{N_c-1} p_j(iT) + T_2 - (T_2-T_1)\sum_{j=0}^{N_c-1} p_j(M_cT) . \tag{28}$$

Therefore, the availability of the maintenance $A_3(M_c, N_c)$ is, by dividing (23) by (28),

$$A_3(M_c, N_c)$$
$$= \frac{T\sum_{i=0}^{\infty}\sum_{j=0}^{N_c-1} p_j(iT) - \sum_{i=0}^{\infty}\sum_{j=0}^{N_c-1}\sum_{k=N-j+1}^{\infty}\frac{k-N+j}{\lambda}P_k(T)}{(T+T_0)\sum_{i=0}^{\infty}\sum_{j=0}^{N_c-1} p_j(iT) + T_2 - (T_2 - T_1)\sum_{j=0}^{N_c-1} p_j(M_cT)}. \quad (29)$$

4.2. *Model 2*

We consider the modified cyclic maintenance :

3)' Failed elements cannot be detected during operation and can be ascertained only according to the diagnosis software executed by a PAR system computer. Failed elements are usually detected at periodic diagnosis. The diagnosis is performed at time interval T and a single diagnosis spends time T_0 until the L-th ($L < M$) diagnosis. Afterward the L-th diagnosis, the diagnosis time interval is switched from T to aT ($0 < a < 1$).

Assumptions 1), 2) and 4) in Section 2.1, remain the same and Assumption 3) is replaced by 3)'.

When the number of failed elements is below N at the M-th diagnosis, the expected effective time until replacement is

$$[LT + (M-L)aT]\sum_{j=0}^{N-1} p_j[LT + (M-L)aT]. \quad (30)$$

When the number of failed elements exceeds N by the L-th diagnosis, the expected effective time until replacement is

$$\sum_{i=1}^{L}\sum_{j=0}^{N-1} p_j[(i-1)T]\sum_{k=N-j}^{\infty}\int_{(i-1)T}^{iT} t dp_k[t-(i-1)T]. \quad (31)$$

When the number of failed elements exceeds N between the L-th and the M-th diagnosis, the expected effective time until replacement is

$$\sum_{i=L+1}^{M}\sum_{j=0}^{N-1} p_j[LT + (i-L-1)aT]$$
$$\times \sum_{k=N-j}^{\infty}\int_{LT+(i-L-1)aT}^{LT+(i-L)aT} t dp_k[t - LT - (i-L-1)aT]. \quad (32)$$

Thus, the total expected effective time until replacement is, from (30), (31) and (32),

$$\begin{aligned}
&T\sum_{i=0}^{L-1}\sum_{j=0}^{N-1}p_j(iT)+aT\sum_{i=L}^{M-1}\sum_{j=0}^{N-1}p_j[LT+(i-L)aT]\\
&-\frac{1}{\lambda}\sum_{i=0}^{L-1}\sum_{j=0}^{N-1}p_j[k-(N-j)]p_k(T)\\
&\quad-\frac{1}{\lambda}\sum_{i=L}^{M-1}\sum_{j=0}^{N-1}p_j[LT+(i-L)aT]\sum_{k=N-j+1}^{\infty}[k-(N-j)]p_k(aT)\,. \qquad (33)
\end{aligned}$$

Similarly, when the number of failed elements is below N at the M-th diagnosis, the expected time between two adjacent regeneration points is

$$[L(T+T_0)+(M-L)(aT+T_0)+T_1]\sum_{j=0}^{N-1}p_j[LT+(M-L)aT]\,. \qquad (34)$$

When the number of failed elements exceeds N by the L-th diagnosis, the expected time between two regeneration points is

$$\sum_{i=1}^{L}\sum_{j=0}^{N-1}p_j[(i-1)T]\sum_{k=N-j}^{\infty}[i(T+T_0)+T_1]p_k(T)\,. \qquad (35)$$

When the number of failed elements exceeds N between the L-th and the M-th diagnosis, the expected time between two regeneration points is

$$\begin{aligned}
&\sum_{i=L+1}^{M}\sum_{j=0}^{N-1}p_j[LT+(i-L-1)aT]\\
&\quad\sum_{k=N-j}^{\infty}[L(T+T_0)+(i-L)(aT+T_0)+T_1]p_k(aT)\,. \qquad (36)
\end{aligned}$$

Thus, the total expected time between two adjacent regeneration points is, from (34), (35) and (36),

$$\begin{aligned}
&(T+T_0)\sum_{i=0}^{L-1}\sum_{j=0}^{N-1}p_j(iT)\\
&\quad+(aT+T_0)\sum_{i=L}^{M-1}\sum_{j=0}^{N-1}p_j[LT+(i-L)aT]+T_1\,. \qquad (37)
\end{aligned}$$

Therefore, the availability of the maintenance $A_4(L,M)$ is, by dividing (33) by (37),

$$A_4(L,M) = \frac{\begin{bmatrix} T\sum_{i=0}^{L-1}\sum_{j=0}^{N-1} p_j(iT) \\ +aT\sum_{i=L}^{M-1}\sum_{j=0}^{N-1} p_j[LT+(i-L)aT] \\ -\frac{1}{\lambda}\sum_{i=0}^{L-1}\sum_{j=0}^{N-1} p_j[k-(N-j)]p_k(T) \\ -\frac{1}{\lambda}\sum_{i=L}^{M-1}\sum_{j=0}^{N-1} p_j[LT+(i-L)aT] \\ \times\sum_{k=N-j+1}^{\infty}[k-(N-j)]p_k(aT) \end{bmatrix}}{\begin{bmatrix} (T+T_0)\sum_{i=0}^{L-1}\sum_{j=0}^{N-1} p_j(iT) \\ +(aT+T_0)\sum_{i=L}^{M-1}\sum_{j=0}^{N-1} p_j[LT+(i-L)aT] \\ +T_1 \end{bmatrix}}. \tag{38}$$

5. Numerical Example

Table 1 gives the optimal number of diagnosis M^* and the optimal managerial number of failed elements N_c^*, and the unavailability $\overline{A}_1(M^*)$ and $\overline{A}_2(N_c^*)$ for $N = 70, 90, 100$, $T = 168, 240, 336\,hours$ (7, 10, 14 *days*), $T_0 = 0.1, 0.5, 1$, $T_1 = 2, 5, 8$ and $\lambda = 0.1, 0.2, 0.3$ */hours*. In all cases in Table 1, $L_1(M)$ is strictly increasing in M.

Table 1 indicates that M^* and N_c^* decrease when N, $1/T$, T_1, and $1/\lambda$ decrease, and the change of T_0 hardly affects M^* and N_c^*. In this calculation, $\overline{A}_1(M^*)$ is always greater than $\overline{A}_2(N_c^*)$. Therefore, we can adjudge in this case that the delayed maintenance is more available than the cyclic one.

6. Conclusions

We have considered cyclic, delayed and two modified maintenances of a PAR which detect failed elements by the periodical diagnosis. Presuming that the element failure occur at a Poisson process, the availability has been derived and the optimal number of diagnosis M^* of cyclic maintenance and the optimal managerial number of failed elements N_c^* of delayed maintenance have been analytically discussed. Comparing the availability numerically, we decide that the delayed maintenance is more available than the cyclic one in this case.

Table 1. Optimal number of diagnosis M^*, optimal managerial number of failed elements N_c^* and unavailabilities $\overline{A}_1(M^*)$ and $\overline{A}_2(N_c^*)$.

N	T	T_0	T_1	λ	M^*	$\overline{A}_1(M^*)$	N_c^*	$\overline{A}_2(N_c^*)$
100	24 × 7	1	8	0.1	5	1.141×10^{-2}	78	0.936×10^{-2}
90	24 × 7	1	8	0.1	4	1.186×10^{-2}	68	1.058×10^{-2}
70	24 × 7	1	8	0.1	3	1.574×10^{-2}	49	1.430×10^{-2}
100	24 × 10	1	8	0.1	3	1.097×10^{-2}	70	0.998×10^{-2}
100	24 × 14	1	8	0.1	2	1.173×10^{-2}	60	1.113×10^{-2}
100	24 × 7	0.5	8	0.1	5	1.144×10^{-2}	78	0.938×10^{-2}
100	24 × 7	0.1	8	0.1	5	1.146×10^{-2}	78	0.941×10^{-2}
100	24 × 7	1	5	0.1	4	0.735×10^{-2}	77	0.593×10^{-2}
100	24 × 7	1	2	0.1	4	0.295×10^{-2}	75	0.242×10^{-2}
100	24 × 7	1	8	0.2	2	2.312×10^{-2}	62	2.143×10^{-2}
100	24 × 7	1	8	0.3	1	4.520×10^{-2}	48	3.830×10^{-2}

Appendix A. Derivation of (3)

The probability that the number of failed elements is below $N-1$ at the $i-1\,(i=1,2,\cdots)$-th diagnosis is

$$\sum_{i=1}^{M}\sum_{j=0}^{N-1} p_j[(i-1)T]\,. \tag{A.1}$$

The probability that the number of failed elements exceeds N from $i-1$-th to i-th diagnosis is

$$\sum_{k=N-j}^{\infty} p_k[iT-(i-1)T] = \sum_{k=N-j}^{\infty}\int_{(i-1)T}^{iT} dp_k[t-(i-1)T]\,. \tag{A.2}$$

From (A.1) and (A.2), the probability that the number of failed elements exceeds N at the i-th disgnosis is

$$\sum_{i=1}^{M}\sum_{j=0}^{N-1} p_j[(i-1)T]\sum_{k=N-j}^{\infty}\int_{(i-1)T}^{iT} dp_k[t-(i-1)T]\,. \tag{A.3}$$

Appendix B. Derivation of (4)

Equation (3) is rewritten as

$$\sum_{i=1}^{M}\sum_{j=0}^{N-1} p_j[(i-1)T]\sum_{k=N-j}^{\infty}[iTp_k(T) - \int_0^T p_k(x)dx]\,. \tag{B.1}$$

The first term of (B.1) is rewritten as

$$
\begin{aligned}
&\sum_{i=1}^{M}\sum_{j=0}^{N-1} p_j[(i-1)T] \sum_{k=N-j}^{\infty} iTp_k(T) \\
&= T\sum_{i=1}^{M}\sum_{j=0}^{N-1} ip_j[(i-1)T][1-\sum_{k=0}^{N-j-1} p_k(T)] \\
&= T\sum_{i=1}^{M}\sum_{j=0}^{N-1} i\{p_j[(i-1)T]-p_j(iT)\} \\
&= T\sum_{j=0}^{N-1}[\sum_{i=0}^{M-1} p_j(iT)-Mp_j(MT)]\,,
\end{aligned}
\tag{B.2}
$$

where we use the following relation

$$
\sum_{j=0}^{N-1} p_j[(i-1)T] \sum_{k=0}^{N-j-1} p_k(T) = \sum_{j=0}^{N-1} p_j(iT)\,. \tag{B.3}
$$

The second term of (B.1) is

$$
\begin{aligned}
&\sum_{i=1}^{M}\sum_{j=0}^{N-1} p_j[(i-1)T] \sum_{k=N-j}^{\infty} \int_0^T p_k(x)dx \\
&= \frac{1}{\lambda}\sum_{i=0}^{M-1}\sum_{j=0}^{N-1} p_j(iT) \sum_{k=N-j}^{\infty}\sum_{l=k+1}^{\infty} p_l(T) \\
&= \frac{1}{\lambda}\sum_{i=0}^{M-1}\sum_{j=0}^{N-1} p_j(iT) \sum_{k=1}^{\infty} kp_{N-j+k}(T) \\
&= \sum_{i=0}^{M-1}\sum_{j=0}^{N-1} p_j(iT) \sum_{k=N-j+1}^{\infty} \frac{k-N+j}{\lambda} p_k(T)\,.
\end{aligned}
\tag{B.4}
$$

Using (B.2) and (B.4), Equation (4) is derived from (2) and (3).

Appendix C. Derivation of (7)

Equation (6) is rewritten as

$$\sum_{i=1}^{M}[i(T+T_0)+T_1]\sum_{j=0}^{N-1}p_j[(i-1)T][1-\sum_{k=0}^{N-j-1}p_k(T)]$$

$$=\sum_{i=1}^{M}[i(T+T_0)+T_1]\{\sum_{j=0}^{N-1}p_j[(i-1)T]-\sum_{j=0}^{N-1}p_j(iT)\}$$

$$=(T+T_0)\sum_{i=1}^{M}\sum_{j=0}^{N-1}\{ip_j[(i-1)T]-ip_j(iT)\}$$

$$+T_1\sum_{i=1}^{M}\sum_{j=0}^{N-1}\{p_j[(i-1)T]-p_j(iT)\}$$

$$=(T+T_0)\sum_{j=0}^{N-1}[\sum_{i=0}^{M-1}p_j(iT)-Mp_j(MT)]$$

$$+T_1-T_1\sum_{j=0}^{N-1}p_j(MT)\,, \qquad \text{(C.1)}$$

where we use the relation (B.3). Using (C.1), Equation (7) is derived from (5) and (6).

Appendix D. Derivation of (14) and (17)

Refering Appendix B. and C., Equations (13) and (16) are rewritten similarly. Equations (14) and (17) are easily derived from (12), (13), (15) and (16) respectively.

References

1. E.Brookner, Phased-array radars, *Scientific American* **252**, 94-102 (1985).
2. E.Brookner, *Practical Phased-Array Antenna Systems*, Artech House, Boston (1991).
3. O.M.Bucci, A.Capozzoli and G.D'elia, Diagnosis of Array Faults from Far-Field Amplitude-Only Data, *IEEE Transaction on Antennas and Propagation* **48**, 647-652 (2000).
4. J.L.Hesse, Maintainability analysis and prototype operations, *Proceedings 1975 Annual Reliability and Maintainability Symposium*, 194-199 (1975).
5. A.H.Hevesh, Maintainability of phased array radar systems, *IEEE Transactions on Reliability* **R-16**, 61-66 (1967).

6. K.Ito and T.Nakagawa, Comparison of cyclic and delayed maintenances for a phased array radar, *Journal of the Operations Research Society of Japan* **47**(1), 51-61 (2004).
7. H.M.Keithley, Maintainability impact on system design of a phased array radar, *Annual New York Conference on Electronic Reliability, 7th* **9**, 1-10 (1966).
8. T.Nakagawa, Modified discrete preventive maintenance policies, *Naval Research Logistics Quarterly* **33**, 703-715 (1986).
9. T.Nakagawa, Maintenance Theory of Reliability, Springer-Verlag, London (2005).
10. M.I.Skolnik, *Introduction to Radar Systems*, McGraw-Hill Publishing Company, New York (1980).
11. M.I.Skolnik, *Radar Handbook*, McGraw-Hill Publishing Company, New York (1990).

OPTIMAL CHECKING TIME OF BACKUP OPERATION FOR A DATABASE SYSTEM

K. NARUSE

Department of Industrial Engineering, Aichi Institute of Technology

1247 Yachigusa, Yakusa-cho, Toyota 470-0392, Japan

S. NAKAGAWA

Institute of Consumer Sciences and Human Life,Kinjo Gakuin University

1723 Omori 2-chome, Moriyama-ku, Nagoya 463-8521, Japan

Y. OKUDA

Department of Industrial Engineering, Aichi Institute of Technology

1247 Yachigusa, Yakusa-cho, Toyota 470-0392, Japan

Abstract- When a failure occurs in the process of a database system, we execute the rollback operation until the latest checking time and make the recovery of database of files. This paper proposes the modified inspection model where the backup is carried out until the latest checking time when some failure was detected. The expected cost until the backup operation is made to the latest checking time is derived, and optimal inspection policies, which minimize it for two cases of periodic and sequential checking times, are analytically discussed. Some further modified models where the operating time is finite and a fault remains hidden are proposed.

1. Introduction

Most units in standby [1, 2] and in storage [3, 4] have to be checked at planned times to detect failures. Barlow and Proschan [5] summarized such inspection policies which minimize the total expected cost until a failure detection. All inspection models have assumed that any failure is known only through checking and summarized in [6]. But, when a failure was detected in the recovery technique of a database system, we execute the rollback operation until the latest checkpointing [7, 8] and reconstruct the consistency of a database. It has been assumed in such models that any failure is always detected immediately, however, there is a loss time or cost associated with the lapsed time of rollback operation between a failure detection and the latest checkpointing.

Further, this model would be applied to the backup policy for hard disks [9, 10]: There is a variety of files in the disk, however, they may be sometimes lost due to human errors or disk failures. To prevent such events, backup files are made at suitable times, which are called a backward time. When failures have occurred, we can make the recovery of files at each backward time.

From the practical viewpoints of database recovery and backup file, we propose the following backup operation model which is one of the modified inspection policies: When a failure was detected, we carry out the backup operation to the latest checking time. In such a model, we do not wish to provide the checks much frequently, and on the other hand, we wish to avoid a long elapsed time between a failure detection and the checking time. It would be an important problem to determine an optimal checking schedule of this model.

By the similar method to that of the usual inspection model [6], we derive the total expected cost until the completion of backup operation after a failure detection, and discuss optimal checking times which minimize it for two cases of periodic and sequential policies. We give numerical examples when failure times of a unit have a Weibull and uniform distributions.

Further, we consider the case where a unit has to be operating for a finite interval. The expected cost is obtained, and an optimal checking time which minimizes it is numerically computed. Finally, the expected cost per unit of time and the availability are also derived. We propose one modified model where a fault occurs and is hidden, and after that, a failure occurs, and obtain the expected cost.

2. Expected Costs

Suppose that the failure time of a unit has a general distribution $F(t)$ with finite mean μ, where $\overline{F}(t) \equiv 1-F(t)$. The checking schedule of a unit is made at successive times $T_k\ (k=1,2,\cdots)$ where $T_0 \equiv 0$. Let c_1 be the cost required for each check. Further, when a failure was detected between T_k and T_{k+1}, we carry out the backup operation to the latest checking time T_k. This incurs a loss cost c_2 per unit of time (Fig.1).

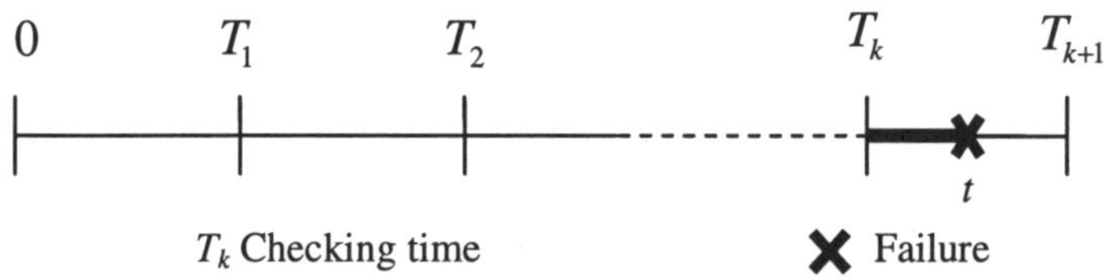

Figure 1. Process of sequential checking time T_k

The total expected cost until a failure is detected and the backup operation is made to the latest checking time is, using the theory of inspection policy [5],

$$\begin{aligned} C_1(T_1, T_2, \cdots) &= \sum_{k=0}^{\infty} \int_{T_k}^{T_{k+1}} \left[kc_1 + c_2 \left(t - T_k \right) \right] dF(t) \\ &= \sum_{k=1}^{\infty} \left[c_1 - c_2 (T_k - T_{k-1}) \right] \overline{F}(T_k) + \mu c_2 . \end{aligned} \tag{1}$$

If a unit is checked at periodic times $kT (k=1,2,\cdots)$ then

$$C_1(T) = (c_1 - c_2 T) \sum_{k=1}^{\infty} \overline{F}(kT) + \mu c_2 . \tag{2}$$

Next, we obtain the expected cost per unit of time for an infinite time span. Since the mean time of backup operation from a failure detection to the latest checking time is

$$\sum_{k=0}^{\infty} \int_{T_k}^{T_{k+1}} (t - T_k) dF(t),$$

the expected cost rate is given by

$$C_2(T_1,T_2,\cdots) = \frac{C_1(T_1,T_2,\cdots)}{\mu + \sum_{k=0}^{\infty} \int_{T_k}^{T_{k+1}} (t - T_k) dF(t)}$$

$$= \frac{c_1 \sum_{k=1}^{\infty} \overline{F}(T_k) - \mu c_2}{2\mu - \sum_{k=1}^{\infty} (T_k - T_{k-1}) \overline{F}(T_k)} + c_2. \tag{3}$$

If a unit is checked at periodic times kT (k=1,2,$\cdots$) then

$$C_2(T) = \frac{c_1 \sum_{k=1}^{\infty} \overline{F}(kT) - \mu c_2}{2\mu - T\sum_{k=1}^{\infty} \overline{F}(kT)} + c_2. \tag{4}$$

3. Optimal Policies

We discuss optimal checking times T_k^* which minimize the expected cost $C_1(T_1,T_2,\cdots)$ in (1). Let $f(t)$ be a density function of $F(t)$, i.e., $f(t) \equiv F'(t)$. Then, differentiating $C_1(T_1,T_2,\cdots)$ with respect to T_k and setting it equal to zero, we have

$$\frac{F(T_{k+1}) - F(T_k)}{f(T_k)} = T_k - T_{k-1} - \frac{c_1}{c_2} \qquad (k = 1,2,\cdots). \tag{5}$$

Thus, we can determine the optimal checking times T_k^*, using Algorithm 1 of [5].

In the periodic inspection case, from (2), we have

$$C_1(0) = \lim_{T \to 0} C_1(T) = \infty,$$
$$C_1(\infty) = \lim_{T \to \infty} C_1(T) = \mu c_2.$$

Hence, we have

$$C_1(\infty)-C_1(\mathrm{T})=(c_2T-c_1)\sum_{k=1}^{\infty}\overline{F}(kT).$$

Thus, there exists an optimal checking time $T_1^*(c_1/c_2<T_1^*<\infty)$ which minimizes $C_1(T)$, and

$$C_1(c_1/c_2)=C_1(\infty)=\mu c_2.$$

Further, differentiating $C_1(T)$ in (2) with respect to T and setting it equal to zero imply

$$T-\frac{\sum_{k=1}^{\infty}\overline{F}(kT)}{\sum_{k=1}^{\infty}k\,f(kT)}=\frac{c_1}{c_2} \tag{6}$$

In the case of $F(t)=1-e^{-\lambda t}$, Equation (6) is

$$T-\frac{1-e^{-\lambda T}}{\lambda}=\frac{c_1}{c_2}. \tag{7}$$

It can be easily seen that the left-hand side of (7) is strictly increasing from 0 to ∞. Thus, there exists a finite and unique T_1^* which satisfies (7), and the resulting cost is

$$C_1(T^*)=c_2T_1^*-c_1. \tag{8}$$

Since $e^a\approx 1+a+a^2/2$ for small $a>0$, the optimal checking time is approximately given by

$$\tilde{T}_1=\sqrt{\frac{2c_1}{\lambda c_2}}, \tag{9}$$

and $T_1^*>\tilde{T}_1$.

We can compute an optimal schedule which minimizes $C_2(T_1,T_2,\cdots)$ in (3), using the algorithm 2 of [5]. When $F(t)=1-e^{-\lambda t}$, Equation (4) is

$$C_2(T)=\frac{c_1e^{-\lambda T}-c_2(1-e^{-\lambda T})/\lambda}{2(1-e^{-\lambda T})/\lambda-Te^{-\lambda T}}+c_2, \tag{10}$$

and

$$C_2(0) = \infty, \qquad C_2(\infty) = \frac{c_2}{2}.$$

Differentiating (10) with respect T and setting it equal to zero, we have

$$\frac{T-(1-e^{-\lambda T})/\lambda}{2-e^{-\lambda T}} = \frac{c_1}{c_2}. \tag{11}$$

It can be easily seen that the left-hand side of (11) is strictly increasing from 0 to ∞. Thus, there exists a finite and unique T_2^* which satisfies (11). By comparing (7) and (11), it can be shown that $T_1^* < T_2^*$, and it is approximately

$$\tilde{T}_2 = \frac{c_1}{c_2} + \sqrt{\left(\frac{c_1}{c_2}\right)^2 + \frac{2c_1}{\lambda c_2}}. \tag{12}$$

4. Finite Interval

Suppose that a unit has to be operating for a finite interval $(0, S]$ $(0 < S < \infty)$, and is replaced at time $S = T_N$ [6]. The other assumptions are the same as those of the previous model. Then, the total expected cost before replacement is

$$C_3(N) = \sum_{k=0}^{N-1} \int_{T_k}^{T_{k+1}} [kc_1 + c_2(t-T_k)]dF(t) + c_1(N-1)\overline{F}(T_N) \quad (N=1,2,\cdots). \tag{13}$$

Putting that $\partial C_3(N)/\partial T_k$, we have (5), and the resulting minimum expected cost is

$$\tilde{C}_3(N) = C_3(N) + c_1 - c_2\int_0^S \overline{F}(t)dt$$

$$= \sum_{k=0}^{N-1} [c_1 - c_2(T_{k+1} - T_k)]\overline{F}(T_k). \tag{14}$$

For example, when N=3, the checking times T_1 and T_2 are given the solution of equations

$$\frac{F(S)-F(T_2)}{f(T_2)}=T_2-T_1-\frac{c_1}{c_2},$$
$$\frac{F(T_2)-F(T_1)}{f(T_1)}=T_1-\frac{c_1}{c_2},$$

and the expected cost is

$$\tilde{C}_3(3)=c_1-c_2T_1+[c_1-c_2(T_2-T_1)]\overline{F}(T_1)+[c_1-c_2(S-T_2)]\overline{F}(T_2).$$

From the above discussion, we compute $T_k(k=1,2,\cdots,N-1)$ which satisfies (5), and substituting them into (14), we obtain the expected cost $\tilde{C}_3(N)$. Next, computing $\tilde{C}_3(N)$ for all $N\geq 1$, we can get the optimal checking number N^* and times $T_k^*(k=1,2,\cdots,N^*)$.

5. Numerical Examples

We compute the optimal checking times numerically when the failure time has a Weibull distribution for the periodic checking time, and when the failure time has a Weibull distribution and a uniform distribution for the sequential checking one.

Suppose that $F(t)=1-\exp[-(\lambda t)^\alpha]$ $(\alpha>1)$. Then, from (6), an optimal time T_1^* is given by a solution of the equation

$$T-\frac{\sum_{k=1}^{\infty} e^{-(\lambda kT)^\alpha}}{\sum_{k=1}^{\infty} \lambda\alpha k(\lambda kT)^{\alpha-1}e^{-(\lambda kT)^\alpha}}=\frac{c_1}{c_2}.$$

Table1 presents the optimal checking times T_1^* for α=1.0, 1.5, 2.0, 3.0 and c_1/c_2=5, 10, 20, 30, 40, 50 when $1/\lambda$=500. Note that when α=1, this corresponds to an exponential case. Approximate times $\tilde{T}_1$ in (9) give a good lower bound for T_1^* for α=1. This indicates that the optimal times become longer as c_1/c_2 are large, however, they are changed little as the values of parameter α increase for α>1.

Table1. Optimal checking times T_1^* to minimize $C_1(T)$ when $F(t)=1-\exp[-(t/500)^{\alpha}]$

α	c_1/c_2					
	5	10	20	30	40	50
1.0	72.42	103.45	148.42	183.81	214.27	241.59
1.5	67.40	95.50	135.53	166.49	192.78	216.11
2.0	66.57	94.14	133.13	163.06	188.28	210.50
2.5	66.59	94.16	133.08	162.89	187.96	210.09
3.0	66.82	94.48	133.57	163.51	188.71	210.87
$\tilde{T}_1$	70.71	100.00	141.42	173.21	200.00	223.61

Table 2 gives the optimal checking schedule $\{T_k^*\}$ which satisfies (5), and $\delta_k \equiv T_{k+1}^* - T_k^*$ when $F(t) = 1 - \exp[-(\lambda t)^2]$ for $c_1/c_2 = 10, 20, 30$. It is roughly seen that the periodic intervals are almost the same value as δ_1, *i.e.*, the interval between the first and second checking times. Each checking time becomes longer as c_1/c_2 are large.

Next, suppose that the failure time is uniformly distributed during $[0, S]$ $(0 < S < \infty)$, *i.e.*,

$$F(t) = \begin{cases} \dfrac{t}{S} & \text{for } 0 \le t \le S, \\ 0 & \text{for } \quad t > S. \end{cases}$$

Then, Equation (5) is rewritten as

$$T_{k+1} - T_k + \frac{c_1}{c_2} = T_k - T_{k-1},$$

which is equal to that of [5]. For example, when S=1000 and c_1/c_2=20, we have that N=10 since $N(N-1)<2c_2S/c_1=100$ [5], and T_k^* = {100, 360, 510, 640, 750, 840, 910, 960, 990, 1000}. These values are a little larger than those in Table 3 for k=2, 3, ⋯, 9 when N=10. It would be trivial in the case of a uniform distribution that the optimal schedule is equal to that of the standard inspection model [6].

Table 2. Optimal checking times T_k^* to minimize $C_1(T_1,T_2,\cdots)$ and $\delta_k=T^*_{k+1}-T^*_k$ when $F(t)=1-\exp[-(t/500)^2]$

k	c_1/c_2=10		c_1/c_2=20		c_1/c_2=30	
	T_k^*	δ_k	T_k^*	δ_k	T_k^*	δ_k
1	145.04	107.24	183.23	136.06	211.09	157.44
2	252.28	91.62	319.29	116.43	368.53	135.06
3	343.90	82.54	435.72	104.90	503.59	121.86
4	426.44	76.49	540.92	97.09	625.45	112.86
5	502.93	72.20	637.71	91.40	738.29	106.13
6	575.13	69.12	729.11	87.15	844.42	100.88
7	644.25	67.03	816.26	84.06	945.30	96.61
8	711.28	65.90	900.32	82.17	1041.91	93.04
9	777.18	65.89	982.49	81.92	1134.95	89.99
10	843.07		1064.41		1224.94	

Table 3 gives the checking times $T_k(k=1,2,\cdots,N)$ and the expected cost $\tilde{C}_3(N)$ for S=1000 and c_1/c_2=20 when $F(t)=1-\exp[-(t/500)^2]$. Comparing $\tilde{C}_3(N)$ for $N=1,2,\cdots,10$ the expected cost is minimum at N=9. That is, the optimal checking number is N^*=9 and optimal checking times are 183.56, 319.94, 436.78, 542.25, 640.21, 732.98, 822.41, 910.45 and 1000. These optimal times are almost the same ones in Table 2 for c_1/c_2=20.

Table 4 gives T_k and the expected cost $\tilde{C}_3(N)$ for S=500 and the same parameters in Table3. In this case, the optimal checking number is N^*=4 and optimal checking times are 174.82, 302.60, 408.84, 500. All checking times in Table 4 are smaller than these in Table 3 for the same N.

Table 3. Checking times T_k and expected cost $\tilde{C}_3(N)$ when S=1000, c_1/c_2=20 and $F(t)=1-\exp[-(t/500)^2]$

k	N									
	1	2	3	4	5	6	7	8	9	10
1	1000	358.24	270.74	231.15	209.68	197.27	190.00	185.83	183.56	182.43
2		1000	518.75	420.95	373.79	347.79	332.93	324.51	319.94	317.68
3			1000	623.89	529.76	483.50	458.25	444.27	436.78	433.10
4				1000	703.78	618.15	576.18	553.93	542.25	536.58
5					1000	769.00	694.31	658.35	640.21	631.54
6						1000	824.00	761.87	732.98	719.61
7							1000	870.82	822.41	801.36
8								1000	910.45	876.46
9									1000	943.53
10										1000
$\tilde{C}_3(N)/c_2$	-17.95	-213.82	-273.18	-299.18	-312.05	-318.61	-321.85	-323.28	-323.73	-323.67

Table 4. Checking times T_k and expected cost $\tilde{C}_3(N)$ for N=1, 2, $\cdots$, 7 when S=500, c_1/c_2=20 and $F(t)=1-\exp[-(t/500)^2]$

k	N						
	1	2	3	4	5	6	7
1	500	263.71	201.84	174.82	162.01	156.55	155.33
2		500	357.26	302.60	277.67	267.21	264.87
3			500	408.84	369.89	353.91	350.37
4				500	443.77	421.41	416.49
5					500	470.32	463.88
6						500	491.91
7							500
$\tilde{C}_3(N)/c_2$	-176.58	-264.10	-280.93	-282.10	-277.75	-271.19	-263.89

6. Concluding Remarks

We have considered the recovery model where we carry out the backup operation to the latest checking time when a failure was detected, and have discussed the optimal policies which minimize the total expected cost or the expected cost rate, using the techniques of inspection policies. Further, we have computed numerically the optimal checking time when a unit should be operating for a finite interval when the failure time has a Weibull distribution. These techniques would be applied to other backup models of a database system [11] and the reliability model with backward time [12].

It is more useful in some systems to adopt the availability than the expected cost as an appropriate objective function. If β_1 is the time for one check then the availability is

$$A(T_1,T_2,\cdots)=\frac{\text{Mean operating time}}{\text{Mean time until the complection of backup operation}}$$

$$=\frac{\mu}{2\mu-\sum_{k=1}^{\infty}(\beta_1+T_k-T_{k-1})\overline{F}(T_k)}. \tag{15}$$

Thus, the optimal policy which maximizes the availability corresponds to the policy which minimizes the expected cost $C(T_1,T_2,\cdots)$ in (1) by replacing $c_1=\beta_1$ and $c_2=1$.

Finally, we consider the following model as one of modified ones: A fault occurs between the j–th and the $(j+1)$–th checking times according to a general distribution $F(t)$ with finite mean μ, and is hidden. After that, a failure occurs between the k–th and the $(k+1)$–th checking times and is detected immediately, and the time from the occurrence of a fault to the failure detection obeys a general distribution $G(x)$ with mean θ. This is called a fault latency [13]. When a failure was detected, we carry out the backup operation to the j–th checking time. Then, the expected cost until a failure is detected and the backup operation is made to the j–th checking time is

$$C_4(T_1,T_2,\cdots)=\sum_{k=1}^{\infty}\sum_{j=0}^{k-1}\int_{T_k}^{T_{k+1}}\left\{\int_{T_j}^{T_{j+1}}\left[kc_1+c_2(x-T_j)\right]dF(t)\right\}dG(x-t)$$
$$+\sum_{j=0}^{\infty}\int_{T_j}^{T_{j+1}}\left\{\int_{T_j}^{x}\left[jc_1+c_2\left(x-T_j\right)\right]dF(t)\right\}dG(x-t)$$
$$=\sum_{j=0}^{\infty}\int_{T_j}^{T_{j+1}}\left\{\sum_{k=j+1}^{\infty}\int_{T_k}^{T_{k+1}}\left[kc_1+c_2\left(x-T_j\right)\right]dG(x-t)\right\}dF(t)$$
$$+\sum_{j=0}^{\infty}\int_{T_j}^{T_{j+1}}\left\{\int_{x}^{T_{j+1}}\left[jc_1+c_2\left(x-T_j\right)\right]dG(x-t)\right\}dF(t)$$
$$=c_1\sum_{j=0}^{\infty}\int_{T_j}^{T_{j+1}}\left[j+\sum_{k=j+1}^{\infty}\bar{G}(T_k-t)\right]dF(t)$$
$$+c_2\sum_{j=0}^{\infty}\int_{T_j}^{T_{j+1}}\left[\theta+t-T_j\right]dF(t)$$
$$=c_1\sum_{k=1}^{\infty}\left[\bar{F}(T_k)+\int_0^{T_k}\bar{G}(T_k-t)\,dF(t)\right]$$
$$+c_2\left\{\mu+\theta-\sum_{k=1}^{\infty}T_k\left[\bar{F}(T_k)-\bar{F}(T_{k+1})\right]\right\}. \qquad (16)$$

If a faults is not hidden and is detected immediately, *i.e.*, $G(x)\equiv 1$ for $x\geq 0$ and $\theta=0$, this corresponds to the previous model and Equation (16) agrees with (1).

Further, suppose that $f(t)$ and $g(x)$ are density functions of $F(t)$ and $G(x)$, respectively. Then, differentiating $C_4(T_1, T_2, \cdots)$ with respect to T_k and setting it equal to zero, we have

$$\frac{(T_k-T_{k-1})f(T_k)-\left[F(T_{k+1})-F(T_k)\right]}{\int_0^{T_k}g(T_k-t)dF(t)}=\frac{c_1}{c_2}. \qquad (17)$$

Thus, using Algorithm 1 of [5], we can compute the optimal checking time numerically for specified distributions $F(t)$ and $G(x)$.

References

1. T.Nakagawa, *J. of Oper Res Soc. of Jpn* **23**, 13(1980).

2. LC.Thomas, IA Jacobs and DI.Gaver, *Commun Statist Stoch Model* **3**, 259 (1987).
3. EC.Martimez, *Proceed of Annual Reliability and Maintainbility Symposium*, 181 (1984).
4. K.Ito and T.Nakagawa, *Comput Math Appl* **24**, 87 (1992).
5. RE.Barlow and F.Proschan, *Mathematical Theory of Reliability*, John Wiley and Sons, New York (1965).
6. T.Nakagawa, *Maintenance Theory of Reliability*, Springer–Verlag, London (2005).
7. A.Reuter, *ACM Trans Database Syst* **9**, 526 (1984).
8. S.Fukumoto, N.Kaio and S.Osaki, *Comput Math Appl* **24**, 63 (1992).
9. H.Sandoh, N.Kaio and H.Kawai, *Reliab Eng Syst Saf* **37**, 29 (1992).
10. H.Sandoh and H.Kawai, *Math Comput Model* **22**, 289 (1995).
11. M.Ohara, M.Arai, S.Fukumoto and K.Iwasaki, *Proceedings of International Workshop on Recent Advances Stochastic Operations Research*, 187 (2005).
12. T.Nakagawa, S.Mizutani and T.Sugiura, *Eleventh ISSAT International Conference on Reliability and Quality in Design*, 219 (2005).
13. KG.Shin and YH.Lee, *IEEE Trans Comput* **C-35**,370 (1986).

ESTIMATING AGE REPLACEMENT POLICIES FROM SMALL SAMPLE DATA*

K. RINSAKA and T. DOHI

Department of Information Engineering,
Graduate School of Engineering,
Hiroshima University, Higashi-Hiroshima, Japan
E-mail: {rinsaka, dohi}@rel.hiroshima-u.ac.jp

In the present paper, we consider the typical age replacement models to minimize the relevant expected costs, and formulate the statistical estimation problems with the complete sample of failure time data. Based on the concept of total time on test statistics, we show that the underlying optimization problems are translated to the graphical ones on the data space. Next, we utilize a kernel density estimator and improve the existing statistical estimation algorithms in terms of convergence speed. Throughout simulation experiments, it can be shown that the developed algorithms are useful especially for the small sample problems, and enable us to estimate the optimal age replacement times with higher accuracy.

1. Introduction

Since Barlow and Proschan[1] published their seminal book, a number of optimal replacement models under uncertainty have been developed in the literature. For the ordinary age replacement problems, Bergman[2] and Bergman and Klefsjö[3,4] developed statistically the non-parametric estimation algorithms based on the total time on test concept to obtain the optimal age replacement times from the complete sample of failure time data. If a sufficient number of failure time data can be obtained, then the resulting estimates of the optimal age replacement times in the references[2,3,4] asymptotically converge to the real but unknown optimal solutions. Hence, the non-parametric estimation algorithms are quite useful to realize a maintenance control with incomplete knowledge on failure time distribution.

*The present research was partially supported by the Ministry of Education, Science, Sports and Culture, Grant-in-Aid for Young Scientists (B); Grant No. 18710145 (2006–2007) and Grant-in-Aid for Scientific Research (B); Grant No. 16310116 (2004–2006).

When the non-parametric estimation is carried out, it is important to obtain the more accurate solution in the situation where only fewer failure data are obtained.

The aim of the present paper is to improve the estimation accuracy of the optimal age replacement times for age replacement problems with a small sample data. More precisely, we propose the non-parametric estimation algorithms based on the kernel density estimator[5,6,7,8,9] for some typical age replacement problems. In Section **2**, we formulate the expected cost per unit time and the expected total discounted cost over an infinite time horizon, and consider these age replacement models; the basic age replacement model, the imperfect age replacement model and the discounted model. In Sections **3** and **4**, we formulate the statistical estimation problems with the complete sample of failure time data. Based on the concept of total time on test statistics, we show that the underlying optimization problems are translated to the graphical ones on the data space. In Section **5**, we utilize a kernel density estimator and improve the statistical estimation algorithms in terms of convergence speed. Throughout simulation experiments, it can be shown that the developed algorithms are useful especially for the small sample problem, and enable us to estimate the optimal age replacement times with higher accuracy.

2. Age Replacement Problems

Under the age replacement policy, a unit is replaced at the failure time or the pre-specified age T (> 0) whichever occurs first. Let $\overline{F} = 1 - F$ denote the survival function of the time to failure of a unit. It is assumed that the lifetime distribution F is continuous and strictly increasing and that the mean time to failure $\mu = \int_0^\infty \overline{F}(t)dt$ is finite. We assume that the unit can be replaced at failure with cost $c + K$ $(c > 0,\ K > 0)$ and a preventive replacement with cost c, where K is an additional cost to the failure replacement.

Model 1: The first model is the basic age replacement problem[1,2], which consists in finding an optimal age $T = T^*$ minimizing the expected cost per unit time in the steady state:

$$C(T) = \frac{(c+K)F(T) + c\overline{F}(T)}{\int_0^T t dF(t) + T\overline{F}(T)} = \frac{c + KF(T)}{\int_0^T \overline{F}(t)dt}. \tag{1}$$

Model 2: The second model considers a more general situation where the preventive maintenance at T is imperfect[3]. Let p $(0 \le p \le 1)$ denote the probability that the preventive maintenance is imperfect. Then the expected cost per unit time in the steady state, $C_p(T)$, is given by

$$C_p(T) = \frac{(c+K)F(T) + [c+p(c+K)]\,\overline{F}(T)}{\int_0^T \overline{F}(t)dt} = \frac{[K-p(c+k)] \times [F(T)+d(p)]}{\int_0^T \overline{F}(t)dt}, \tag{2}$$

where

$$d(p) = \frac{c/(c+K)+p}{K/(c+K)-p}. \tag{3}$$

Model 3: The third model justifies the present value of expected cost over an infinite time horizon by taking account of discounting[4]. Let us define the discount factor α (> 0) to represent the net present value of the total expected cost over an infinite time horizon $C_\alpha(T)$. Then, we have

$$C_\alpha(T) = \frac{ce^{-\alpha T}\overline{F}(T) + (c+K)\int_0^T e^{-\alpha t}dF(t)}{1-e^{-\alpha T}\overline{F}(T) - \int_0^T e^{-\alpha t}dF(t)} = \frac{c+K\left[1-e^{-\alpha T}\overline{F}(T)\right] - \alpha(c+K)\int_0^T e^{-\alpha t}\overline{F}(t)dt}{\alpha\int_0^T e^{-\alpha t}\overline{F}(t)dt}. \tag{4}$$

3. The TTT Concept

To derive the optimal age replacement time on the graph, we define the equilibrium distribution or equivalently the scaled total time on test (TTT) transform[10] of the lifetime distribution function $F(t)$ by

$$\phi(u) = \frac{1}{\mu}\int_0^{F^{-1}(u)} \overline{F}(t)dt. \tag{5}$$

Since $F(t)$ is a nondecreasing function, there always exists its inverse function:

$$F^{-1}(u) = \inf\{t; F(t) \geq u\}, \quad 0 \leq u \leq 1. \tag{6}$$

Because the expected costs per unit time given in Eqs.(1) and (2) are represented by the scaled TTT transform of $F(t)$, the following result can be easily obtained[2,3].

Theorem 1: Obtaining the optimal age replacement time which minimizes the expected cost per unit time for Model i $(= 1, 2)$ can be reduced to the following maximization problem:

$$\max_{0 \leq u \leq 1} \frac{\phi(u)}{u + \eta_i}, \tag{7}$$

where

$$\eta_1 = c/K, \quad \eta_2 = d(p). \tag{8}$$

Theorem 1 can be obtained by transforming $C(T)$ and $C_p(T)$ to the functions of u by means of $u = F(t)$. If the lifetime distribution $F(t)$ is known, then the optimal age replacement times can be obtained from Theorem 1 by $T^* = F^{-1}(u^*)$, where $u^*(0 \leq u^* \leq 1)$ is given by the x coordinate value u^* for the point of the curve with the largest slope among the line pieces drawn from the point $(-\eta_i, 0)$ $(-\infty < -\eta_i < 0)$ on a two-dimensional plane to the curve $(u, \phi(u)) \in [0, 1] \times [0, 1]$.

Next, consider the discounting problem. Following Bergman and Klefsjö[4], we define the modified scaled total time on test transform of the lifetime distribution by

$$\phi_\alpha(u) = \frac{1}{\mu_\alpha} \int_0^{F_\alpha^{-1}(u)} \overline{F}_\alpha(t)dt, \tag{9}$$

$$F_\alpha(t) = 1 - e^{-\alpha t}\overline{F}(t), \tag{10}$$

$$\mu_\alpha = \int_0^\infty \overline{F}_\alpha(t)dt, \tag{11}$$

$$F_\alpha^{-1}(u) = \inf\left\{x : F_\alpha(x) \geq u\right\}. \tag{12}$$

Since the expected total discounted cost in Eq.(4) is represented as a function of $\phi_\alpha(u)$ and u, the following result can be obtained[4].

Theorem 2: Obtaining the optimal age replacement time which minimizes the expected total discounted cost over an infinite time horizon for Model 3 can be reduced to the following maximization problem:

$$\max_{0 \leq u \leq 1} \frac{\phi_\alpha(u)}{u + c/K}. \tag{13}$$

4. An Empirical Method

Next, we consider the case where the failure time distribution $F(t)$ is unknown. It is assumed that the order statistics $x_1, x_2, \cdots, x_n$ for n (> 0) failure time data are observed and that they are the complete data without truncation from $F(t)$. Let us define the estimate of the failure occurrence time distribution $F(t)$ by the empirical distribution function:

$$F_n(x) = \begin{cases} j/n & \text{for } x_j \leq x \leq x_{j+1} \\ 1 & \text{for } x_n \leq x. \end{cases} \tag{14}$$

As an estimate of the scaled TTT transform based on the empirical distribution function, we define the following scaled total time on test statistics:

$$\phi_{n,j} = \psi_j / \psi_n, \tag{15}$$

where the function

$$\psi_j = \sum_{k=1}^{j} (n-k+1)(x_k - x_{k-1}), \quad j = 1, 2, \cdots, n; \quad \psi_0 = 0 \tag{16}$$

is called the TTT statistics[10]. Plot the point sequence $(j/n,\ \phi_{n,j})$ $(j = 0, 1, 2, \cdots, n)$ on the two-dimensional plane. By connecting the points, the scaled TTT plot is obtained. Since $(j/n,\ \phi_{n,j})$ $(j = 0, 1, 2, \cdots, n)$ is a nonparametric estimate of $(u, \phi(u))$, $u \in [0, 1]$, the following theorem on the optimal age replacement time is obtained by direct application of the result in Theorem 1.

Theorem 3: Suppose that the order statistics $x_1 \leq x_2 \leq \cdots \leq x_n$ of n complete data on the failure time are observed in Model i $(= 1, 2)$. The nonparametric estimate $\hat{T}$ of the optimal age replacement time minimizing the expected cost per unit time is given by x_{j^*}, where

$$j^* = \left\{ j \,\middle|\, \max_{0 \leq j \leq n} \frac{\phi_{n,j}}{j/n + \eta_i} \right\}. \tag{17}$$

Next, define the modified scaled TTT statistics based on this sample by

$$\phi_{n,j,\alpha} = \psi_{j,\alpha} / \psi_{n,\alpha}, \tag{18}$$

where

$$\psi_{j,\alpha} = \sum_{k=1}^{j} (n-k+1)(x_k - x_{k-1}) e^{-\alpha x_k}, \quad j = 1, 2, \cdots, n; \quad \psi_{0,\alpha} = 0. \tag{19}$$

The similar but somewhat different statistics in Eq.(18) was proposed by Bergman and Klefsjö[4]. Plotting the point $(j/n,\ \phi_{n,j,\alpha})$ $(j = 0, 1, 2, \cdots, n)$ and connecting them by line segments yield the modified curve for the discounting problem.

Theorem 4: Suppose that the order statistics $x_1 \le x_2 \le \cdots \le x_n$ of n of complete data on the failure time are observed in Model 3. The nonparametric estimate $\hat{T}$ of the optimal age replacement time minimizing the expected total discounted cost over an infinite time horizon is given by x_{j^*}, where

$$j^* = \left\{ j | \max_{0 \le j \le n} \frac{\phi_{n,j,\alpha}}{j/n + c/K} \right\}. \tag{20}$$

5. Kernel Method

In this section, we propose the kernel density estimation to obtain the optimal age replacement time from the small sample data. Suppose that the lifetime data $x_1, x_2, \cdots, x_n$ are the sample from a probability density function f. Define the kernel density estimator[5,6,7,8,9] by

$$\hat{f}(y) = \frac{1}{nh} \sum_{i=1}^{n} \Phi\left(\frac{y - x_i}{h}\right), \tag{21}$$

where h (> 0) is the window width, and is often called the smoothing parameter or bandwidth. The function Φ is called the kernel function which satisfies the conditions:

$$\int_{-\infty}^{\infty} \Phi(t)dt = 1, \quad \int_{-\infty}^{\infty} t\Phi(t)dt = 0, \quad \int_{-\infty}^{\infty} t^2\Phi(t)dt = r^2 \neq 0. \tag{22}$$

In many cases, the kernel function Φ will be selected as a symmetric probability density function. In Table 1, we give the typical examples of the kernel function. Since the kernel estimator is a sum of 'bumps' placed at the observations, the kernel function Φ determines the shape of the bumps while the window width h determines their width.

In order to estimate the optimal age replacement time from the failure time data, we define the estimate of the scaled TTT transform by

$$\phi_k(u) = \frac{1}{\hat{\mu}_n} \int_0^{\hat{F}^{-1}(u)} \overline{\hat{F}}(t)dt, \tag{23}$$

where

$$\hat{\mu}_n = \sum_{j=1}^{n} x_j/n, \tag{24}$$

Table 1. Examples of the kernel function[8].

Kernel	$\Phi(t)$
Rectangular	$\frac{1}{2}$ for $\|t\| < 1$, 0 otherwise
Gaussian	$\frac{1}{\sqrt{2\pi}} e^{-(1/2)t^2}$
Triangular	$1 - \|t\|$ for $\|t\| < 1$, 0 otherwise
Biweight	$\frac{15}{16}\left(1 - t^2\right)^2$ for $\|t\| < 1$, 0 otherwise
Epanechnikov	$\frac{3}{4}\left(1 - \frac{1}{5}t^2\right)/\sqrt{5}$ for $\|t\| < \sqrt{5}$, 0 otherwise

and

$$\hat{F}(t) = \int_0^t \hat{f}(s) ds. \tag{25}$$

The following theorem on the optimal age replacement time is easily obtained from the analogy to Theorem 1.

Theorem 5: Suppose that n complete data $x_1, x_2, \cdots, x_n$ on the failure time are observed in Model i $(= 1, 2)$. The nonparametric estimate $\hat{T}$ of the optimal age replacement time minimizing the expected cost per unit time is given by $T^* = \hat{F}^{-1}(u^*)$ satisfying:

$$\max_{0 \le u \le 1} \frac{\phi_k(u)}{u + \eta_i}. \tag{26}$$

Next, define the estimator of the modified scaled TTT transform by

$$\phi_{k,\alpha}(u) = \frac{1}{\hat{\mu}_{n,\alpha}} \int_0^{\hat{F}_\alpha^{-1}(u)} \overline{\hat{F}}_\alpha(t) dt, \tag{27}$$

where,

$$\hat{\mu}_{n,\alpha} = \sum_{j=1}^{n} \frac{(n - j + 1)(x_j - x_{j-1}) e^{-\alpha x_j}}{n}. \tag{28}$$

Theorem 6: Suppose that n complete data $x_1, x_2, \cdots, x_n$ on the failure time are observed in Model 3. The nonparametric estimate $\hat{T}$ of the optimal age replacement time minimizing the expected total discounted cost is given by $T^* = \hat{F}^{-1}(u^*)$ satisfying the following:

$$\max_{0 \le u \le 1} \frac{\phi_{k,\alpha}(u)}{u + c/K}. \tag{29}$$

When we utilize the kernel method, the problem of choosing the design parameter h is of crucial importance. In this paper, we apply two methods for choosing an ideal value of the smoothing parameter; likelihood cross-validation, and reference to a standard distribution.

The former is based on the likelihood function to judge the goodness-of-fit of a statistical model[9,11,12]. In the basic algorithm, an arbitrary data x_k is removed from the sample, and the appropriate density estimate at the point x_k from the remaining $n-1$ sample is calculated by

$$\hat{f}_{nk}(x_k) = \frac{1}{(n-1)h} \sum_{\substack{j=1 \\ j \neq k}}^{n} \Phi\left(\frac{x_k - x_j}{h}\right). \tag{30}$$

Next we choose the ideal value of h so as to satisfy the following the maximum likelihood criterion:

$$\max_{h \geq 0} L(h) = \frac{1}{n} \sum_{k=1}^{n} \log \hat{f}_{nk}(x_k). \tag{31}$$

On the other hand, when the Gaussian kernel function

$$\Phi(t) = \frac{1}{\sqrt{2\pi}} e^{-(1/2)t^2} \tag{32}$$

is assumed for the kernel estimation, the reference to standard distribution can be applied to select the ideal smoothing parameter. We select the ideal window width so as to minimize the mean integrated square error (MISE)[6]:

$$\mathrm{MISE}(\tilde{f}) = E \int_{-\infty}^{\infty} \left\{\tilde{f}(x) - f(x)\right\}^2 dx, \tag{33}$$

where $\tilde{f}$ means the kernel density estimator of the underlying density f. The MISE is the most widely used measure on the global accuracy of $\tilde{f}$ as an estimator of f, and it can be approximated as

$$\begin{aligned}
\mathrm{MISE}(\tilde{f}) &= \int_{-\infty}^{\infty} E\left\{\tilde{f}(x) - f(x)\right\}^2 dx \\
&= \int_{-\infty}^{\infty} \left\{E\tilde{f}(x) - f(x)\right\}^2 dx + \int_{-\infty}^{\infty} \mathrm{Var}\tilde{f}(x) dx \\
&\approx \frac{1}{4} h^4 r^2 \int_{-\infty}^{\infty} f''(x)^2 dx + n^{-1} h^{-1} \int_{-\infty}^{\infty} \Phi(t)^2 dt.
\end{aligned} \tag{34}$$

It can be shown that the ideal value of the window width[5], from the viewpoint of minimizing Eq.(34) is given by

$$h_{ideal} = r^{-2/5} \left\{\int_{-\infty}^{\infty} \Phi(t)^2 dt\right\}^{1/5} \left\{\int_{-\infty}^{\infty} f''(x)^2 dx\right\}^{-1/5} n^{-1/5}. \tag{35}$$

The most tractable approach is to assume the normal distribution with density φ and variance σ^2 to assign a value to the term $\int f''(x)^2 dx$ in Eq.(35) for the ideal window width. This yields

$$\int_{-\infty}^{\infty} f''(x)^2 dx = \sigma^{-5} \int_{-\infty}^{\infty} \varphi''(x)^2 dx = \frac{3}{8}\varphi^{-1/2}\sigma^{-5}. \tag{36}$$

If the Gaussian kernel in Eq.(32) is used, then the window width obtained from Eq.(35) is given by

$$\begin{aligned} h_{ideal} &= (4\pi)^{-1/10}\frac{3}{8}\pi^{-1/2}\sigma n^{-1/5} \\ &= \left(\frac{4}{3}\right)^{1/5} \sigma n^{-1/5} \approx 1.06\sigma n^{-1/5}. \end{aligned} \tag{37}$$

We can see from Eq.(37) that the ideal window width becomes small as the number of observed data increases. Note that the ideal window size in Eq.(37) can be found very easily compared with the likelihood cross-validation in Eq.(31).

6. Simulation Experiments

Of our interest in this section is the investigation of asymptotic properties and convergence speed of estimates proposed in previous sections. Suppose that the lifetime obeys the Weibull distribution:

$$F(x) = 1 - e^{-(x/\theta)^\gamma} \tag{38}$$

with shape parameter $\gamma = 2.0$ and scale parameter $\theta = 0.2$. The other parameters are fixed as $c = 1$, $K = 9$, $p = 0.2$, $\alpha = 0.1$. Under these assumptions, the optimal age replacement times for Model 1, Model 2 and Model 3 can be derived as $T^* = 0.0673$, $T^* = 0.1358$ and $T^* = 0.0674$, respectively.

Let us consider an estimation of the optimal age replacement time minimizing the expected cost per unit time when the failure time data are already observed. It is assumed that the observed data consist of 30 pseudo random numbers generated from the Weibull distribution in Eq.(38). For the 30 pseudo random numbers, we determine the window size as $h^* = 0.0698$ for the Epanechnikov's kernel by solving the maximization problem in Eq.(31) where the Epanechnikov's kernel function[13] is given by

$$\Phi(t) = \begin{cases} \frac{3}{4}\left(1 - \frac{1}{5}t^2\right)/\sqrt{5} & \text{for } |t| < \sqrt{5}, \\ 0 & \text{otherwise}. \end{cases} \tag{39}$$

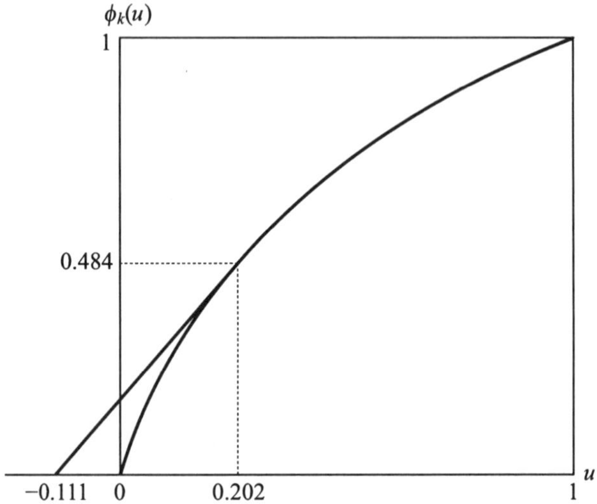

Figure 1. Estimation of the optimal age replacement time based on the kernel density estimation (Model 1).

In Fig. 1, we present an estimation example of the optimal age replacement time minimizing the expected cost per unit time in Model 1 based on the kernel density estimation from 30 data. The point with the steepest slope among the line segments drawn from $(-\eta_1, 0) = (-0.111, 0)$ to the scaled TTT plot $\phi_k(u)$ is $u^* = 0.202$. Hence, the optimal age replacement time can be estimated as $\hat{T}^* = 0.0877$.

Next, let us study the asymptotic behavior of two nonparametric estimation algorithms, namely, the empirical distribution and the kernel density estimation. Monte Carlo simulations are carried out with pseudo random numbers based on the Weibull distribution in Eq.(38), in order to investigate the convergence toward the real optimal solution. Figures 2 to 4 show the asymptotic behavior of the optimal age replacement times for Model 1, Model 2 and Model 3. Here, the asymptotic behavior in Figs. 2–4 were obtained from the same data. It can be seen that the results by the Epanechnikov kernel with the likelihood cross-validation are quite similar to ones by the Gaussian kernel with the reference to the standard distribution for Model 1 and Model 3. It is found from these figures that the results converge to the real optimal solutions when the number of failure time data is close to 20. The convergence speed of the estimates based on the empirical distribution in Fig. 3 is faster than that of the kernel density, since the

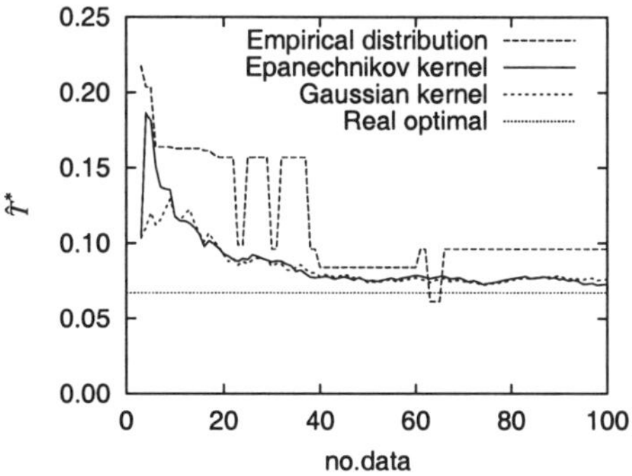

Figure 2. Asymptotic behavior of estimates of the optimal age replacement time (Model 1).

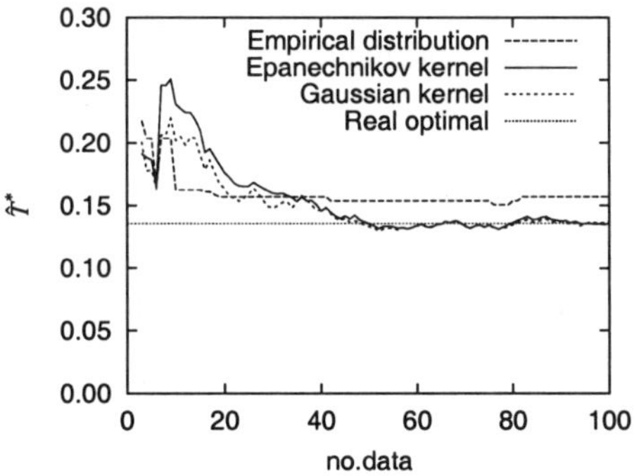

Figure 3. Asymptotic behavior of estimates of the optimal age replacement time (Model 2).

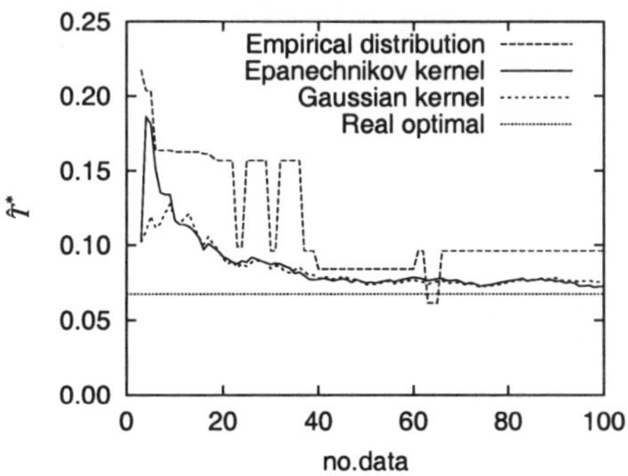

Figure 4. Asymptotic behavior of estimates of the optimal age replacement time (Model 3).

6th observed failure time data is close to the real optimal age replacement time. However we cannot recognize the remarkable improvements in the accuracy of estimates based on the empirical distribution after the 6th observation. In the empirical distribution, when the failure time data close to the real optimal solution is observed, the estimation may function satisfactorily, and vice versa. This is because the estimate of the optimal solution is given by one of n points $x_{(j)}$, $j = 1, \cdots, n$. In the kernel methods, on the other hand, even if the failure time close to the real optimal solution is not observed, an accurate estimate may be obtained, since the kernel method can evaluate the density function continuously.

Finally, we investigate the convergence speed of the kernel method. Figures 5 to 7 show the relative absolute error average (RAEA) of estimates of the optimal age replacement times, where the Monte Carlo simulations are carried out 1,000 times. For a small sample problem, we can observe that the convergence speed of the optimal age replacement time estimated by the kernel density estimation is faster than that by the empirical distribution. Especially, the estimation algorithm based on the Gaussian kernel with the reference to standard distribution provides a very quick convergence speed. From these results, we conclude that the statistical algorithm based on the kernel density estimation can be recommended to estimate the optimal age replacement time, especially for the small sample problem.

7. Concluding Remarks

In the present paper, we have considered the typical age replacement models and developed the statistical estimation algorithms with the complete sample of failure time data. The non-parametric estimation algorithms based on the kernel density estimation have been proposed to improve the estimation accuracy for small sample of failure time data. The determination of the window width has been quite important in the kernel density estimation. In this paper, two methods for choosing the smoothing parameter have been applied; namely, the likelihood cross-validation, and the reference to standard distribution. Throughout simulation experiments, it has been shown that the proposed algorithm based on the kernel density estimation had higher estimation accuracy than the empirical distribution, and faster convergence speed to the theoretical optimal replacement time. Especially, the Gaussian kernel with the reference to standard distribution has provided the very nice convergence speed.

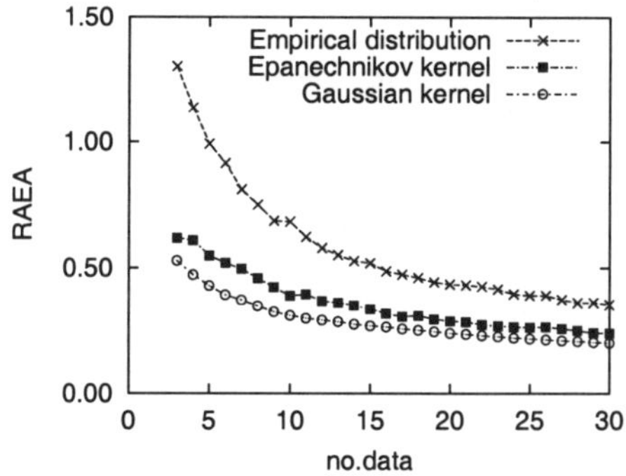

Figure 5. Relative absolute error average of estimates of the optimal age replacement time (Model 1).

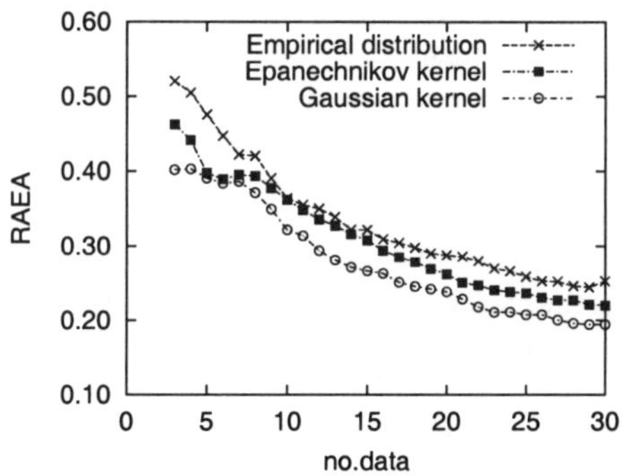

Figure 6. Relative absolute error average of estimates of the optimal age replacement time (Model 2).

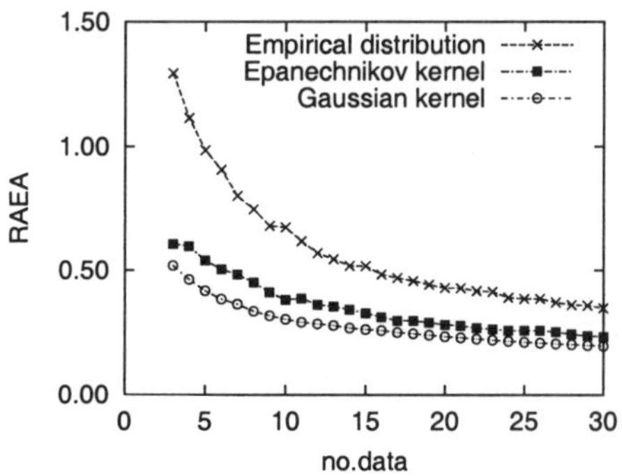

Figure 7. Relative absolute error average of estimates of the optimal age replacement time (Model 3).

References

1. R.E Barlow and F. Proschan, *Mathematical Theory of Reliability*, John Wiley & Sons, New York (1965).
2. B. Bergman, On age replacement and the total time on test concept, *Scandinavian Journal of Statistics*, **6**,161 (1979).
3. B. Bergman and B. Klefsjö, A graphical method applicable to age-replacement problems, *IEEE Transactions on Reliability*, **R-31**, 478 (1982).
4. B. Bergman and B. Klefsjö, TTT transforms and age replacements with discounted costs, *Naval Research Logistics Quarterly*, **30**, 631 (1983).
5. E. Parzen, On the estimation of a probability density function and the mode, *Annals of Mathematical Statistics*, **33**, 1065 (1962).
6. M. Rosenblatt, Remarks on some nonparametric estimates of a density function, *Annals of Mathematical Statistics*, **27**, 832 (1956).
7. T. Cacoullos, Estimation of a multivariate density, *Annals of the Institute of Statistical Mathematics*, **18**, 178 (1966).
8. B.W. Silverman, *Density Estimation for Statistics and Data Analysis*, Chapman and Hall, London (1986).
9. A.J. Izenman, Recent developments in nonparametric density estimation, *Journal of American Statistical Association*, **86**, 205 (1991).
10. R.E. Barlow and R. Campo, Total time on test processes and applications to failure data, in *Reliability and Fault Tree Analysis*, eds. R.E. Barlow, J. Fussell and N.D. Singpurwalla, 451, SIAM, Philadelphia (1975).
11. R.P.W. Duin, On the choice of smoothing parameters for Parzen estimators of probability density functions, *IEEE Trnasactions on Computer* **C-25**, 1175 (1976).
12. J.D.F. Habbema, J. Hermans, and K. van der Broek, A stepwise discrimination program using density estimation, *Proceedings of Computational Statistics* (ed. by G. Bruckman), 100, Physica Verlag, Vienna (1974).
13. V.A. Epanechnikov, Nonparametric estimation of a multidimensional probability density, *Theory of Probability and Its Applications*, **14**, 153 (1969).

PART C

Finance

STOCK REPURCHASE POLICY WITH TRANSACTION COSTS UNDER JUMP RISKS*

HIROMICHI GOKO

Bank of Japan
2-1-1 Nihonbashi-Hongokucho Chuo-ku, Tokyo, 103-8660, JAPAN
E-mail: hiromichi.gouko@boj.or.jp

MASAMITSU OHNISHI

Graduate School of Economics, Osaka University
1-7 Machikaneyama Toyonaka, Osaka, 560-0043, JAPAN
E-mail: ohnishi@econ.osaka-u.ac.jp

MOTOH TSUJIMURA†

Faculty of Economics, Ryukoku University
67 Fukakusa Tsukamoto-cho Fushimi-ku, Kyoto, 612-8577, JAPAN
E-mail: tsujimura@econ.ryukoku.ac.jp

We examine a stock repurchase policy with fixed and proportional transaction costs under jump risks. The firm's problem is to maximize the expected total discounted stock repurchases. To solve the problem, we formulate it as a stochastic impulse control problem, and then approach it using quasi-variational inequalities (QVI). Then, we prove that the value function is a solution to the QVI and that the QVI policy is optimal. Furthermore, we present the results of numerical examples and conduct comparative-static analysis. The amount of the ith stock repurchase is increasing in the fixed and proportional transaction costs. An increase in the fixed and proportional transaction costs lengthens the expected interval of stock repurchase time. Unfortunately, the effect of jump risk on the stock repurchase policy is ambiguous.

*The second and third authors were partially supported by daiwa securities group inc. The third author was also partially supported by the ministry of education, culture, sports, science, and technology under a grant-in-aid for young scientists (b), 157170113. The authors would like to thank two anonymous referees for helpful comments.
†Corresponding author.

1. Introduction

Firms mainly distribute cash flows to shareholders in the form of dividends or stock repurchases. Stock repurchasing has become an important method of distributing cash flows to shareholders. Firms repurchase stock for the following reasons: to distribute cash flow (see, for example, Jensen[7] and Stephens and Weisbach[13]); to announce a firm's manager's belief that the firm's stock is undervalued (see, for example, Vermaelen[16] and Stephens and Weisbach[13]); to avoid unwanted takeover attempts (see, for example, Bagwell[1]); and to counter the dilution effects of employee and management stock options (see, for example, Fenn and Liang[4]). Refer to Dittmar[3] for the reasons for stock repurchases. Jagannathan, Stephens and Weisbach[6] investigate the firm's decision between distributing cash flows in the form of dividends and stock repurchases. See also Guay and Harford[5].

We assume that a firm's accumulated net revenues are governed by a jump-diffusion and that the firm distributes cash flow as dividends and by repurchasing stock. We assume that dividends represent a stable cash distribution to stockholders, so that the firm constantly pays out the same amount of dividends in each dividend period. On the other hand, we assume that stock repurchases represent temporary cash distribution to stockholders. Thus, we concentrate on how the firm repurchases stock and examine an optimal stock repurchase policy. In this context, we assume that when the firm repurchases stock, it incurs both fixed and proportional transaction costs. For example, the fixed transaction costs could be associated with the firm's decision making, while the proportional transaction costs might be taxes. We also assume that the firm's net revenue jumps because of, for example, business conditions and lawsuits. Furthermore, we assume that the firm goes bankrupt when the cash reserve falls to zero. Then, the firm's problem is to maximize the expected total discounted amount of stock repurchases. To solve this problem, we formulate it as an impulse control problem. Then, we prove that a policy, which is derived from quasi-variational inequalities (QVI), is an optimal stock repurchase policy for the firm's problem and show that a function that satisfies the QVI coincides with the value function of the firm's problem. Furthermore, we present the numerical results of simulations for the amount of stock repurchased, the expected interval of stock repurchase time, and the value function. We can then obtain comparative-static results. The amount of the ith stock repurchase is increasing in the fixed and proportional transaction costs. An increase in the fixed and proportional transaction costs lengthens the expected interval of stock repurchase time. Unfortunately, the jump risk has an ambiguous effect on the stock repurchase policy.

In Jeanblanc-Picqué and Shiryaev[8] and Ohnishi and Tsujimura[10], optimal dividend problems are examined by using stochastic impulse control. We extend the cash reserve process of Jeanblanc-Picqué and Shiryaev[8] and Ohnishi and Tsujimura[10] by dealing with a jump-diffusion process. Furthermore, in this paper, we consider positive and negative jumps. Takashima and Tsujimura[14] investigate a dividend and stock repurchase policy by using combined stochastic control: absolutely continuous and impulse control. By contrast, we concentrate on a stock repurchase policy and consider the jump risks, which we examine by using stochastic impulse control.

This paper is organized as follows. In the next section, we describe the firm's problem. In Section 3, we analyze the problem. In Section 4, we present the results of numerical examples and the comparative statics. Section 5 concludes the paper.

2. The Model

We assume that a firm's accumulated cash reserves X_t jump at the random times $\mathcal{T}_1, \cdots, \mathcal{T}_n, \cdots$, and that the associated relative changes in the value of these reserves at a jump time are given by $Y_1, \cdots, Y_n, \cdots$. We assume that $\mathcal{T}_n$ is the nth jump time of a Poisson process $N := (N_t)_{t\geq 0}$, which has intensity $\lambda \in (0,1)$, and that the sequence $Y = (Y_n)_{n\geq 1}$ comprises independent, identically distributed random variables that take values in $(-\infty, \infty)$ and have the distribution F. We assume that Y has a finite mean:

$$\mathbb{E}[Y] = \int_{-\infty}^{\infty} y dF(y) < \infty. \tag{1}$$

For $t \in [\mathcal{T}_n, \mathcal{T}_{n+1})$, the process of the firm's accumulated cash reserves $X := (X_t)_{t\geq 0}$ follows a Brownian motion with drift, as in Radner and Shepp[11]:[a]

$$dX_t = \mu dt + \sigma dW_t, \tag{2}$$

where $\mu \in \mathbb{R}$ and $\sigma \in \mathbb{R}\backslash\{0\}$ are constants. $W := (W_t)_{t\geq 0}$ is a Brownian motion process on a filtered probability space $(\Omega, \mathcal{F}, \mathbb{P}; (\mathcal{F}_t)_{t\geq 0})$, which

[a]Radner and Shepp[11] rigorously explain why the cash reserve is governed by a Brownian motion with drift in Section 1. Furthermore, it is shown that if the accumulated net revenue follows a geometric Brownian motion, the model defined in this paper yields an empty problem, and that all the profits of the firm are drawn at once.

satisfies the usual conditions. $(\mathcal{F}_t)_{t\geq 0}$ is generated by a Brownian motion process W, a Poisson process N, and the sequence Y. We assume that W, N, and Y are mutually independent. At $t = \mathcal{T}_n$, the jump is given by $X_{\mathcal{T}_n} - X_{\mathcal{T}_n-} = X_{\mathcal{T}_n-}Y_n$, so that

$$X_{\mathcal{T}_n} = X_{\mathcal{T}_n-}(1 + Y_n). \tag{3}$$

See, for example, Section 7.2 of Lamberton and Lapeyre[9] and Subsection 3.1.1 of Runggaldier[12]. By using Ito's formula to obtain the solution to Eq. (2) and by using a recursive argument based on Eq. (3), at the generic time t, X_t is given by

$$X_t = \left(X_0 + \int_0^t \mu ds + \int_0^t \sigma dW_s\right) \prod_{n=1}^{N_t}(1 + Y_n). \tag{4}$$

Clearly, Eq. (4) is the solution of

$$dX_t = \mu dt + \sigma dW_t + X_{t-}Y_t dN_t. \tag{5}$$

Let ξ_i be the amount of the ith stock repurchased. Let τ_i be the ith stock repurchase time such that $\tau_i \to +\infty$ as $i \to +\infty$ a.s. We set $\tau_0- := 0$ and $\tau_{i+1}- = \tau_i$ if $\tau_{i+1} = \tau_i$. The values of τ_i and ξ_i correspond, respectively, to the stopping times and the impulses in impulse control theory. A stock repurchase policy is defined as the following double sequences:

$$v := \{(\tau_i, \xi_i)\}_{i\geq 0}. \tag{6}$$

If the stock repurchase policy v is given by (6), then the dynamics of the cash reserve process, $X^{x,v} := (X_t^{x,v})_{t\geq 0}$, is given by

$$\begin{cases} dX_t^{x,v} = \mu dt + \sigma dW_t + X_{t-}^{x,v} Y_t dN_t, & \tau_i \leq t < \tau_{i+1} \leq T, \quad i \geq 0; \\ X_{\tau_i}^{x,v} = X_{\tau_i-}^{x,v} - \xi_i; & X_{0-}^{x,v} = x, \end{cases} \tag{7}$$

where T represents a bankruptcy time that is defined by

$$T = \inf\{t > 0; X^{x,v} \leq 0\}. \tag{8}$$

We assume that when the firm goes bankrupt, the illiquid assets of the firm have no salvage value.

Definition 2.1. (Admissible Stock Repurchase Policy). *A stock repurchase policy v is admissible if the following conditions are satisfied:*

$$0 \leq \tau_i \leq \tau_{i+1}, \quad a.s. \quad i \geq 0, \tag{9}$$

$$\tau_i \textit{ is an } (\mathcal{F}_t)_{t\geq 0}\textit{-stopping time}, \quad i \geq 0, \tag{10}$$

$$\xi_i \textit{ is } \mathcal{F}_{\tau_i}\textit{-measurable}, \quad i \geq 0, \tag{11}$$

$$P\left[\lim_{i\to\infty} \tau_i \leq \tilde{T}\right] = 0, \quad \forall \tilde{T} \in [0, \infty), \tag{12}$$

Condition (12) means that the stock repurchase policy will only occur finitely before a terminal time $\tilde{T}$. Let $\mathcal{V}$ denote the set of admissible stock repurchase policies.

Let $K : \mathbb{R}_+ \to \mathbb{R}$ represent the net stock repurchases defined by

$$K(\xi) := k_1 \xi - k_0, \tag{13}$$

where $(1 - k_1) \in (0, 1)$ is the parameter for the proportional transaction cost and $k_0 \in \mathbb{R}_{++}$ is the fixed transaction cost. Note that $K(\xi)$ satisfies superadditivity with respect to ξ:

$$K(\xi + \xi') \geq K(\xi) + K(\xi'), \quad \xi, \xi' \in \mathbb{R}_+. \tag{14}$$

This implies that reasonable $(\mathcal{F}_t)_{t\geq 0}$-stopping times are strictly increasing sequences; that is, $0 = \tau_0 < \tau_1 < \tau_2 < \cdots < \tau_i < \cdots < T$. The expected total discounted stock repurchases function associated with the stock repurchase policy v is defined by

$$J(x; v) = \mathbb{E}\left[\sum_{i=1}^{\infty} e^{-r\tau_i} K(\xi_i) 1_{\{\tau_i < T\}}\right], \tag{15}$$

where $r \in \mathbb{R}_{++}$ is a discount rate. We assume that

$$J(0; v) = 0. \tag{16}$$

Therefore, the firm's problem is to maximize Eq. (15) over $v \in \mathcal{V}$:

$$V(x) = \sup_{v \in \mathcal{V}} J(x; v) = J(x; v^*), \tag{17}$$

where V is the value function of the agent's problem and v^* is the optimal stock repurchase policy.

3. Analysis

In the previous section, we formulated the firm's problem as a stochastic impulse control problem. In this section, we prove that a policy, which is a QVI policy, is an optimal stock repurchase policy for the problem (17).

Let $\phi : \mathbb{R}_+ \to \mathbb{R}$ be a function. Let $\mathcal{M}$ denote the stock repurchase operator on the space of functions ϕ defined by

$$\mathcal{M}\phi(x) = \sup_{\xi \in \mathcal{Z}(x,\xi)} \{\phi(\eta(x,\xi)) + K(\xi)\}, \tag{18}$$

where $\mathcal{Z}(x,\xi)$ is the feasible set defined by $\mathcal{Z}(x,\xi) = \{\xi; \eta(x,\xi) \geq 0\}$, and $\eta(x,\xi)$ is the new cash reserve value following the repurchase of stock. This operator means that the firm repurchases stock in the current period. We assume that ϕ is a twice continuously differentiable function on $\mathbb{R}$, C^2, except on the boundary of the considered region. The integrodifferential operator $\mathcal{L}$ of X^x is defined as

$$\mathcal{L}\phi(x) = \frac{1}{2}\sigma^2\phi''(x) + \mu\phi'(x) - r\phi(x) + \lambda\left[\int_{-\infty}^{\infty} \phi((1+y)x)dF(y) - \phi(x)\right]. \tag{19}$$

This operator is well defined provided that

$$\int_{-\infty}^{\infty} \phi((1+y)x)dF(y) (= \mathbb{E}[\phi((1+Y)x])) \tag{20}$$

Since ϕ is not necessarily C^2 in the whole region, we must apply the generalized Dynkin formula for ϕ. The formula is applicable if ϕ is stochastically C^2. That is, we use the generalized Dynkin formula on the set that has a Green measure of $X^{x,v}$ of zero. The Green measure of $X^{x,v}$ in a domain is the expected total occupation measure defined by

$$\mathbb{G}(\Xi; x, v) := \mathbb{E}\left[\int_0^{\infty} X_t^{x,v} 1_{\Xi} dt\right], \tag{21}$$

where $\Xi \subset \mathfrak{B}(\mathbb{R})$ is a Borel set. ϕ is stochastically C^2 with respect to $X^{x,v}$ if $\mathcal{L}\phi(x)$ is well defined pointwise for almost all x with respect to the Green measure $\mathbb{G}(\cdot; x, v)$. The following equality, which is the generalized Dynkin formula, is used in the proof of Theorem 3.1.

$$\mathbb{E}[e^{-r\theta}\phi(X_\theta^{x,v})] = \mathbb{E}[e^{-r\tau_i}\phi(X_{\tau_i}^{x,v})] + \mathbb{E}\left[\int_{\tau_i}^{\theta} e^{-rt}\mathcal{L}\phi(X_t^{x,v})dt\right]. \tag{22}$$

This is defined for all i, and for all bounded stopping times θ such that $\tau_i \leq \theta \leq \tau_{i+1}$.

Definition 3.1. (QVI). *The following relations are the QVI for the problem given by Eq. (17):*

$$\mathcal{L}\phi(x) \leq 0, \quad \textit{for a.a. } x \textit{ w.r.t. } \mathbb{G}(\cdot, x; v),\ \forall v \in \mathcal{V}; \tag{23}$$

$$\phi(x) \geq \mathcal{M}\phi(x); \tag{24}$$

$$[\mathcal{L}\phi(x)][\phi(x) - \mathcal{M}\phi(x)] = 0, \quad \textit{for a.a. } x \textit{ w.r.t. } \mathbb{G}(\cdot, x; v),\ \forall v \in \mathcal{V}. \tag{25}$$

Definition 3.2. (QVI policy). *Let ϕ be a solution of the QVI. Then, the following stock repurchase policy, $\hat{v} = \{(\hat{\tau}_i, \hat{\xi}_i)\}_{i\geq 0}$, is termed a QVI policy:*

$$(\hat{\tau}_0, \hat{\xi}_0) = (0, 0); \tag{26}$$

$$\hat{\tau}_i = \inf\{t \geq \hat{\tau}_{i-1}; X_t^{x,\hat{v}} \notin H\}; \tag{27}$$

$$\hat{\xi}_i = \arg\max\{\phi(\eta(X_{\hat{\tau}_i -}^{x,\hat{v}}, \xi_i)) + K(\xi_i); \xi_i \in \mathcal{Z}\} \textit{ if } \hat{\tau}_i < T. \tag{28}$$

In this context, H is the continuation region defined by

$$H := \{x; \phi(x) > \mathcal{M}\phi(x)\}, \tag{29}$$

and $X_t^{x,\hat{v}}$ results from applying the stock repurchase policy $\hat{v} = \{(\hat{\tau}_i, \hat{\xi}_i)\}_{i\geq 0}$.

We now prove that a QVI policy is an optimal stock repurchase policy by using the following theorem. See also Brekke and Øksendal[2] and Ohnishi and Tsujimura[10].

Theorem 3.1. *Suppose that Eq. (1) holds.*

(I) Let the function ϕ be a solution of the QVI that satisfies the following conditions:

$$\phi \textit{ is stochastically } C^2 \textit{ w.r.t. } X^{x,v}; \tag{30}$$

$$\lim_{t\to\infty} e^{-rt}\phi(X_t^{x,v}) = 0, \quad \textit{a.s.} \quad \forall v \in \mathcal{V}; \tag{31}$$

$$\textit{the family } \{\phi(X_\tau^{x,v})\}_{\tau<\infty} \textit{ is uniformly integrable w.r.t. } P \quad \forall v \in \mathcal{V}. \tag{32}$$

Then, we obtain

$$\phi(x) \geq J(x; v), \quad \forall v \in \mathcal{V}. \tag{33}$$

(II) From (23)–(25) and (29), we have

$$\mathcal{L}\phi(x) = 0, \quad x \in H. \tag{34}$$

Suppose that $\hat{v} \in \mathcal{V}$; that is, assume that the QVI policy is admissible. Then, we obtain

$$\phi(x) = J(x; \hat{v}). \tag{35}$$

Hence, we have

$$\phi(x) = V(x) = J(x; \hat{v}), \tag{36}$$

and, therefore, $\hat{v}$ is optimal.

Proof

(I) Assume that ϕ satisfies Eqs. (30) through (32). Choose $v \in \mathcal{V}$. Let us define $\theta_{i+1} := \tau_i \vee (\tau_{i+1} \wedge s)$ for any $s \in \mathbb{R}_+$. Then, from the generalized Dynkin formula (22) and Ineq. (23), we obtain

$$\mathbb{E}[e^{-r\theta_{i+1}}\phi(X^{x,v}_{\theta_{i+1}})] \leq \mathbb{E}[e^{-r\tau_i}\phi(X^{x,v}_{\tau_i})]. \tag{37}$$

By taking $\lim_{s\to\infty}$ and by using the dominated convergence theorem, we obtain

$$\mathbb{E}[e^{-r\tau_{i+1}}\phi(X^{x,v}_{\tau_{i+1}})] \leq \mathbb{E}[e^{-r\tau_i}\phi(X^{x,v}_{\tau_i})]. \tag{38}$$

For all $\tau_{i+1} < T$, following the repurchase of stock, the cash reserve jumps immediately from $X^{x,v}_{\tau_{i+1}-}$ to a new cash reserve level $X^{x,v}_{\tau_{i+1}-} - \xi_i \; (= X^{x,v}_{\tau_{i+1}})$. At $\tau_{i+1} = T$, the cash reserve falls to zero and the firm goes bankrupt. That is,

$$\begin{cases} \phi(X^{x,v}_{\tau_{i+1}}) \geq \mathcal{M}\phi(X^{x,v}_{\tau_{i+1}}) \\ \qquad\qquad = \phi(X^{x,v}_{\tau_{i+1}-} - \xi_{i+1}) + K(\xi_{i+1}), & \text{if } \tau_{i+1} < T, \\ \phi(X^{x,v}_{\tau_{i+1}}) = 0, & \text{if } \tau_{i+1} = T. \end{cases} \tag{39}$$

From (38) and (39), it follows that

$$\begin{aligned} \mathbb{E}[e^{-r\tau_{i+1}}\phi(X^{x,v}_{\tau_{i+1}-} - \xi_{i+1})1_{\{\tau_{i+1}<T\}} + e^{-r\tau_{i+1}}K(\xi_{i+1})1_{\{\tau_{i+1}<T\}}] \\ \leq \mathbb{E}[e^{-r\tau_i}\phi(X^{x,v}_{\tau_i})]. \end{aligned} \tag{40}$$

Summing from $i = 0$ to $i = m - 1$ yields

$$\mathbb{E}[e^{-r\tau_m}\phi(X^{x,v}_{\tau_m-} - \xi_m)1_{\{\tau_m<T\}} + \sum_{i=1}^{m} e^{-r\tau_i}K(\xi_i)1_{\{\tau_i<T\}}] \leq \phi(x). \tag{41}$$

Taking $\lim_{m\to\infty}$ and using (31), (32), and the dominated convergence theorem yields

$$\phi(x) \geq \mathbb{E}\left[\sum_{i=1}^{\infty} e^{-r\tau_i} K(\xi_i) 1_{\{\tau_i<T\}}\right]. \tag{42}$$

Therefore, Eq. (33) is proven.

(II) Assume that Eq. (34) holds and that $\hat{v} = \{(\hat{\tau}_i, \hat{\xi}_i)\}_{i\geq 1}$ is a QVI policy. Repeat the argument in part (I) for $v = \hat{v}$. Then, Ineqs. (37) through (41) become equalities. Thus, we obtain

$$\phi(x) = \mathbb{E}\left[\sum_{i=1}^{\infty} e^{-r\hat{\tau}_i} K(\hat{\xi}_i) 1_{\{\hat{\tau}_i<T\}}\right]. \tag{43}$$

Hence, we obtain Eq. (35). Combining Eq. (35) with Ineq. (33) yields

$$\phi(x) \geq \sup_{v\in\mathcal{V}} J(x; v) \geq J(x; \hat{v}) = \phi(x). \tag{44}$$

Therefore, $\phi(x) = V(x)$ and $v^* = \hat{v}$ is optimal; that is, a solution of the QVI is the value function and the QVI policy is optimal. □

4. Numerical Examples and Comparative Statics

In this section, we present a comparative-static analysis for the amount of stock repurchased, the expected interval of the stock repurchase time, and the value function. Unfortunately, we do not have an analytical solution to $\mathcal{L}\phi(x) = 0$ that is satisfied in the continuation region H. Thus, we obtain these solutions by using a numerical method. The comparative-static analysis yields an economic implication. When the firm decides to repurchase stock with transaction costs under uncertainty, these results are useful.

To find the value function $V(x)$, we perform a simulation of Eq. (5). Since we formulate the firm's problem as the impulse control problem, it is reasonable to assume that, under an optimal stock repurchase policy, the firm repurchases stock whenever the cash reserve reaches a threshold, $\bar{x}$, so that it instantaneously falls to another level, β. Ohnishi and Tsujimura[10] show that such policy is indeed optimal under an appropriate set of sufficient conditions in the context of a dividend policy. Takashima and Tsujimura[14] also show that such a policy is optimal in the context of a dividend and stock repurchase policy. The expected total discounted stock repurchases function, Eq. (15), is evaluated at grid points defined by $(ih_{\bar{x}}, jh_{\beta})$, where i and $j(1 \leq i, j \leq Q,\ i > j)$ are integers. We obtain M paths from Euler's method for every combination of β and $\bar{x}$ in order to find the value function defined by (17). Unless otherwise stated, we set $r = 0.1$, $\sigma = 0.3$, $\mu = 0.05$, $k_1 = 0.7$, $k_0 = 0.1$, and $\lambda = 0.05$. We assume that F is the standard normal cumulative distribution function.

The results of numerical examples are shown in Tables 1–6. Note that, in these tables, the initial value is given by $x = 0.5$, and $V(x)|_{x=0.5}$ is calculated by using a function of either the r, μ, σ, k_1, k_0 or λ parameters , with $h = 0.02$ and $M = 2{,}000{,}000$. We set $1 \leq i \leq 5$, $39 \leq j \leq 64$ for Table 1; we set $1 \leq i \leq 5$, $40 \leq j \leq 57$ for Table 2; we set $1 \leq i \leq 5$, $44 \leq j \leq 57$ for Table 3; we set $1 \leq i \leq 4$, $45 \leq j \leq 58$ for Table 4; we set $1 \leq i \leq 3$, $42 \leq j \leq 56$ for Table 5; and we set $1 \leq i \leq 3$, $45 \leq j \leq 55$ for Table 6. Moreover, these tables show the optimal amount of stock repurchased and the expected interval of the stock repurchase time, as defined by $\mathbb{E}[\tau_{\beta,\bar{x}}] = \mathbb{E}[\tau_{i+1} - \tau_i]$. The results for the numerical examples in Tables 1–6 illustrate the comparative-static analysis. First, from Tables 1–6, the comparative-static results for the amount of the ith stock repurchase, ξ_i, are as follows. *The amount of the ith stock repurchase is increasing in the drift parameter, the diffusion parameter, the parameter for the proportional transaction cost, and the fixed transaction cost. On the other hand, ξ_i is decreasing in the discount rate.* Next, the comparative-static analysis of the expected interval of the stock repurchase time, $\mathbb{E}[\tau_{\beta,\bar{x}}]$, yields the following results. *An increase in the drift parameter, the proportional-transaction-cost parameter, or the fixed transaction cost lengthens the expected interval of the stock repurchase time, whereas increases in the discount rate and the diffusion parameter shorten it.* The intensity of the Poisson process has an ambiguous effect on ξ_i and $\mathbb{E}[\tau_{\beta,\bar{x}}]$.

In this paper, we assume that stock repurchases represent temporary cash distribution to stockholders. In this context, stockholders would like to avoid transaction costs. Thus, firms repurchase more stock when fixed or proportional transaction costs rise. Furthermore, the optimal expected interval of the stock repurchase time lengthens.

5. Conclusion

In this paper, we examined an optimal stock repurchase policy with transaction costs under jump risks. We assumed that the firm's cash reserve is governed by a jump-diffusion process, and that, when the firm repurchases stock, it incurs both fixed and proportional transaction costs. Further,

Table 1. The results of the numerical example for r.

r	β	$\bar{x}$	ξ_i	$V(x)$	$\mathbb{E}[\tau_{\beta,\bar{x}}]$
0.070000	0.060000	1.240000	1.180000	0.343539	4.605722
0.100000	0.020000	1.000000	0.980000	0.310766	2.811643
0.130000	0.020000	0.840000	0.820000	0.292536	1.807374

Notes: This table illustrates the comparative-static effects of the discount rate. Equation (15) is evaluated at grid points defined by $(ih_{\bar{x}}, jh_\beta)$, where $1 \leq i \leq 5$, $39 \leq j \leq 64$, $h = 0.02$. We obtained $M = 2,000,000$ paths by using Euler's method for every combination of β and $\bar{x}$ in order to find the value function defined by (17).

Table 2. The results of the numerical example for μ.

μ	β	$\bar{x}$	ξ_i	$V(x)$	$\mathbb{E}[\tau_{\beta,\bar{x}}]$
0.035000	0.020000	0.880000	0.860000	0.291858	2.059807
0.050000	0.020000	0.980000	0.960000	0.310742	2.680953
0.065000	0.080000	1.120000	1.040000	0.335980	3.677407

Notes: This table illustrates the comparative-static effects of the drift parameter. Equation (15) is evaluated at grid points defined by $(ih_{\bar{x}}, jh_\beta)$, where $1 \leq i \leq 5$, $40 \leq j \leq 57$, $h = 0.02$.

Table 3. The results of the numerical example for σ.

σ	β	$\bar{x}$	ξ_i	$V(x)$	$\mathbb{E}[\tau_{\beta,\bar{x}}]$
0.210000	0.080000	0.900000	0.820000	0.308109	4.175046
0.300000	0.020000	1.020000	1.000000	0.310772	2.949139
0.390000	0.020000	1.100000	1.080000	0.315439	2.202687

Notes: This table illustrates the comparative-static effects of the diffusion parameter. Equation (15) is evaluated at grid points defined by $(ih_{\bar{x}}, jh_\beta)$, where $1 \leq i \leq 5$, $44 \leq j \leq 57$, $h = 0.02$.

we supposed that the firm goes bankrupt when its cash reserves fall to zero. The firm's problem is to maximize the expected total discounted stock repurchases. To solve this problem, we formulated it as a stochastic impulse control problem, and then approached this problem by using quasi-variational inequalities (QVI). Then, we proved that the value function is a solution to the QVI and that the associated QVI policy is optimal. Furthermore, we presented numerical examples and comparative statics.

Table 4. The results of the numerical example for k_1.

k_1	β	$\bar{x}$	ξ_i	$V(x)$	$\mathbb{E}[\tau_{\beta,\bar{x}}]$
0.490000	0.020000	1.100000	1.080000	0.203387	3.505510
0.700000	0.020000	0.980000	0.960000	0.310810	2.681492
0.910000	0.060000	0.920000	0.860000	0.419925	2.346563

Notes: This table illustrates the comparative-static effects of the parameter for the proportional transaction cost. Note that, to be precise, $1 - k_1(\in (0,1))$ is the proportional-transaction-cost parameter. Equation (15) is evaluated at grid points defined by $(ih_{\bar{x}}, jh_\beta)$, where $1 \le i \le 4$, $45 \le j \le 58$, $h = 0.02$.

Table 5. The results of the numerical example for k_0.

k_0	β	$\bar{x}$	ξ_i	$V(x)$	$\mathbb{E}[\tau_{\beta,\bar{x}}]$
0.070000	0.040000	0.920000	0.880000	0.327068	2.320092
0.100000	0.020000	1.020000	1.000000	0.310864	2.947472
0.130000	0.020000	1.100000	1.080000	0.296103	3.510047

Notes: This table illustrates the comparative-static effects of the fixed-transaction-cost parameter. Equation (15) is evaluated at grid points defined by $(ih_{\bar{x}}, jh_\beta)$, where $1 \le i \le 3$, $42 \le j \le 56$, $h = 0.02$.

Table 6. The results of the numerical example for λ.

λ	β	$\bar{x}$	ξ_i	$V(x)$	$\mathbb{E}[\tau_{\beta,\bar{x}}]$
0.035000	0.020000	1.000000	0.980000	0.311148	2.813397
0.050000	0.020000	1.000000	0.980000	0.310828	2.812030
0.065000	0.020000	1.020000	1.000000	0.310858	2.951523

Notes: This table illustrates the comparative-static effects of the intensity of jump risks. The expected total discounted stock repurchase function, Eq.(15), is evaluated at grid points defined by $(ih_{\bar{x}}, jh_\beta)$, where $1 \le i \le 3$, $45 \le j \le 55$, $h = 0.02$.

To conclude this paper, we discuss problems remaining for further research. One problem is to determine the effect of the jump risk on the stock repurchase policy. The other is to prove that a solution to the QVI exists and is unique. Furthermore, although applying impulse control theory to stock repurchases has interesting extensions, we do not discuss the effect of stock repurchases on the firm's value or on its share price. In addition, the problem of a random amount of stock repurchases remains. In fact, because firms do not repurchase stock as they plan to, the amount repurchased is a random variable. Finally, it is interesting to study the time taken to repurchase stock because firms do not necessarily immediately repurchase stock as planned.

References

1. L. S. Bagwell, Share Repurchase and Takeover Deterrence, *RAND Journal of Economics*, 22, pp. 72–88 (1991).
2. K. A. Brekke and B. Øksendal, A Verification Theorem for Combined Stochastic Control and Impulse Control, in: Decreusefond, L. et al. (Eds.), *Stochastic Analysis and Related Topics*, Vol. 6. Bikhäuser, Basel. pp. 211–220 (1998).
3. A. K. Dittmar, Why Do Firms Repurchase Stock?, *Journal of Business*, 73, pp. 331–355 (2000).
4. G. W. Fenn and N. Liang, Good News and Bad News about Share Repurchases, *Finance and Economics Discussion Series*, Board of Government of the Federal Reserve, 1998-4 (1997).
5. W. Guay and J. Harford, The Cash-Flow Permanence and Information Content of Dividend Increases versus Repurchases, *Journal of Financial Economics*, 57, pp. 385–415 (2000).
6. M. Jagannathan, C. P. Stephens and M. S. Weisbach, Financial Flexibility and the Choice between Dividends and Stock Repurchases, *Journal of Financial Economics*, 57, pp. 355–384 (2000).
7. M. C. Jensen, Agency Costs of Free Cash Flow, Corporate Finance, and Takeovers, *American Economic Review*, 76, pp. 323–329 (1986).
8. M. Jeanblanc-Picqué and A. N. Shiryaev, Optimization of the Flow of Dividends, *Russian Mathematical Surveys*, 50, pp. 257–277 (1995).
9. D. Lamberton and B. Lapeyre, *Introduction to Stochastic Calculus Applied to Finance*, Chapman & Hall, Cornwall (1996).
10. M. Ohnishi and M. Tsujimura, Optimal Dividend Policy with Transaction Costs under a Brownian Cash Reserve, *Discussion Papers in Economics and Business*, Osaka University, 02-07 (2002).
11. R. Radner and L. Shepp, Risk vs. Profit Potential: A Model for Corporate Strategy, *Journal of Economic Dynamics and Control*, 20, pp. 1373–1393 (1996).
12. W. J. Runggaldier, Jump-Diffusion Models, in: Rachev, S. T. (Ed.) *Handbook of Heavy Tailed Distributions in Finance*, North Holland, Amsterdam, pp. 169–209 (2001).
13. C. P. Stephens and M. S. Weisbach, Actual Share Reacquisitions in Open-Market Repurchase Programs, *Journal of Finance*, 53, pp. 313–333 (1998).
14. R. Takashima and M. Tsujimura, Dividend and Stock Repurchase Policy with Transaction Costs, Working Paper (2006).
15. M. Tsujimura, Impulse Control and its Applications to Mathematical Finance, *Osaka Economic Papers*, 53, pp. 199–220 (2003).

16. T. Vermaelen, Common Stock Repurchases and Market Signalling, *Journal of Financial Economics*, 9, pp. 139–183 (1981).

THE PRICING OF PERPETUAL GAME PUT OPTIONS AND OPTIMAL BOUNDARIES *

ATSUO SUZUKI

Nanzan University,
27 Seirei-cho,
Aichi, Japan
E-mail: a2o@nanzan-u.ac.jp

KATSUSHIGE SAWAKI

Nanzan University,
27 Seirei-cho,
Aichi, Japan
E-mail: sawaki@nanzan-u.ac.jp

In this paper we deal with the pricing model of game options introduced by Kifer which is a contract in that the seller and the buyer have both the rights to cancel and to exercise it at any time, respectively. First, we discuss the pricing of perpetual game put options by applying the first hitting time approach of a Brownian motion, when the stock pays dividends continuously. Secondly, we investigate properties of optimal boundaries of the seller and the buyer. Finally, some numerical results are presented to demonstrate analytical properties of the value function.

1. Introduction

Game option introduced by Kifer[1] is a contract that the seller and the buyer have the rights to cancel and exercise the contract at any time, respectively. It is represented as an optimal stopping problem between the seller and the buyer, and is analysed by Dynkin Game. Also it is thought as callable American option.

If the buyer exercise, the buyer receives the profit. If the seller cancels, the seller has the obligation to pay extra cost. It is interpreted as the penalty for the cancel. Because of the cancel, the price of game option is

*This work is supported in part by the Grant-in-Aid for Scientific Research (No.16651090) by Ministry of Education, Culture, Sports, Science and Technology.

cheaper than American option. If it is too large, it is natural that the seller does not cancels. Then game option is reduced to American option.

Merton[2] solved various type of perpetual options. Since perpetual option price do not depend on the time, it is solved by ordinary differential equation. Karatzas and Shreve [3] summarizes the value function of perpetual American options and their analytical properties.

On the other hand, Kyprianou[4] gave the value function, which is called the perpetual Israeli put option, and its optimal boundaries when the stock pays no dividend. Furthermore, from the fact that boundary is a constant if the maturity is infinity, Suzuki and Sawaki [5] formulated it as double barrier option with the rebate.

In this paper we study the pricing of perpetual game put option and its optimal boundaries and explore properties of the value function, when the stock price pays dividend.

In the next section, we present the pricing formula of game option that the maturity is finite. We define perpetual game put option and the optimal stopping times of the buyer and the seller. In Sec. 3, we state the main result Theorem 3.1. We calculate the value function in continuous region using the first hitting time approach of a Brownian motion. And we get the optimal boundaries of the seller and the buyer. They are saddle point of the value function. The boundary of the buyer is the interval. If the stock price is in the interval, the buyer exercise immediately. On the other hand, that of the seller is the point. Furthermore, we verify that it holds the smooth fit condition. In Sec. 4, we illustrate the optimal boundary of the buyer and the value function.

2. Preliminary

We consider the Black-Scholes model. Let B_t be the bond price at time t which is given by

$$dB_t = rB_t dt,\ B_0 > 0,\ r > 0,$$

where r is the riskless interest rate. Let S_t be the stock price at time t which satisfies the stochastic differential equation

$$dS_t = \mu S_t dt + \kappa S_t dW_t, \tag{1}$$

where μ and $\kappa > 0$ are constants. W_t is a standard Brownian motion on a probability space $(\Omega, \mathcal{F}, \{\mathcal{F}_t\}_{0\le t\le T}, P)$. Define the process

$$Z_t = \exp\left\{-\frac{1}{2}\left(\frac{\mu - r + d}{\kappa}\right)^2 t - \frac{\mu - r + d}{\kappa} W_t\right\},$$

where d is dividend rate and constant. We define the risk neutral measure $\tilde{P}$ which is given by

$$\tilde{P}(A) = E[Z_T 1_A], \ A \in \mathcal{F}_T.$$

Under $\tilde{P}$, define the process $\tilde{W}_t$ by

$$\tilde{W}_t = W_t + \frac{\mu - r + d}{\kappa} t, \tag{2}$$

which is a standard Brownian motion by Girsanov's theorem. Substituing Eq. (2) into Eq. (1), we get

$$dS_t = (r - d)S_t dt + \kappa S_t d\tilde{W}_t. \tag{3}$$

Solving the above stochastic differential equation with $S_0 = x$, we express the stock price as $S_t(x)$ by emphasizing on the dependence of the initial stock price.

$$S_t(x) = xH(t),$$

where

$$H(t) = \exp\left\{\left(r - d - \frac{\kappa^2}{2}\right) t + \kappa \tilde{W}_t\right\}.$$

Consider the game option introduced by Kifer[1], which is a contract that the buyer and the seller have both the rights to exercise and to cancel it at any time, respectively. To formulate the pricing model of the game option based on a coupled stopping game, denote $\mathcal{T}_{t,T}$ the set of all stopping times with values in $[t, T]$. Let σ and τ be the stopping times of the seller and the buyer, respectively. If the buyer exercises, the seller must pay the buyer Y_τ. If the seller cancels the contract, the seller pays the buyer $X_\sigma (X_t > Y_t)$. If $\sigma = \tau$, the seller pays the buyer to X_τ. Let $\delta = X_t - Y_t$. δ can be interpreted as the penalty for the cancel. If the penalty is large enough, it is optimal for the seller not to cancel the contract. Then the game option is reduced to the American option. The payoff function of the game option is given by

$$R(\sigma, \tau) = X_\sigma 1_{\{\sigma < \tau\}} + Y_\tau 1_{\{\tau \le \sigma\}}.$$

Let $V_T(t, x)$ be the no arbitrage price of the game option with finite maturity T. Then, $V_T(t, x)$ is given by the following theorem. Kifer[1] proved for the finite maturity.

Theorem 2.1. ***(Kifer)*** *Define*

$$J_t^x(\sigma, \tau) = \tilde{E}[e^{-r(\sigma \wedge \tau - t)} R(\sigma, \tau) | S_t = x].$$

Then $V_T(t,x)$ is represented by

$$V_T(t,x) = \inf_{\sigma\in\mathcal{T}_{tT}} \sup_{\tau\in\mathcal{T}_{tT}} J_t^x(\sigma,\tau)$$
$$= \sup_{\tau\in\mathcal{T}_{tT}} \inf_{\sigma\in\mathcal{T}_{tT}} J_t^x(\sigma,\tau).$$

And the optimal stopping times are

$$\hat{\sigma}_t = \inf\{t \in [0,T) \mid V_T(t,S_t) = X_t\} \wedge T,$$
$$\hat{\tau}_t = \inf\{t \in [0,T) \mid V_T(t,S_t) = Y_t\} \wedge T.$$

3. Pricing and Optimal Boundaries

In this section we consider the valuation of the perpetual game put by using the first hitting time approach. It is obvious that $V(x) \le V^{ap}(x)$, where $V^{ap}(x)$ is the value function of American put option.

Lemma 3.1. *Let $V^{ap}(x)$ be the value function of the perpetual American put option. Let $\delta^* = V^{ap}(K)$ evaluated at $x = K$. If $\delta \ge \delta^*$, the perpetual game put option becomes the perpetual American put option.*

Proof. Consider the function

$$U(x) = V^{ap}(x) - (K-x)^+ - \delta.$$

Since $U'(x) = \frac{dV^{ap}}{dx} + 1 \ge 0$ for $x < K$ and $U'(x) = \frac{dV^{ap}}{dx} < 0$ for $x > K$, $U(x)$ attains the maximum $V^{ap}(K) - \delta$ at $x = K$. We have $V^{ap}(K) - \delta < 0$ for $\delta \ge \delta^*$, and then obtain $U(x) < 0$, *i.e.* $V^{ap}(x) < (K-x)^+ + \delta$. By $V(x) \ge V^{ap}(x)$, We get

$$V(x) < (K-x)^+ + \delta.$$

Therefore the seller never cancels the contract for $\delta \ge \delta^*$. □

Theorem 2.1 can easily be extended into the case of the perpetual game option with the infinite maturity. The value function $V(x)$ of the perpetual put game options is defined by

$$V(x) = \inf_\sigma \sup_\tau J^x(\sigma,\tau),$$

where

$$J^x(\sigma,\tau) = \tilde{E}[e^{-r(\sigma\wedge\tau)}\{((K-S_\sigma(x))^+ + \delta)1_{\{\sigma<\tau\}} + (K-S_\tau(x))^+ 1_{\{\tau<\sigma\}}\}].$$

Let σ_a^x *and* τ_b^x be the first times the process $S_t(x)$ hits the seller's critical stock price a and the buyer's critical stock price b respectively. *i.e.*,

$$\sigma_a^x = \inf\{t > 0 \mid S_t(x) = a\}$$
$$\tau_b^x = \inf\{t > 0 \mid S_t(x) = b\}.$$

For $0 < b < x < a < \infty$, we consider the function

$$\begin{aligned} &V^x(a,b) \\ &= \tilde{E}[e^{-r(\sigma_a^x \wedge \tau_b^x)}((K - S_{\sigma_a^x}(x))^+ + \delta)1_{\{\sigma_a^x < \tau_b^x\}} + (K - S_{\tau_b^x}(x))^+ 1_{\{\tau_b^x \le \sigma_a^x\}}] \\ &= ((K-a)^+ + \delta)\tilde{E}[e^{-r\sigma_a^x} 1_{\{\sigma_a^x < \tau_b^x\}}] + (K-b)^+ \tilde{E}[e^{-r\tau_b^x} 1_{\{\tau_b^x \le \sigma_a^x\}}]. \end{aligned} \tag{4}$$

We prepare the following lemma to compute the conditional expectations in Eq. (4).

Lemma 3.2. *Set* $\nu = \frac{r-d}{\kappa} - \frac{1}{2}\kappa$, $\gamma_1 = \frac{1}{\kappa}(\sqrt{\nu^2 + 2r} + \nu)$, *and* $\gamma_2 = \frac{1}{\kappa}(\sqrt{\nu^2 + 2r} - \nu)$. *Then we have*

$$\tilde{E}[e^{-r\sigma_a^x} 1_{\{\sigma_a^x < \tau_b^x\}}] = \frac{\left(\dfrac{b}{x}\right)^{\gamma_1} - \left(\dfrac{x}{b}\right)^{\gamma_2}}{\left(\dfrac{b}{a}\right)^{\gamma_1} - \left(\dfrac{a}{b}\right)^{\gamma_2}}, \tag{5}$$

$$\tilde{E}[e^{-r\tau_b^x} 1_{\{\tau_b^x < \sigma_a^x\}}] = \frac{\left(\dfrac{a}{x}\right)^{\gamma_1} - \left(\dfrac{x}{a}\right)^{\gamma_2}}{\left(\dfrac{a}{b}\right)^{\gamma_1} - \left(\dfrac{b}{a}\right)^{\gamma_2}}. \tag{6}$$

Proof. First we prove Eq. (5). Define

$$L_t = \exp\left(-\frac{1}{2}\nu^2 t - \nu \tilde{W}_t\right).$$

We define $\hat{P}$ as $d\hat{P} = L_T d\tilde{P}$. By Girsanov's theorem, $\hat{W}_t \equiv \tilde{W}_t + \nu t$ is a standard Brownian motion under the probability measure $\hat{P}$. Define the first time that the process $\hat{W}_t$ hits λ or ρ by T_λ or T_ρ, respectively as follows;

$$T_\lambda = \inf\{t > 0 \mid \hat{W}_t = \lambda\}$$
$$T_\rho = \inf\{t > 0 \mid \hat{W}_t = \rho\}.$$

Since we obtain $\log S_t(x) = \log x + \kappa \hat{W}_t$ from $S_t(x) = x\exp(\kappa \hat{W}_t)$, we have

$$\sigma_a^x = T_\lambda, \quad a.s., \quad \lambda = \frac{1}{\kappa}\log\frac{a}{x},$$
$$\tau_b^x = T_\rho, \quad a.s., \quad \rho = \frac{1}{\kappa}\log\frac{b}{x}.$$

And

$$\begin{aligned}L_{T_\lambda}^{-1} &= \exp\left(\frac{1}{2}\nu^2 T_\lambda + \nu\tilde{W}_{T_\lambda}\right)\\ &= \exp\left(-\frac{1}{2}\nu^2 T_\lambda + \nu\hat{W}_{T_\lambda}\right)\\ &= \exp\left(-\frac{1}{2}\nu^2 T_\lambda + \nu\lambda\right).\end{aligned}$$

Therefore,

$$\begin{aligned}\tilde{E}[e^{-r\sigma_a^x}1_{\{\sigma_a^x<\tau_b^x\}}] &= \tilde{E}[e^{-rT_\lambda}1_{\{T_\lambda<T_\rho\}}]\\ &= \hat{E}\left[\exp\left(-\frac{1}{2}\nu^2 T_\lambda + \nu\lambda\right)e^{-rT_\lambda}1_{\{T_\lambda<T_\rho\}}\right]\\ &= e^{\nu\lambda}\hat{E}\left[\exp\left\{-\left(r+\frac{1}{2}\nu^2\right)T_\lambda\right\}1_{\{T_\lambda<T_\rho\}}\right].\end{aligned}$$

It is easy to see from Karatzas and Shreve[6] (Exercise 8.11, p.100) that by using the equation

$$\hat{E}\left[\exp\left\{-\left(r+\frac{1}{2}\nu^2\right)T_\lambda\right\}1_{\{T_\lambda<T_\rho\}}\right] = \frac{\sinh\rho\sqrt{\nu^2+2r}}{\sinh(\rho-\lambda)\sqrt{\nu^2+2r}},$$

we have

$$\begin{aligned}\tilde{E}[e^{-r\sigma_a^x}1_{\{\sigma_a^x<\tau_b^x\}}] &= \frac{\sinh\rho\sqrt{\nu^2+2r}}{\sinh(\rho-\lambda)\sqrt{\nu^2+2r}}\frac{e^{\nu\rho}}{e^{\nu(\rho-\lambda)}}\\ &= \frac{e^{\kappa\rho\gamma_1}-e^{-\kappa\rho\gamma_2}}{e^{\kappa(\rho-\lambda)\gamma_1}-e^{-\kappa(\rho-\lambda)\gamma_2}}\\ &= \frac{\left(\frac{b}{x}\right)^{\gamma_1}-\left(\frac{x}{b}\right)^{\gamma_2}}{\left(\frac{b}{a}\right)^{\gamma_1}-\left(\frac{a}{b}\right)^{\gamma_2}}.\end{aligned}$$

The proof for Eq. (6) follows similarly. □

From the above lemma, for $0 < b < x < a < \infty$, $V^x(a,b)$ is given by

$$V^x(a,b) = ((K-a)^+ + \delta)\frac{\left(\frac{x}{b}\right)^{\gamma_2}-\left(\frac{b}{x}\right)^{\gamma_1}}{\left(\frac{a}{b}\right)^{\gamma_2}-\left(\frac{b}{a}\right)^{\gamma_1}} + (K-b)^+\frac{\left(\frac{a}{x}\right)^{\gamma_1}-\left(\frac{x}{a}\right)^{\gamma_2}}{\left(\frac{a}{b}\right)^{\gamma_1}-\left(\frac{b}{a}\right)^{\gamma_2}}. \quad (7)$$

To prove the main theorem, we need the following lemma.

Lemma 3.3.

$$\frac{\partial V^x}{\partial b}(K,b)$$
$$= \frac{\left(\frac{K}{x}\right)^{\gamma_1} - \left(\frac{x}{K}\right)^{\gamma_2}}{\left(\left(\frac{K}{b}\right)^{\gamma_1} - \left(\frac{b}{K}\right)^{\gamma_2}\right)^2} \left(\frac{K}{b}\right)^{\gamma_1+1} \left\{ -(\gamma_1+\gamma_2)\varepsilon \left(\frac{K}{b}\right)^{-\gamma_2} \right.$$
$$\left. -(\gamma_1+1)\left(\frac{b}{K}\right) + \gamma_1 + (1-\gamma_2)\left(\frac{b}{K}\right)^{\gamma_1+\gamma_2+1} + \gamma_2 \left(\frac{b}{K}\right)^{\gamma_1+\gamma_2} \right\} \quad (8)$$

Proof. By Eq. (7), $V^x(K,b)$ is given by

$$V^x(K,b) = \delta \frac{\left(\frac{b}{x}\right)^{\gamma_1} - \left(\frac{x}{b}\right)^{\gamma_2}}{\left(\frac{b}{K}\right)^{\gamma_1} - \left(\frac{K}{b}\right)^{\gamma_2}} + (K-b)\frac{\left(\frac{x}{K}\right)^{\gamma_2} - \left(\frac{K}{x}\right)^{\gamma_1}}{\left(\frac{b}{K}\right)^{\gamma_2} - \left(\frac{K}{b}\right)^{\gamma_1}}, \; b < x < K. \quad (9)$$

First we derivative the first term.

$$\frac{\partial}{\partial b}\left(\frac{\left(\frac{x}{b}\right)^{\gamma_2} - \left(\frac{b}{x}\right)^{\gamma_1}}{\left(\frac{K}{b}\right)^{\gamma_2} - \left(\frac{b}{K}\right)^{\gamma_1}}\right) = \frac{(\gamma_1+\gamma_2)\frac{1}{b}\left(\frac{b}{K}\right)^{\gamma_1-1}\left(\frac{K}{b}\right)^{\gamma_2-1}\left\{\left(\frac{x}{K}\right)^{\gamma_2} - \left(\frac{K}{x}\right)^{\gamma_1}\right\}}{\left(\left(\frac{K}{b}\right)^{\gamma_2} - \left(\frac{b}{K}\right)^{\gamma_1}\right)^2}.$$

Next,

$$\frac{\partial}{\partial b}\left(\frac{K-b}{\left(\frac{K}{b}\right)^{\gamma_1} - \left(\frac{b}{K}\right)^{\gamma_2}}\right)$$
$$= \frac{-\left(\left(\frac{K}{b}\right)^{\gamma_1} - \left(\frac{b}{K}\right)^{\gamma_2}\right) - (K-b)\left(-\gamma_1\left(\frac{K}{b}\right)^{\gamma_1}\frac{1}{b} - \gamma_2\left(\frac{b}{K}\right)^{\gamma_2-1}\frac{1}{K}\right)}{\left(\left(\frac{K}{b}\right)^{\gamma_1} - \left(\frac{b}{K}\right)^{\gamma_2}\right)^2}$$
$$= \frac{1}{\left(\left(\frac{K}{b}\right)^{\gamma_1} - \left(\frac{b}{K}\right)^{\gamma_2}\right)^2}\left\{-(\gamma_1+1)\frac{a}{K}\left(\frac{K}{b}\right)^{\gamma_1} + \gamma_1\left(\frac{K}{b}\right)^{\gamma_1+1}\right.$$
$$\left. +(1-\gamma_2)\left(\frac{b}{K}\right)^{\gamma_2} + \gamma_2\left(\frac{b}{K}\right)^{\gamma_2-1}\right\}. \quad (10)$$

Therefore, we get

$$\begin{aligned}&\delta\frac{\partial}{\partial b}\left(\frac{\left(\frac{x}{b}\right)^{\gamma_2}-\left(\frac{b}{x}\right)^{\gamma_1}}{\left(\frac{K}{b}\right)^{\gamma_2}-\left(\frac{b}{K}\right)^{\gamma_1}}\right)\\&=\frac{\left(\left(\frac{x}{K}\right)^{\gamma_2}-\left(\frac{K}{x}\right)^{\gamma_1}\right)}{\left(\left(\frac{K}{b}\right)^{\gamma_2}-\left(\frac{b}{K}\right)^{\gamma_1}\right)^2}(\gamma_1+\gamma_2)\\&\quad\times\left\{\frac{K+\delta}{b}\left(\frac{b}{K}\right)^{\gamma_1-1}\left(\frac{K}{b}\right)^{\gamma_2-1}-\left(\frac{b}{K}\right)^{\gamma_1-1}\left(\frac{K}{b}\right)^{\gamma_2}\right\}.\end{aligned}$$

Using $\left(\left(\frac{K}{b}\right)^{\gamma_2}-\left(\frac{b}{K}\right)^{\gamma_1}\right)\left(\frac{a}{b}\right)^{\gamma_1-1}\left(\frac{b}{a}\right)^{\gamma_2-1}=\left(\frac{K}{b}\right)^{\gamma_1}-\left(\frac{b}{K}\right)^{\gamma_2}$, the right hand side of the above equation equals

$$\frac{\left(\frac{x}{K}\right)^{\gamma_2}-\left(\frac{K}{x}\right)^{\gamma_1}}{\left(\left(\frac{K}{b}\right)^{\gamma_1}-\left(\frac{b}{K}\right)^{\gamma_2}\right)^2}(\gamma_1+\gamma_2)\varepsilon\left(\frac{K}{b}\right)^{\gamma_1-\gamma_2+1}. \tag{11}$$

By multiplying $\left(\frac{K}{x}\right)^{\gamma_1}-\left(\frac{x}{K}\right)^{\gamma_2}$ for Eq. (10) and adding Eq. (11), we obtain Eq. (8). □

Theorem 3.1. *The value function $V(x)$ of the perpetual game put option is represented by*

$$V(x)=\begin{cases}K-x & 0\le x\le b^*,\\ \delta\dfrac{\left(\dfrac{x}{b^*}\right)^{\gamma_2}-\left(\dfrac{b^*}{x}\right)^{\gamma_1}}{\left(\dfrac{K}{b^*}\right)^{\gamma_2}-\left(\dfrac{b^*}{K}\right)^{\gamma_1}}+(K-b^*)\dfrac{\left(\dfrac{K}{x}\right)^{\gamma_1}-\left(\dfrac{x}{K}\right)^{\gamma_2}}{\left(\dfrac{K}{b^*}\right)^{\gamma_1}-\left(\dfrac{b^*}{K}\right)^{\gamma_2}} & b^*<x<K,\\ \delta\left(\dfrac{K}{x}\right)^{\gamma_1} & x\ge K.\end{cases}$$

And the optimal boundary of the buyer $b^*=z_1K$. z_1 is the solution in $(0,1)$ to the equation $g(z)=0$, where $g(z)$ satisfies the equation

$$g(z)=(1-\gamma_2)z^{\gamma_1+\gamma_2+1}+\gamma_2z^{\gamma_1+\gamma_2}-\varepsilon(\gamma_1+\gamma_2)z^{\gamma_2}-(\gamma_1+1)z+\gamma_1.$$

Furthermore, $V(x)$ holds

$$\frac{dV}{dx}(b^*-)=\frac{dV}{dx}(b^*+)=-1. \tag{12}$$

Proof. First, consider $0<x<K$. Because the seller wants to minimize the payment to the buyer, the seller never cancels. For $x\ge K$ the buyer never exercise since payoff of the buyer $(K-x)^+$ equals 0. Therefore only

the seller selects an optimal stopping time. We suppose that the seller cancels the contract if $S_t(x) = a$.

$$\delta E[e^{-r\sigma_a^x}] = \delta \left(\frac{a}{x}\right)^{\gamma_1}. \tag{13}$$

To minimize Eq. (13), we differentiate $u(x,a)$ with respect to a.

$$\frac{\partial u(x,a)}{\partial a} = \frac{\delta\gamma_1}{x}\left(\frac{a}{x}\right)^{\gamma_1-1} > 0.$$

Because $u(x,a)$ is a monotonically increasing function with respect to a, it takes the minimum at $a = K$. Therefore it is optimal for the seller to cancel at $a = K$.

From Lemma 3.3, we have

$$g(z) = (1-\gamma_2)z^{\gamma_1+\gamma_2+1} + \gamma_2 z^{\gamma_1+\gamma_2} - \varepsilon(\gamma_1+\gamma_2)z^{\gamma_2} - (\gamma_1+1)z + \gamma_1,$$

where,$z = b/K, \varepsilon = \delta/K$. Since $g(0) = \gamma_1 > 0$ and $g(1) = -\varepsilon(\gamma_1+\gamma_2) < 0$, the equation $g(z) = 0$ has the solution z_0 in $(0,1)$.

Finally, we prove Eq. (12). Since $V(x) = K - x$ in $0 \le x < b$, it holds $\frac{dV}{dx}(b^*-) = -1$. Consider $b \le x < K$. We compute $\frac{dV}{dx}$.

$$\begin{aligned}
&\frac{dV}{dx}(x)\\
&= (K-b)\frac{-\gamma_1\left(\frac{K}{x}\right)^{\gamma_1}\frac{1}{x} - \gamma_2\left(\frac{x}{K}\right)^{\gamma_2-1}\frac{1}{K}}{\left(\frac{K}{b}\right)^{\gamma_1} - \left(\frac{b}{K}\right)^{\gamma_2}} + \delta\frac{\gamma_2\left(\frac{x}{b}\right)^{\gamma_2-1}\frac{1}{b} + \gamma_1\left(\frac{b}{x}\right)^{\gamma_1}\frac{1}{x}}{\left(\frac{K}{b}\right)^{\gamma_2} - \left(\frac{b}{K}\right)^{\gamma_1}}\\
&= \frac{1}{\left(\frac{K}{b}\right)^{\gamma_1} - \left(\frac{b}{K}\right)^{\gamma_2}}\left\{-\gamma_1\left(\frac{K}{x}\right)^{\gamma_1+1} - \gamma_2\left(\frac{x}{K}\right)^{\gamma_2-1} + \gamma_1\left(\frac{K}{x}\right)^{\gamma_1}\frac{b}{x}\right.\\
&\qquad \left. +\gamma_2\left(\frac{x}{K}\right)^{\gamma_2-1}\frac{b}{K} + \varepsilon\gamma_2\left(\frac{K}{b}\right)^{\gamma_1}\left(\frac{x}{K}\right)^{\gamma_2-1} + \varepsilon\gamma_1\frac{b}{x}\left(\frac{K}{x}\right)^{\gamma_1}\left(\frac{b}{K}\right)^{\gamma_2-1}\right\}\\
&= \frac{1}{1-\left(\frac{b}{K}\right)^{\gamma_1+\gamma_2}}\left\{-\gamma_1\left(\frac{b}{x}\right)^{\gamma_1+1}\frac{K}{b} - \gamma_2\left(\frac{x}{K}\right)^{\gamma_2-1}\left(\frac{b}{K}\right)^{\gamma_1} + \gamma_1\left(\frac{b}{x}\right)^{\gamma_1+1}\right.\\
&\qquad \left. +\gamma_2\left(\frac{x}{K}\right)^{\gamma_2-1}\left(\frac{b}{K}\right)^{\gamma_1+1} + \varepsilon\gamma_2\left(\frac{x}{K}\right)^{\gamma_2-1} + \varepsilon\gamma_1\left(\frac{b}{x}\right)^{\gamma_1+1}\left(\frac{b}{K}\right)^{\gamma_2-1}\right\}\\
&= \frac{1}{1-\left(\frac{b}{K}\right)^{\gamma_1+\gamma_2}}\left\{\gamma_1\left(\varepsilon\left(\frac{b}{K}\right)^{\gamma_2-1} + 1 - \frac{K}{b}\right)\left(\frac{b}{x}\right)^{\gamma_1+1} + \varepsilon\gamma_2\left(\frac{x}{K}\right)^{\gamma_2-1}\right.\\
&\qquad \left. -\gamma_2\left(1-\frac{b}{K}\right)\left(\frac{b}{K}\right)^{\gamma_1}\left(\frac{x}{K}\right)^{\gamma_2-1}\right\}.
\end{aligned}$$

Therefore, we have

$$\frac{dV}{dx}(b+)+1=\frac{1}{1-\left(\frac{b}{K}\right)^{\gamma_1+\gamma_2}}\left\{\gamma_1\left(\varepsilon\left(\frac{b}{K}\right)^{\gamma_2-1}+1-\frac{b}{K}\right)+\varepsilon\gamma_2\left(\frac{b}{K}\right)^{\gamma_2-1}\right.$$

$$\left.-\gamma_2\left(1-\frac{b}{K}\right)\left(\frac{b}{K}\right)^{\gamma_2-1}\left(\frac{b}{K}\right)^{\gamma_1}+1-\left(\frac{b}{K}\right)^{\gamma_1+\gamma_2}\right\}$$

$$=\frac{1}{1-\left(\frac{b}{K}\right)^{\gamma_1+\gamma_2}}\frac{K}{b}\left\{\gamma_1\left(\varepsilon\left(\frac{b}{K}\right)^{\gamma_2}+\frac{b}{K}-1\right)+\varepsilon\gamma_2\left(\frac{b}{K}\right)^{\gamma_2}\right.$$

$$\left.-\gamma_2\left(1-\frac{b}{K}\right)\left(\frac{b}{K}\right)^{\gamma_1+\gamma_2}+\frac{b}{K}-\left(\frac{b}{K}\right)^{\gamma_1+\gamma_2+1}\right\}$$

$$=\frac{1}{1-\left(\frac{b}{K}\right)^{\gamma_1+\gamma_2}}\frac{K}{b}\left\{(\gamma_2-1)\left(\frac{b}{K}\right)^{\gamma_1+\gamma_2+1}+\varepsilon(\gamma_1+\gamma_2)\left(\frac{b}{K}\right)^{\gamma_2}\right.$$

$$\left.-\gamma_2\left(\frac{b}{K}\right)^{\gamma_1+\gamma_2}+(\gamma_1+1)\frac{b}{K}-\gamma_1\right\}$$

$$=\frac{1}{1-\left(\frac{b}{K}\right)^{\gamma_1+\gamma_2}}\frac{K}{b}g\left(\frac{b}{K}\right)$$

$$=0. \qquad \square$$

4. Numerical Examples

In this section we provide numerical examples to evaluate the optimal boundary of the buyer and the value function of the perpetual game put option. We set the exercise price $K = 100$, volatility $\kappa = 0.3$, interest rate $r = 0.1$ and dividend $d = 0.09$.

Figure 1 shows the optimal exercise boundary as the penalty δ increases from 5 up to 25. Figure 1 reveals that b^* is a monotonically decreasing function with respect to δ. From Lemma 3.1, $\delta^* = 22.63$. Then we know that the perpetual game put is reduced to the perpetual American put.

Figure 2 demonstrates the value function $V(x), (\delta = 10)$ expressed by continuous line curve and perpetual American put $V^{ap}(x)$ with dividend by dashed line curve. We know that $V(x)$ satisfies $(K-x)^+ \leq V(x) \leq (K-x)^+ + \delta$ and is not differentiable at $x = K$. It is important to recognize that $V(x)$ is a decreasing convex function in x.

Figure 3 show that $V(x)$ is increasing in δ. When δ equals δ^*, the value function of the perpetual game put coincide with one of perpetual American put. Figure 4 shows that the price of the perpetual game put option with dividend dominates the one without dividend.

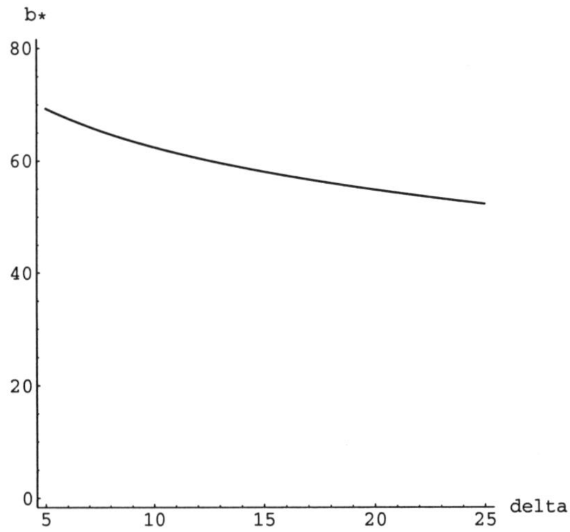

Figure 1. Optimal exercise boundary of the buyer

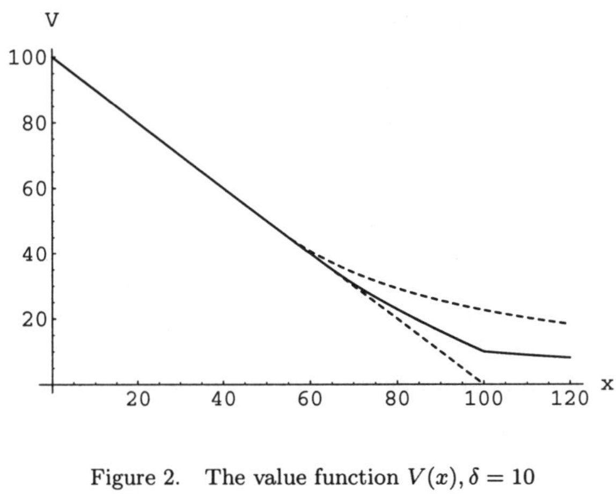

Figure 2. The value function $V(x), \delta = 10$

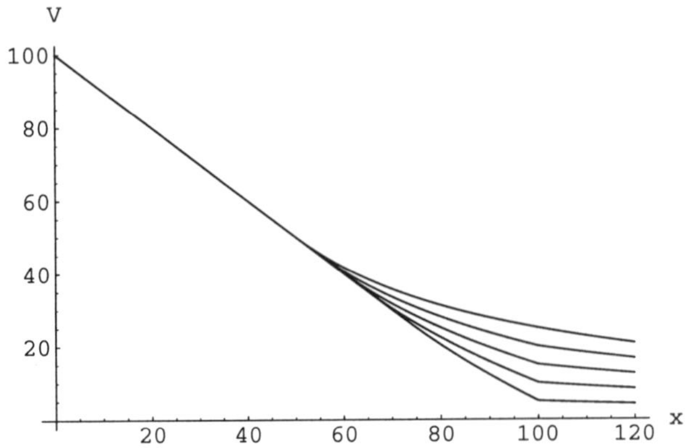

Figure 3. The value function $V(x)$, $\delta = 5, 10, 15, 20, \delta^*$

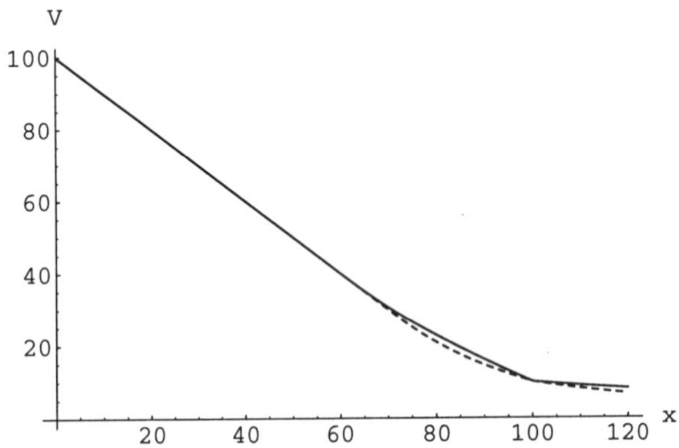

Figure 4. The value function $V(x)$, real line: dividend, dashed line: no dividend

5. Conclusion

In this paper we have studied the pricing model of the perpetual game put option and the optimal boundaries by means of a coupled stopping game based on the first hitting approach of a Brownian motion. We also explored some analytical properties of the value function and the optimal boundary of the perpetual game put option which are useful to provide an approximation of the finite lived game option. Furthermore, numerical examples are presented to illustrate the optimal boundary and the value function in Figure 1 and 2. Further investigation is left for future research.

References

1. Y. Kifer, Game options, *Finance and Stochastics* **4**, pp. 443–463, (2000).
2. R. C. Merton, Theory of rational option pricing, *Bell Journal of Economics and Management Science* **4**, pp. 141–183, (1973).
3. I. Karatzas and S.E. Shreve, *Methods of Mathematical Finance*, Springer, (1998).
4. A. E. Kyprianou, Some calculations for Israeli options, *Finance and Stochastics* **4**, pp. 73–86, (2004).
5. A. Suzuki and K. Sawaki, The Pricing of Callable Perpetual American Options, *Transactions of the Operations Research Society of Japan (in Japanese)* **49**, pp. 19–31, (2006).
6. I. Karatzas and S.E. Shreve, *Brownian Motion and Stochastic Calculus*, Second Edition, Springer, (1991).

ON THE VALUATION AND OPTIMAL BOUNDARIES OF CONVERTIBLE BONDS WITH CALL NOTICE PERIODS*

K. YAGI

Nanzan University,
27 Seirei, Seto, Aichi, 489-0863, Japan,
JSPS Research Fellow,
E-mail: d04mm003@nanzan-u.ac.jp

K. SAWAKI

Nanzan University,
27 Seirei, Seto, Aichi, 489-0863, Japan,
E-mail: sawaki@nanzan-u.ac.jp

In this paper we present a valuation model of callable convertible bonds with call notice periods in a setting of optimal stopping problem between the issuer (firm) and the holder (investor). The convertible bond holder can convert the bond into the underlying stock at any time. On the other hand, when the issuer wants to call (s)he must give an advance notice of calling the bond after a certain period. We analyze the pricing of callable convertible bonds with call notice periods. Furthermore, we explore the analytical properties of optimal conversion and call notice boundaries by the holder and the issuer, respectively. The value of convertible bonds and the optimal critical prices are examined numerically by using the finite difference method.

1. Introduction

A callable convertible bond is a hybrid security in that it enables investors to choose the best maturities suited to their portfolios and the issuer can redeem in whole or in part at its option before maturity. It is well known that the value of such a callable convertible bond should be enough to the amount subtracted the callable discount from the sum of the values of the bond, the European option and conversion premium. Yagi and Sawaki[1] has established the analytical decomposition of the valuation of the callable con-

*This work is supported in part by the Grant-in-Aid for Scientific Research (No. 16651090) of the Japan Ministry of Education, Science, Sports, and Culture.

vertible bond, based on the game option given by Kifer[2]. Seko and Sawaki[3] also considers the valuation of callable American options in which the optimal boundaries for the holder and the issuer are investigated. Brennan and Schwartz[4] and Ingersoll[5] have paved a way of pricing the convertible bond as pioneer works. McConell and Schwartz[6] analyzes an example of such typical securities, called LYON.

In this paper we extend the previous work[1] into the valuation of convertibles with call notice periods. Most convertible bonds are callable. When the convertible is called by the issuer, the holder has been put on notice that (s)he has only a short period to switch the convertible bond into common stock before the maturity. (S)He can either convert during this call notice period or accept the call and give up the convertible bond for the call price in cash offered by the issuer. Hence, when the call comes out, holders must decide whether it is in their optimal decision making to accept the call or to convert into common stock. Grau, Forsyth and Vetzal[7] analyzes the value of callable convertible bonds with call notice periods by using credit risk models given by Tsiveriotis and Fernandes[8] and Ayache, Forsyth and Vetzal[9], and provides numerical valuation by using a finite difference method. Dai and Kwok[10] also studies, in aid of variational inequalities, theoretical characterization of issuer's optimal call policy and holder's conversion policy of callable convertible securities with call notice requirement in which zero default risk and zero coupon payment are assumed.

We present a valuation model for callable convertible bonds with call notice periods in the following section. In particular, we explore analytical properties of optimal conversion and call notice boundaries for convertible bonds in Sec. 3. Also, we investigate how the call notice period gives influence the value of the convertible bond and the optimal conversion and call policies for the holder and the issuer, respectively. Furthermore, in Sec. 4 the value of convertible bonds and the optimal critical prices are examined numerically by using the finite difference method.

2. The Pricing Model of Convertible Bonds

In this section we present the pricing model of callable convertible bonds (here after, abbreviated by CB) with call notice periods.

We consider the Black-Scholes' economy consisting of a riskless asset and a risky stock. The riskless asset price B_t with the interest rate r is given by

$$dB_t = rB_t dt, \quad B_0 > 0, \ r > 0. \tag{1}$$

The stochastic differential equation of the stock price S_t under the risk-neutral measure $\tilde{P}$ can be written as the well known geometric Brownian motion

$$dS_t = (r - \delta)S_t dt + \kappa S_t d\tilde{Z}_t \tag{2}$$

where $\delta(< r)$ and κ are the constant rate of dividend payments and the volatility for the stock price S_t, respectively, and $\tilde{Z}_t$ is the standard Brownian motion defined on the probability space $(\Omega, \mathcal{F}, \{\mathcal{F}_t\}_{t\geq 0}, \tilde{P})$.

We assume that the CB is convertible for the investor at any time. Letting a be the number of stocks converted from the CB, the conversion value $C(t, S)$ is given by

$$C(t, S) = aS \quad \text{for all } t. \tag{3}$$

Let $V(t, S; T_0)$ be the value of a callable CB with the face value F, a call notice period T_0 and the maturity $T(> T_0)$ and $\overline{V}(t, S; F, T)$ the value of non-callable CB with the face value F and the maturity T. At the maturity the investor can either convert the CB or receive the face value. The terminal payoff at the maturity T is

$$V(T, S; T_0) = \overline{V}(T, S; F, T) = \max(aS, F). \tag{4}$$

Let τ be a conversion time (a stopping time) by the investor and $\mathcal{T}_{t,T}$ the set of stopping times with respect to the filtration $\{\mathcal{F}_t; 0 \leq t \leq T\}$. From Yagi and Sawaki[1], the value of non-callable CB is given by the following proposition.

Proposition 2.1. *The value of non-callable CB $\overline{V}(t, s; F, T)$ can be represented by*

$$\overline{V}(t, s; F, T) = \sup_{\tau \in \mathcal{T}_{t,T}} I_t^s(\tau), \tag{5}$$

$$I_t^s(\tau) = \tilde{E}\left[e^{-r(T-t)} \max(aS_T, F) 1_{\{\tau = T\}} + e^{-r(\tau - t)} aS_\tau 1_{\{\tau < T\}} \,\middle|\, S_t = s \right] \tag{6}$$

where "sup" is taken in the sense of the essential sup *with respect to the measure $\tilde{P}$. Moreover, the optimal stopping time τ_t^{**} for non-callable CB is determined by*

$$\tau_t^{**} = \inf\left\{ \tau \in [t, T) \mid \overline{V}(\tau, S_\tau; F, T) = aS_\tau \right\} \wedge T. \tag{7}$$

Hence, we have $\overline{V}(t, s; F, T) = I_t^s(\tau_t^{**})$.

The proof for the non-callable CB immediately follows from Ingersoll[5] or Yagi and Sawaki[1] because the value of non-callable CB can be rewritten as the sum of the bond price and the price of American call option with continuous dividend payments.

For callable CB if we assume no call notice period, the investor can immediately make a choice between selling the CB back to the issuer at a call price and converting the CB when the issuer calls the CB. Let $X(\geq F)$ be the call price. Then, the payoff in call is the maximum of the conversion value aS and the call price X; $\max(aS, X)$. If the call notice period is $T_0(>0)$, the investor is provided with an option whether to convert the CB in the notice period or to sell the CB back to the issuer at the call price after the notice period when the issuer gives an advance notice of calling the CB. In the other words, the callable CB should be changed to a non-callable CB with the face value replaced by X and the remaining period replaced by T_0 when the issuer notifies the call. Hence, when the issuer has notified of calling at time t, the value of the callable CB $V(t, S; T_0)$ may turn out equal to the one of the non-callable CB $\overline{V}(t, S; X, t+T_0)$ with the face value X and the maturity $t + T_0$. Letting σ be a call notice time (a stopping time) by the issuer, we can obtain the following theorem which is a revised version of the pricing model for a game option given by Kifer[2].

Theorem 2.1. *Define*

$$\begin{aligned} J_t^s(\sigma,\tau) = \tilde{E}\Big[& e^{-r(T-t)}\max(aS_T, F)1_{\{\sigma=\tau=T\}} \\ & + e^{-r(\tau-t)}aS_\tau 1_{\{\tau\leq\sigma<T\}} \\ & + e^{-r(\sigma-t)}\overline{V}(\sigma, S_\sigma; X, \sigma+T_0)1_{\{\sigma<\tau<T\}}\Big| S_t = s\Big]. \end{aligned} \tag{8}$$

Suppose that for all $s > 0$

$$\tilde{E}\left[\sup_{0\leq\tau\leq T} e^{-r\tau}aS_\tau \;\middle|\; S_0 = s\right] < \infty.$$

There is a unique no-arbitrage price process $V(t, s; T_0)$ of the CB in the sense of Black-Scholes model such that $V(t, s; T_0)$ can be represented by the right continuous process

$$\begin{aligned} V(t,s;T_0) &= \inf_{\sigma\in\mathcal{T}_{t,T}} \sup_{\tau\in\mathcal{T}_{t,T}} J_t^s(\sigma,\tau) \\ &= \sup_{\tau\in\mathcal{T}_{t,T}} \inf_{\sigma\in\mathcal{T}_{t,T}} J_t^s(\sigma,\tau) \end{aligned} \tag{9}$$

where "sup" and "inf" are taken in the sense of the essential sup *and the* essential inf *with respect to the measure $\tilde{P}$.*

Corollary 2.1. *The optimal stopping times* τ_t^* *and* σ_t^* *for the investor and the issuer, respectively, are given by*

$$\begin{aligned}\tau_t^* &= \inf\{\tau \in [t,T) \mid V(\tau, S_\tau; T_0) = aS_\tau\} \wedge T,\\ \sigma_t^* &= \inf\{\sigma \in [t,T) \mid V(\sigma, S_\sigma; T_0) = \overline{V}(\sigma, S_\sigma; X, \sigma + T_0)\} \wedge T \quad (10)\end{aligned}$$

and hence $V(t,s;T_0) = J_t^s(\sigma_t^*, \tau_t^*)$ *for all* t *and* s.

The proof in Theorem 2.1 and Corollary 2.1 for the value of callable CB with call notice periods should be referred to Kifer[2] after putting $Y_t = aS_t$ and $X_t = \overline{V}(t, S_t; X, t+T_0)$ as the payoff function of the investor and issuer, respectively.

Remark 2.1. Note that the issuer actually stops at time $t = \sigma + T_0$ where σ is the call notice time. The issuer never stops after time $t \geq T - T_0$.

When the issuer notifies to call the CB, the callable CB with the call notice period T_0 should be replaced by the non-callable CB with the remaining time T_0. If the remaining time ending the maturity is shorter than T_0, the issuer gives no advance notice of calling. Hence, the call notice time should be selected in the period until $T - T_0$.

3. Optimal Policies and Boundaries for the Issuer and the Investor

To value the callable CB we must identify the policies for the issuer and the investor to follow. In according to the spirit of Brennan and Schwartz[4] we assume that it is optimal for the investor to maximize the value of the CB and for the issuer to minimize the value of CB at each stopping time.

- **Optimal conversion policy**
  It is optimal for the investor never to convert the CB when its value is greater than the conversion value given by (3) because the investor seeks to maximize the value of the CB.
- **Optimal call notice policy**
  It is optimal for the issuer never to notify the investor of calling the CB when its value is less than the value of non-callable CB with face value X with maturity $t + T_0$ because the issuer wishes to minimize the value of the CB.

Let $\overline{s}$ be the optimal conversion boundary defined by $\overline{s} = \inf\{s | \overline{V}(t,s;X,t+T_0) = as\}$. Then, we have the following proposition 3.1.

Proposition 3.1. *Define* $T_0^* = (1/r)\log(X/F)$ *for* $X > F$. *If* $T_0 \leq T_0^*$, *then the inequality*

$$\overline{s} \geq \frac{F}{a} \tag{11}$$

holds and the value of the callable CB with the call notice period satisfies the inequalities

$$aS \leq V(t, S; T_0) \leq \overline{V}(t, S; X, t + T_0) \tag{12}$$

for all S *and* t.

Proof. If $T_0 \leq T_0^*$, we have

$$\begin{aligned} a\overline{s} &= \overline{V}(t, \overline{s}; X, t + T_0) \\ &\geq V(t, \overline{s}; T_0) \\ &\geq Xe^{-rT_0} \\ &\geq Xe^{-rT_0^*} \\ &= Xe^{-\log(X/F)} \\ &= F \end{aligned}$$

for $\overline{s}$ satisfying $a\overline{s} = \overline{V}(t, \overline{s}; X, t + T_0)$.

If $T_0 > T_0^*$, then we have

$$\begin{aligned} \lim_{t \to T^-} V(t, 0; T_0) &= F \\ &= Xe^{-\log(X/F)} \\ &= Xe^{-rT_0^*} \\ &> Xe^{-rT_0} \\ &= \lim_{t \to T^-} \overline{V}(t, 0; X, t + T_0) \end{aligned}$$

from $\lim_{S \to 0+} V(t, S; T_0) = Fe^{-r(T-t)}$ and $\lim_{S \to 0+} \overline{V}(t, S; X, t + T_0) = Xe^{-rT_0}$ for all t. Hence there exists a pair of t and S such that $V(t, S; T_0) > \overline{V}(t, S; X, t+T_0)$ for some t and S. From optimal conversion and call notice policies, if the value of callable CB with call notice period does not satisfy the Ineqs. (12), then there exists an arbitrage opportunity. Therefore, we must have the call notice period T_0 small enough to satisfy $T_0 \leq T_0^*$. □

Remark 3.1. If $T_0 > T_0^*$, then there exists an arbitrage opportunity for the investor.

Remark 3.2. It is important from a practical point of view to show that the call notice period must be shorter than T_0^* in Proposition 3.1.

For the non-callable CB the following inequality must hold;

$$\overline{V}(t, S; X, t+T_0) \geq aS, \quad \text{for } 0 \leq t \leq T. \tag{13}$$

Proposition 3.1 insists that the optimal conversion stock price should be bigger than or equal to F/a which is the stock price converted by CB. It is also easy to show that

$$\lim_{s \to \overline{s}} \frac{\partial \overline{V}}{\partial s} = a$$

that guarantees the existence of $\overline{s}$ satisfying $a\overline{s} = \overline{V}(t, \overline{s}; X, t+T_0)$.

Let $\mathcal{S}^i$ and $\mathcal{S}^f$ be the stopping regions of the callable CB with call notice periods T_0 for the investor and the firm, respectively, and $\mathcal{C}$ the continuation region for the both of them as follows;

$$\begin{aligned}
\mathcal{S}^i &= \{(t,s) \mid V(t,s;T_0) = as\} \\
\mathcal{S}^f &= \{(t,s) \mid V(t,s;T_0) = \overline{V}(t,s;X,t+T_0)\} \\
\mathcal{C} &= \{(t,s) \mid as < V(t,s;T_0) < \overline{V}(t,s;X,t+T_0)\}.
\end{aligned}$$

Now, define $V(t,s;0) = V_0(t,s)$ as the value of callable CB with no call notice period ($T_0 = 0$). The investor must immediately surrender the CB for redemption or convert when their claim is called by the issuer. The value of callable CB with no call notice periods must satisfy

$$aS \leq V_0(t,S) \leq \max(aS, X), \quad \text{for } 0 \leq t \leq T. \tag{14}$$

The stopping regions $\hat{\mathcal{S}}^i$ and $\hat{\mathcal{S}}^f$ for the investor and the firm, respectively, and the continuation region $\hat{\mathcal{C}}$ for the both of them, suited to the callable CB with no call notice period are

$$\begin{aligned}
\hat{\mathcal{S}}^i &= \{(t,s) \mid V_0(t,s) = as\} \\
\hat{\mathcal{S}}^f &= \{(t,s) \mid V_0(t,s) = \max(as, X)\} \\
\hat{\mathcal{C}} &= \{(t,s) \mid as < V_0(t,s) < \max(as, X)\}.
\end{aligned}$$

For each t define

$$\mathcal{S}_t^i = \{s \mid (t,s) \in \mathcal{S}^i\}, \quad \mathcal{S}_t^f = \{s \mid (t,s) \in \mathcal{S}^f\}, \quad \mathcal{C}_t = \{s \mid (t,s) \in \mathcal{C}\}$$

associated with the callable CB with call notice period, and similarly

$$\hat{\mathcal{S}}_t^i = \{s \mid (t,s) \in \hat{\mathcal{S}}^i\}, \quad \hat{\mathcal{S}}_t^f = \{s \mid (t,s) \in \hat{\mathcal{S}}^f\}, \quad \hat{\mathcal{C}}_t = \{s \mid (t,s) \in \hat{\mathcal{C}}\}$$

associated with the callable CB without call notice period.

Proposition 3.2. *The value of CB with the call notice period is more than or equal to the value of CB with no call notice period, that is,*

$$V_0(t,s) \leq V(t,s;T_0) \quad \textit{for all } t \textit{ and } s \tag{15}$$

and moreover,

$$(\hat{\mathcal{S}}_t^i \cup \hat{\mathcal{S}}_t^f) \supseteq (\mathcal{S}_t^i \cup \mathcal{S}_t^f) \quad \textit{for all } t. \tag{16}$$

Proof. When the issuer calls the callable CB, the investor has only a notice period to exchange the callable CB with call notice period for the non-callable CB which is maximized by the investor at the notice period. Hence it is clear to hold Ineq. (15). Equation (16) follows from Eqs. (14) and (15). □

The optimal conversion boundary for the investor can be defined as the graph of $s_t^i \equiv \inf\{s |\ s \in \mathcal{S}_t^i\}$. Similarly, the optimal call notice boundary for the issuer is the graph of $s_t^f \equiv \inf\{s |\ s \in \mathcal{S}_t^f\}$, $t \in [0,T]$. Set $s_t^* = \min(s_t^i, s_t^f)$ and similarly $\hat{s}_t^i$ the optimal conversion boundary and $\hat{s}_t^f$ the optimal call boundary for the CB with no call notice periods. Set $\hat{s}_t^* = \min(\hat{s}_t^f, \hat{s}_t^i)$.

Theorem 3.1. *The following relationships hold for the CB with call notice periods,*

(i) $\mathcal{S}_t^i = [s_t^i, \infty)$, $\mathcal{S}_t^f = [s_t^f, \infty)$, $C_t = [0, s_t^*)$

(ii) *(a) The optimal conversion boundary*
$\hat{s}_t^i \leq s_t^i \leq \overline{s}$

(b) The optimal call notice boundary
$\hat{s}_t^f \leq s_t^f \leq \overline{s}$

(c) Interaction of the optimal boundaries
$\hat{s}_t^* \leq s_t^* \leq \overline{s}$.

Proof. Property (i) follows from the definitions of $\mathcal{S}_t^i$, $\mathcal{S}_t^f$ and C_t. Inequalities $\hat{s}_t^i \leq s_t^i$ in property (ii) (a) and $\hat{s}_t^f \leq s_t^f$ in (b) can easily be proved from Ineq. (15). Suppose that $s_t^i > \overline{s}$ for some t. $V(t,\overline{s};T_0) > \overline{V}(t,\overline{s};X,t+T_0)$ holds, which contradict Ineq. (12). Hence $s_t^i \leq \overline{s}$ holds for any t. Similarly, $s_t^f \leq \overline{s}$ also holds. Property (c) follows from properties (a) and (b). □

4. Numerical Examples

In this section we present the numerical valuation of the CB with call notice periods, and the optimal conversion and call notice boundaries through a simple finite difference method. In particular, we illustrate how the prices and optimal boundaries of callable CB with call notice differ from them of non-callable CB.

Table 1. Data 1

Face value	F	100
Call price	X	120
Conversion number	a	2
Interest rate	r	0.04
Dividend rate	δ	0.02
Volatility	κ	0.3

Table 1 shows the data we use to evaluate the values of the CB and the optimal boundaries in Figs. 1, 2, 3 and 4. In this case, we have $T_0^* = 4.56$. In Fig. 1 the value of CB with maturity $T = 5$ is drawn as a function of the stock prices for several different call notice periods $T_0 = 1.0, 0.5, 0.1$ and 0.0. In either cases, $T_0 < T_0^* = 4.56$. We may observe the fact that the value of CB with call notice is not less than the ones with no call notice. Since CB has several properties for both option and bond, the value of CB doesn't have monotonicity with respect to t. However, we can numerically show that the value of CB is monotonically increasing with respect T_0 for $0 \leq T_0 \leq T_0^*$ from Fig. 1. In Fig. 2 optimal conversion boundaries have been illustrated for the same call notice periods as the ones used in Fig. 1. Figure 2 depicts a result of Theorem 3.1 (ii) (a) that optimal conversion boundaries with call notice are always higher than the ones with no call notice. Figure 3 demonstrates numerically Theorem 3.1 (ii) (b), in that the call notice boundaries is not less than the call boundary for the firm and not bigger than the optimal conversion boundary of the non-callable CB. Actually, either of the investor or the firm stops the CB, that is, when the stock price hits the minimum value of the conversion and call notice boundaries s_t^*, the CB is stopped. Therefore, we illustrate the minimum of the optimal conversion and call notice boundaries in Fig. 4. We may recognize Theorem 3.1 (ii) (c) in Fig. 4. When the call notice period T_0 is 0.5, the minimum of two boundaries s_t^* is the call notice boundary s_t^f for $12 \leq t \leq 17$ and s_t^* is the conversion boundary s_t^i except the period between $t = 12$ and $t = 17$. In Fig. 2 when the optimal conversion boundary is more than the call boundary, the conversion boundary is flat. Similarly, we have the same feature in Fig. 3.

Table 2. Data 2

Face value	F	100
Call price	X	110
Conversion number	a	1
Interest rate	r	0.04
Dividend rate	δ	0.02
Volatility	κ	0.3
Maturity	T	30
Call notice period	T_0	29

In Fig. 5, the value of CB also is illustrated for the data given by Table 2. Then, $\overline{s} = 99.36 < 100 = F/a$ and $T_0^* = 2.38 < 29 = T_0$. This shows that if Ineq. (11) in Proposition 3.1 does not hold, then $T_0 > T_0^*$.

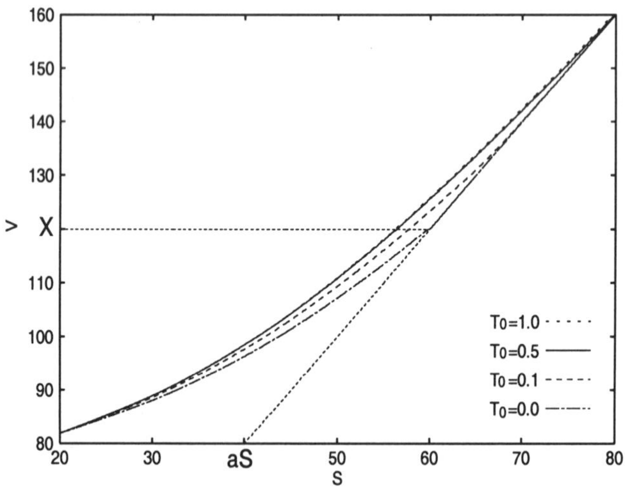

Figure 1. The value of CB
$F = 100, X = 120, a = 2, r = 0.04, \delta = 0.02, \kappa = 0.3, T = 5$

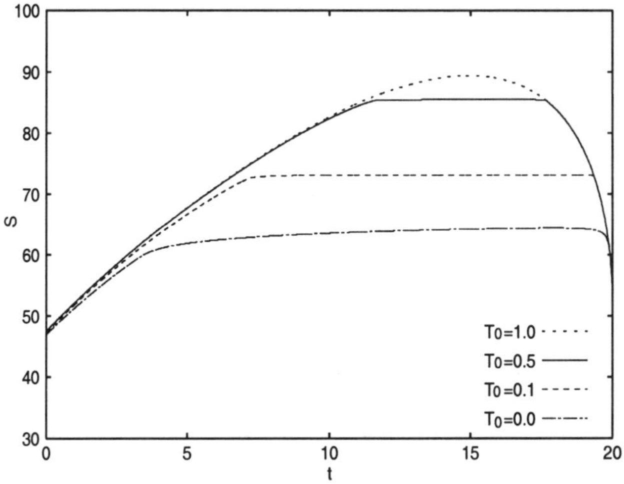

Figure 2. Optimal conversion boundaries for the investor
$F = 100, X = 120, a = 2, r = 0.04, \delta = 0.02, \kappa = 0.3$

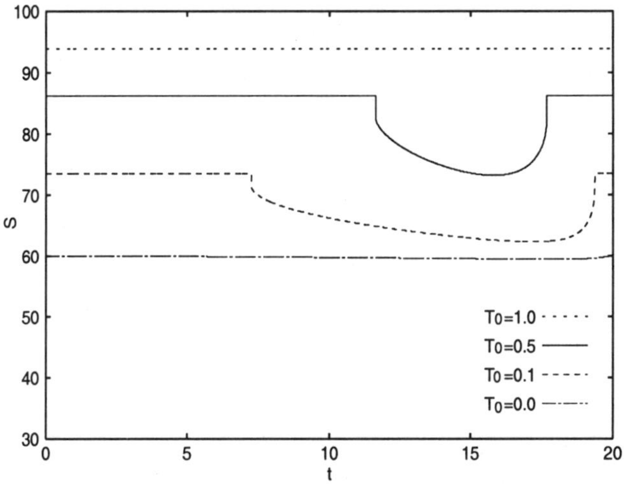

Figure 3. Optimal call notice boundaries for the firm
$F = 100, X = 120, a = 2, r = 0.04, \delta = 0.02, \kappa = 0.3$

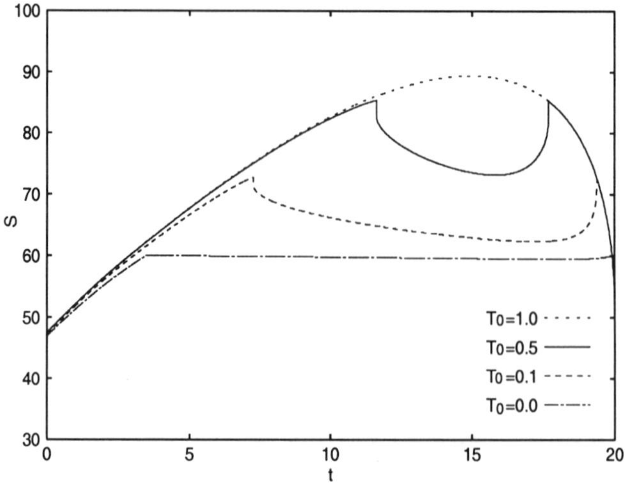

Figure 4. Interaction of the boundaries
$F = 100, X = 120, a = 2, r = 0.04, \delta = 0.02, \kappa = 0.3$

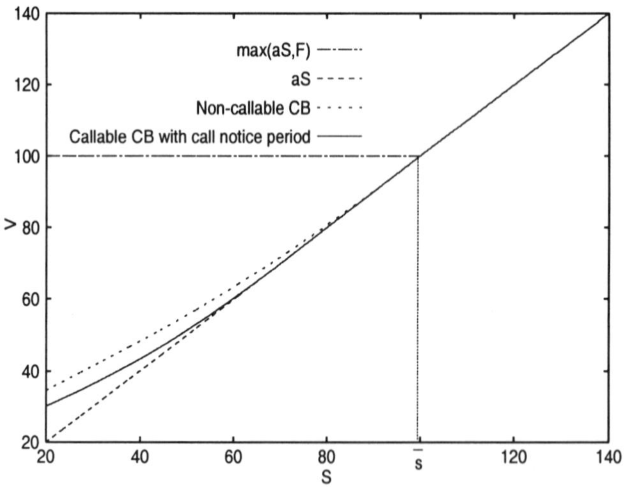

Figure 5. The value of CB
$F = 100, X = 110, a = 1, r = 0.04, \delta = 0.02, \kappa = 0.3, T = 30, T_0 = 29$

5. Conclusion

A callable CB gives its holder the right to swap the bond for stock and its issuer the right to force conversion by calling the bond when the market price of the callable CB exceeds the call price. In this paper we have studied the valuation model of the callable CB with the call notice period and the optimal conversion and call notice policies of the CB by using a coupled stopping game between the investor and the issuer. We have shown that the value of CB with the call notice is less than the one without call notice and both the optimal conversion and call notice boundaries with the call notice are more than the ones without call notice. Furthermore, we have provided the optimal conversion and call notice boundaries numerically computed by the finite difference method.

In this paper we emphasis the impact of call notice periods upon the value and optimal boundaries of callable CB. Such a call notice requirement is likely to be inherent in many callable securities issued by risky companies. It is of interest to study other hybrid securities with different provisions like structured bonds with different requirements of revised prices for conversion or of payments with foreign currencies and so on. We leave further investigation for future research.

Acknowledgements

We thank the referees for their useful comments and suggestions which are much helpful to improve the final version of the paper.

References

1. K. Yagi and K. Sawaki, The valuation and optimal strategies of callable convertible bonds, *Pacific Journal of Optimization*, **1**(2), 375-386, (2005).
2. Y. Kifer, Game options, *Finance and Stochastics*, **4**, 443-463, (2000).
3. S. Seko and K. Sawaki, The valuation of callable contingent claims, Presented at the 3rd World Congress, Bachelier Finance Society, (2004).
4. M.J. Brennan and E.S. Schwartz, Convertible bonds: Valuation and optimal strategies for call and conversion, *Journal of Finance*, **32**, 1699-1715, (1977).
5. J.E. Ingersoll, A contingent-claims valuation of convertible securities, *Journal of Financial Economics*, **4**, 289-322, (1977).
6. J.J. McConnell and E.S. Schwartz, The origin of LYONs: A case study in financial innovation, *Journal of Applied Corporate Finance* **4**(4), 40-47, (1992).
7. A.J. Grau, P.A. Forsyth and K.R. Vetzal, Convertible bonds with call notice periods, working paper, University of Waterloo, (2003).
8. K. Tsiveriotis and C. Fernandes, Valuing convertible bonds with credit risk, *Journal of Fixed Income*, **8**(2), 95-102, (1998).

9. E. Ayache, P.A. Forsyth and K.R. Vetzal, Next generation models for convertible bonds with credit risk, *Wilmott magazine*, pp. 68-77, (2002).
10. M. Dai and Y.K. Kwok, Optimal policies of call with notice period requirement for American warrants and convertible bonds, *Preprint*, (2005).

PART D

Performance Evaluation

AN EFFICIENT APPROACH TO ANALYZE FINITE BUFFER QUEUES WITH GENERALIZED PARETO INTERARRIVAL TIMES AND EXPONENTIAL SERVICE*

F. FERREIRA

Department of Mathematics and CEMAT
University of Trás-os-Montes e Alto Douro
Quinta dos Prados, 5001-911 Vila Real, Portugal
E-mail: mmferrei@utad.pt

A. PACHECO

Department of Mathematics and CEMAT
Instituto Superior Técnico, Technical University of Lisbon
Av. Rovisco Pais, 1049-001 Lisboa, Portugal
E-mail: apacheco@math.ist.utl.pt

The finding that heavy tailed Pareto distributions are adequate to model Internet packet interarrival times has motivated recent work aimed at developing efficient methods to evaluate or accurately approximate steady-state performance measures of Pareto/$M/\cdot/\cdot$ queues, overcoming or getting around certain problems that arise on their analysis using classical methods. In this work we give an overview of some of these methods and compare them with an approach we have recently proposed, which combines embedding, uniformization and stochastic ordering techniques and is suited to solve queues with Generalized-Pareto interarrival time distributions.

1. Introduction

Motivated by the relevance of Pareto/$M/\cdot/\cdot$ queueing systems in Internet applications (see, e.g., [6, 9, 10, 20] and references therein) and, at the same time, the difficulties that arise to treat these systems by standard methods, we apply a methodology we have recently proposed in [3, 4] to study $GP/M/\cdot/\cdot$ queues, where GP denotes the Generalized Pareto dis-

*This work was partially supported by *Programa Operacional "Ciência, Tecnologia, Inovação"* (POCTI) of the *Fundação para a Ciência e a Tecnologia* (FCT), cofinanced by the European Community fund FEDER, and the projects POSC/EIA/60061/2004 and Euro-NGI.

tribution as presented in [12] and next defined. The approach along with the obtained results are compared with others provided in the literature.

It is well known the importance of a good fitting of the interarrival time distributions to study $GI/M/\cdot/\cdot$ queues, as different interarrival time distributions can lead to significatively different queueing performances even if the first two moments are matched (see, e.g., [3, 4, 20]). This motivated also our interest in $GP/M/\cdot/\cdot$ queues, for which we are not aware of previous works. The Generalized Pareto distribution, apart from unifying the most common Pareto-Type distributions (used in queues), has yet more freedom to fit data.

The Generalized Pareto distribution, $GP(\kappa,\beta,\theta)$, $\theta,\beta,\kappa \in (0,\infty)$, has respectively density and distribution functions

$$a(x;\kappa,\beta,\theta) = \frac{\Gamma(\kappa+\beta)\theta^{\kappa}x^{\beta-1}}{\Gamma(\kappa)\Gamma(\beta)(\theta+x)^{\kappa+\beta}}$$

and

$$A(x;\kappa,\beta,\theta) = \frac{\Gamma(\kappa+\beta)}{\Gamma(\kappa)\Gamma(\beta)}\int_0^{\frac{x}{\theta+x}} y^{\beta-1}(1-y)^{\kappa-1}\,\mathrm{d}y$$

for $x > 0$, whereas the corresponding associated Shifted Generalized Pareto distribution, $SGP(\kappa,\beta,\theta,\gamma)$, $\theta,\beta,\kappa \in (0,\infty)$ and $\gamma \in \mathbb{R}$, has respectively density and distribution functions

$$a_S(x;\kappa,\beta,\theta,\gamma) = a(x-\gamma;\kappa,\beta,\theta)$$

and

$$A_S(x;\kappa,\beta,\theta,\gamma) = A(x-\gamma;\kappa,\beta,\theta)$$

for $x > \gamma$. Note that, in general, the Generalized Pareto and the Shifted Generalized Pareto distribution functions do not possess explicit inverses.

Generalized Pareto and Shifted Generalized Pareto distribution functions generalize the most commonly used Pareto-type distributions in queueing applications: the two-parameter Pareto distribution, $P(\kappa,\theta) = GP(\kappa,1,\theta)$, and its shifted version, $SP(\kappa,\theta,\gamma) = SGP(\kappa,1,\theta,\gamma)$, which have distributions functions

$$A(x;\kappa,1,\theta) = 1 - \left(\frac{\theta}{\theta+x}\right)^{\kappa},\ x > 0$$

and

$$A_S(x;\kappa,1,\theta) = 1 - \left(\frac{\theta}{\theta+x-\gamma}\right)^{\kappa},\ x > \gamma$$

respectively, along with their one-parameter particular cases, $P(\kappa) = P(\kappa, 1)$ and $SP(\kappa, \gamma) = SP(\kappa, 1, \gamma)$. In these distributions κ is a shape parameter (or tail index) which affects the thickness of the tail (the smaller is κ the heavier the tail becomes), and θ and γ are scale and a location parameters, respectively.

Traditional methods to study the (continuous-time) state process (the number of customers in the system) of $GI/M/\cdot/\cdot$ queues are based on the Laplace transform (LT) of the interarrival time distribution (see, e.g., [7, 10]) or on the one-step transition probability matrix of the state of the system embedded at customer arrival epochs (see, e.g., [15]).

As happens in general for heavy-tailed distributions, Pareto-type distributions do not possess finite moments of high integer orders. In fact, the n-th moment of the $GP(\kappa, \beta, \theta)$ distribution is finite only if $n < \kappa$, being given by $\theta^n \prod_{i=1}^{n} (\beta + n - i)/(\kappa - i)$. Moreover, Pareto-type distributions do not possess explicit analytic LTs, rendering impossible the direct use of the standard methods based on LTs to analyze $P/M/\cdot/\cdot$ queues, where P stands for "Pareto". In addition, difficulties arise also with the embedded Markov chain approach, as no simple expressions exists for the one-step transition probability matrix of the state of a $P/M/\cdot/\cdot$ queue embedded at customer arrival epochs. To get around these problems, researchers have been working in several directions, namely:

(i) resorting to simulation;
(ii) proposing methods to approximate the Pareto distribution itself by other distributions suited to be used in standard methods;
(iii) approximating the LT of the Pareto distribution followed by the use of standard methods based on LTs; and
(iv) approximating the one-step transition probability matrix of embedded discrete time Markov chains, used further to derive results in continuous time.

Good solutions are available in the literature to compute LTs of Pareto distributions [1, 8, 9], which may be used in the analysis of $P/M/s$ (delay) and $P/M/s/s$ (loss) queues using standard queueing methods based on LTs [6, 10]. However, for general finite buffer $GI/M/s/c$ systems, no simple LT based solutions are known for the customer prearrival and continuous time steady-state distributions. In addition, there is in general no way to analyze a $GI/M/s/c$ system from its corresponding delay or loss system. Alternative solutions propose to approximate the Pareto distribution by other distributions [2].

Efficient methods to deal with finite buffer queues are required, in particular for queues exhibiting tail raising effects (i.e., such that its steady-state probabilities increase in the tail) whose steady-state distribution cannot be properly approximated by truncation and normalization of the steady-state distribution of the corresponding delay system. Realizing this problem, Kim [14] proposed a model to approximate the steady-state distribution of $P/M/1/c$ queues.

A traditional approach to study the (continuous time) state process of $GI/M/s/c$ systems is to use information on the one-step transition probability matrix of the state of the system embedded at customers arrival epochs. If this transition probability matrix is known, then, among other approaches, it is possible to derive the associated steady-state distribution as a function of the LT of the interarrival time distribution and its derivatives ([13], Theorem 4.1) and obtain the continuous time steady-state distribution from Takács's relation ([22], see equation (4) below). However, as mentioned before, the one-step transition probability matrix of the state of a $P/M/s/c$ system embedded at arrival epochs does not possess a closed form expression and numerical procedures are needed for its computation.

In this line of work, the authors proposed a method to analyze $GI/M/s/c$ queues [3], further extended to $GI^X/M/s/c$ queues [4], which approximates, with arbitrarily chosen accuracy level, the transition probability matrix of the state of the system embedded at arrival epochs, provided that it is possible to compute in an efficient way the mixed-Poisson probabilities associated to the customer interarrival time distribution. The method combines Markov chain embedding with uniformization, for which it requires the evaluation of mixed-Poisson probabilities. In addition, it uses stochastic ordering as a way to bound the errors of the computed distributions and performance measures. As the Shifted Pareto mixed-Poisson distribution can be computed using simple stable recursions, the method can be used for $SP^X/M/s/c$ systems [3, 4]. In this paper, we show that the method is also suited to analyze $GP^X/M/s/c$ and $SGP^X/M/s/c$ queues by exhibiting efficient recursions to evaluate Generalized Pareto and Shifted Generalized Pareto mixed-Poisson distributions.

The paper will proceed as follows. In Section 2 we present an overview of our method and other methods aimed at overcoming the difficulties that arise when analyzing queues with Pareto customer interarrival times. Due to space constraints and aiming to show the efficiency of our method when confronted with the others presented, we focus only on queues with single arrivals and refer to [4] for the extension to batch arrival queues. In

Section 3 we show how to efficiently compute the Generalized Pareto and Shifted Generalized Pareto mixed-Poisson distributions. Finally in Section 4 we present some numerical results.

2. Some Approaches to Study P/M/ · /· Queues

In this section we give an overview of approaches to obtain the steady-state distribution (of the number of customers in the system) in continuous time, p, and at customer prearrivals, π, in $GI/M/s/c$ systems with Pareto-type customer interarrival time distribution. We let $A(\cdot)$ denote the customer interarrival time distribution function, $1/\lambda$ its mean, and $L(\cdot)$ its Laplace transform. Moreover, we let μ denote the customer service rate and $\rho = \lambda/(s\mu)$ denote the (offered) traffic intensity of the queueing system.

2.1. *Simulation*

Simulation studies for $P/M/\cdot/\cdot$ queues may be found, e.g., in [5, 9]. Although being simple and not leading to problems, the simulation of these queues is not an attractive method as it requires long running times in order to achieve convergence. Thus, it tends to be used only as an instrument to validate other approaches.

2.2. *Approximating the Pareto distribution*

Feldmann and Whitt [2] proposed a recursive procedure to fit hyperexponential distributions to distributions with decreasing failure rate, which applies, in particular, to fitting Pareto-type distributions. Using this approach, $GI/M/\cdot/\cdot$ queues are approximated by $H_k/M/\cdot/\cdot$, which can be treated by standard methods based on LTs [10, 22] or by matrix analytic methods [17]. However, despite the good accuracy the method can provide, the fitting procedure can get complicated as it has many degrees of freedom to chose the parameters involved, namely to decide the best number of exponentials to be used in the fitting and the same number of special points that are used in the fitting procedure. Thus, alternative approaches should be investigated.

2.3. *Approximating the Laplace transform of the Pareto distribution*

Many different approaches are available to approximate the LT of Pareto distributions. Gordon [9] expressed the LT of the $P(\kappa, \theta)$ distribution as a power series

$$L(t) = \kappa\Gamma(-\kappa)\left[(\theta t)^{\kappa}e^{\theta t} - \sum_{n=0}^{\infty}\frac{(\theta t)^n}{\Gamma(n-\kappa+1)}\right] \qquad (1)$$

which is convergent for $1 < \kappa < 2$. Later, Abate and Whitt [1] suggested the use of continued-fraction expansions to evaluate the LT of completely monotone density functions and expressed, in particular, the LT of the $P(\kappa)$ distribution by the continued-fraction expansion

$$L(t) = 1 - t_1 \circ t_2 \circ t_3 \circ \ldots (0)$$

with

$$t_n(t) = \frac{a_n}{b_n + t}$$

where $a_1 = t$ and $a_{2n+1} = n$, $a_{2n} = n + \kappa - 1$, $b_{2n-1} = t$, and $b_{2n} = 1$, for $n \in \mathbb{N}_+$. The truncation of the previous LT expansions leads to efficient numerical approximations for the LT of the $P(\kappa)$ distribution.

Another efficient approach to approximate LTs that has been used recently is the so called Transformed Approximation Method (TAM), which can be applied to any continuous distribution. TAM's first version, due to Harris and Marchal [11], approximates the LT of a distribution by

$$L(t) \simeq \frac{1}{N}\sum_{i=1}^{N} e^{-tx_i},$$

where the points x_i are the quantiles of probability $i/(N+1)$. The method was later improved [8] to allow freedom in the choice of the quantiles x_i, using the approximation

$$L(t) \simeq \sum_{i=1}^{N} p_i e^{-tx_i},$$

where the points x_i are the quantiles of previously arbitrarily selected increasing probabilities, and

$$p_1 = \frac{A(x_1) + A(x_2)}{2}, \qquad p_N = 1 - \frac{A(x_{N-1}) + A(x_N)}{2}$$

and

$$p_i = \frac{A(x_{i+1}) - A(x_{i-1})}{2}, \quad \text{for } 2 \le i \le N-1.$$

The method is specially useful for distribution functions $A(\cdot)$ that are easily invertible, which is the case of the $\mathrm{P}(\kappa, \theta)$ and $\mathrm{SP}(\kappa, \theta)$ distributions, whose quantile of probability p is given by

$$x = \theta(1-p)^{-1/\kappa} - \theta \quad \text{and} \quad x = \theta(1-p)^{-1/\kappa}$$

respectively. However the method is not particularly fast when the distribution functions are not easily invertible – as it is the case in general for the Generalized Pareto distribution – since an additional search procedure to compute quantiles is required.

In this work we are interested in finite buffer queueing systems, but recall that once we get an approximation $\bar{L}(\cdot)$ of the LT $L(\cdot)$ of the Pareto distribution, the $P/M/s/\infty$ system may be approximated by standard methods based on the geometric parameter $\sigma = L(s\mu(1-\sigma))$ (see, e.g., [10]) by approximating the geometric parameter using $\bar{L}(\cdot)$ instead of $L(\cdot)$. As the customer prearrival steady-state probabilities of a finite buffer $P/M/s/s$ system are given as follows [22]

$$\pi_n = \sum_{k=n}^{s} (-1)^{k-n} \binom{k}{n} C_k \frac{\sum_{l=k}^{s} \binom{s}{l} C_l^{-1}}{\sum_{l=0}^{s} \binom{s}{l} C_l^{-1}}, \quad n = 0, 1, \ldots, s \tag{2}$$

with

$$C_0 = 1 \quad \text{and} \quad C_k = \prod_{l=1}^{k} \frac{L(l\mu)}{1 - L(l\mu)}, \; 1 \le k \le s,$$

Fischer *et al.* [6] proposed to use as approximations of the customer prearrival steady-state probabilities the values $\pi_n^\star$ given by (2) with the C_k replaced by $\prod_{l=1}^{k}[\bar{L}(l\mu)/(1-\bar{L}(l\mu))]$, $1 \le k \le s$. The continuous time steady-state probability vector (p_n) is then approximated by $(p_n^\star)$, where

$$p_n^\star = \frac{\lambda}{n\mu}\pi_{n-1}^\star, \; n = 1, 2, \ldots, s \quad \text{and} \quad p_0^\star = 1 - \sum_{j=1}^{s} p_j^\star.$$

Taking into account the heavy-tailed nature of the $P(\kappa, \theta)$ distribution with $1 < \kappa < 2$ and the tail raising effect it produces on the continuous time steady-state distribution of the associated $P/M/1/c$ system, Koh and Kim [14] proposed approximations to the corresponding steady-state blocking probability and the continuous time steady-state distribution. The interpretation of the steps that are described in [14] leads to the conclusion that Koh and Kim's method consists of the following four steps:

1. Approximate the LT $L(\cdot)$ of the $P(\kappa,\theta)$ distribution by $\bar{L}(\cdot)$, where
$$\bar{L}(t) = \kappa\Gamma(-\kappa)\left[(\theta t)^{\kappa}e^{\theta t} - \sum_{n=0}^{N}\frac{(\theta t)^n}{\Gamma(n-\kappa+1)}\right]$$
where N is such that
$$\kappa\Gamma(-\kappa)\frac{[(\kappa-1)/\rho]^N}{\Gamma(N-\kappa+1)} \le \varepsilon$$
for an appropriately chosen small positive value ε. That is, $\bar{L}(\cdot)$ is obtained from $L(\cdot)$ by truncating at N the power series in (1).
2. Approximate the geometric parameter σ of the corresponding $P(\kappa,\theta)/M/1$ queue, satisfying $\sigma = L(\mu(1-\sigma))$, by the solution $\bar{\sigma}$ of the equation $\sigma = \bar{L}(\mu(1-\sigma))$ on the interval $(0,1)$.
3. Approximate the steady-state blocking probability by
$$\bar{P}_b = \left(\frac{1-\bar{\sigma}}{1-\bar{\sigma}^{c+1}}\right)^a \bar{\sigma}^c$$
with
$$a = \frac{\log(1+\mu\frac{d\bar{L}(t)}{dt}|_{t=\mu(1-\bar{\sigma})})}{\log(1-\bar{\sigma})}.$$
4. Finally, approximate the steady-state distribution of the number of customers in the system (p_n) by $(p_n^\star)$ by finding a solution u of the equation $\sum_{n=0}^{c} p_n^\star = 1$, where
$$p_0^\star = 1-(1-\bar{P}_b)\rho \quad \text{and} \quad p_n^\star = b_n + \frac{u(\bar{P}_b - b_c)/2}{c-n+u}, \; 1 \le n \le c$$
with $b_n = \frac{p_0^\star\rho(1-\bar{\sigma})}{1-\rho}\bar{\sigma}^{n-1}$, for $1 \le n \le c$.

This method is asymptotically exact and provides good results for moderate/large buffer sizes, specially for κ close to 2, but it does not guarantee a given approximation error. For an idea of what the method's approximation error might be either simulation or alternative methods whose approximation error may be controlled need to be used. Moreover, the method applies only to $P(\kappa,\theta)$ interarrival times and cannot be used for multiserver systems. Steps 1-3 of the method, which deal with the loss probability approximation, are carried out fast. However, the complexity of the root finding problem that needs to be solved in step 4, to obtain the value u and in order to compute the approximation of the steady-state distribution of the number of customers in the system, grows fast with the queue capacity.

2.4. *Combining embedding with uniformization*

In this section we briefly present the method we proposed in [3] to analyze $GI/M/s/c$ queues, whose extension to batch arrivals can be found in [4]. Let $Y(t)$ denote the state of a $GI/M/s/c$ system at time t and $\bar{Y}_n$ denote the corresponding state at the prearrival of the n-th customer, i.e., $\bar{Y}_n = Y(T_n^-)$ with T_n denoting the arrival epoch of the n-th customer. As the continuous time state process, $Y = (Y(t))$, is a Markov regenerative process (MRGP, see [15]) associated to the renewal sequence $(T_n)_{n\in\mathbb{N}_+}$ of customer arrival epochs, information on Y may be extracted from the analysis of the DTMC $\bar{Y} = (\bar{Y}_n)$, the customer prearrival state process.

The analysis of $\bar{Y}$ takes into account that in-between two consecutive customer arrivals the state process Y evolves as a pure-death process with infinitesimal generator matrix $Q = (q_{ij})_{i,j=0,1,\ldots,c}$, such that

$$q_{i,i-1} = -q_{ii} = \mu \min(s,i), \quad i = 1, 2, \ldots, c$$

with all other entries null. As $\max |q_{ii}| = s\mu < \infty$, this death process is uniformized with rate $s\mu$, whose associated uniformized embedded transition probability matrix is $\hat{P} = I + Q/s\mu$ (see, e.g., [15]). Then, the computation of the transition probability matrix $\bar{P}$ of $\bar{Y}$ involves the powers of $\hat{P}$ along with the computation of mixed-Poisson probabilities with parameter $s\mu$. Namely,

$$\bar{P} = \Delta\left(\sum_{m\in\mathbb{N}} \alpha_m \hat{P}^m\right)$$

where $[\Delta(V)]_{ij} = [V]_{\min(c,i+1)j}$, for $V = [v_{ij}]_{i,j=0,1,\ldots,c}$, and

$$\alpha_m \equiv \alpha_m^A(s\mu) = \int_0^{+\infty} e^{-s\mu t}\frac{(s\mu t)^m}{m!}\, A(\mathrm{d}t) \tag{3}$$

denotes the m-th mixed-Poisson probability with mixture distribution (the customer interarrival time distribution) function $A(\cdot)$ and rate $s\mu$. Thus, α_m is the probability that exactly m renewals take place in the uniformizing Poisson process between two consecutive customer arrivals to the system. In detail, we proceed as follows:

1. Evaluate the mixed-Poisson probabilities with structural distribution function $A(\cdot)$ and rate $s\mu$ in (3).
2. For a sufficiently large positive integer N, compute the matrices

$$\underline{M}^{(N)} = \Delta\left(\sum_{m\leq N} \alpha_m \hat{P}^m + \sum_{m>N} \alpha_m U\right)$$

and

$$\overline{M}^{(N)} = \Delta \left(\sum_{m \le N} \alpha_m \hat{P}^m + \sum_{m > N} \alpha_m \hat{P}^N \right)$$

with $U = (u_{ij}) = (\delta_{j0})$, along with their associated stationary probability vectors $\underline{\pi}^{(N)}$ and $\overline{\pi}^{(N)}$.

3. Approximate the customer prearrival steady-state probability vector π by

$$\pi^\star = \frac{\underline{\pi}^{(N)} + \overline{\pi}^{(N)}}{2}$$

and the continuous time steady-state probability vector p by $p^\star$, where

$$p_n^\star = \frac{\lambda}{\mu \min(s,n)} \pi_{n-1}^\star,\ n = 1, 2, \ldots, c \quad \text{and} \quad p_0^\star = 1 - \sum_{j=1}^{c} p_j^\star. \tag{4}$$

The matrices $\underline{M}^{(N)}$ and $\overline{M}^{(N)}$ computed in step 2 are such that

$$\underline{M}^{(N)} \le_K \bar{P} \le_K \overline{M}^{(N)}$$

with $\le_K$ denoting the Kalmykov ordering of matrices (see, e.g., [19, 21]), and $\underline{M}^{(N)}$ and $\overline{M}^{(N)}$ converge to $\bar{P}$ as N tends to infinity. In addition, their associated stationary probability vectors $\underline{\pi}^{(N)}$ and $\overline{\pi}^{(N)}$ are such that

$$\underline{\pi}^{(N)} \le_{st} \pi \le_{st} \overline{\pi}^{(N)}$$

with $\le_{st}$ denoting the usual stochastic ordering of probability vectors (see, e.g., [19, 21]), and $\underline{\pi}^{(N)}$ and $\overline{\pi}^{(N)}$ converge to π as N tends to infinity. Moreover, it follows that

$$\|p^\star - p\|_\infty \le \varepsilon$$

whenever

$$\|\overline{\pi}^{(N)} - \underline{\pi}^{(N)}\|_\infty < \varepsilon/(c+1).$$

These facts make it possible to compute the customer prearrival and continuous time steady-state distributions, π and p, in a controlled way with any desired degree of accuracy.

A very special case is the single server system, $s = 1$, for which exact solutions are obtained using only the first c mixed-Poisson probabilities without need to resort to stochastic bounding [3].

The described method is numerically stable as it involves only additions and multiplications of nonnegative numbers, and it is fast, as the special block-structure of the transition probability matrix $\hat{P}$ can be explored in order to compute its powers [3]. It is easy to implement and depends on the interarrival time distribution only through its mixed-Poisson probabilities. In addition, the mixed-Poisson probabilities can be computed in a fast recursive way for a large range of distributions [16, 23, 24], which, as presented in the next section, includes Generalized Pareto distributions.

3. Generalized Pareto Mixed-Poisson Probabilities

We recall that, for $m \in \mathbb{N}$, the m-th mixed-Poisson probability with mixing distribution function $A(\cdot)$ defined on (x_0, x_1), $0 \leq x_0 < x_1 \leq \infty$, and rate q, is given by $\alpha_m^A(q) = \int_{x_0}^{x_1} e^{-qt} \frac{(qt)^m}{m!} A(\mathrm{d}t)$. In general, it is not possible to get simple closed form expressions for the mixed-Poisson probabilities. However, with the arrival of faster computers, this task constitutes no problem for a large range of mixing distributions, specially since Willmot [24] proposed a general method for the derivation of simple recursive formulas for the evaluation of mixed-Poisson probabilities with rate 1 provided that the probability density function $a(\cdot)$ associated to $A(\cdot)$ is such that

$$\frac{d \log a(x)}{dx} = \frac{\sum_{n=0}^{k} \eta_n \, x^n}{\sum_{n=0}^{k} \psi_n \, x^n} = \frac{\eta(x)}{\psi(x)}$$

for some polynomials $\eta(x)$ and $\psi(x)$ of degree $k > 0$. Following [24], we provide the corresponding method for the mixed-Poisson probabilities with general rate q, $q > 0$. Namely, letting

$$\phi(x) = \sum_{n=0}^{k} \phi_n x^n = \eta(x) + \frac{d\psi(x)}{dx} - q\psi(x)$$

the approach of Willmot leads to the following recursion formula

$$\sum_{n=-1}^{k} \frac{m\psi_{n+1} + \phi_n}{q^{n+m}} \, (n+m)! \, \alpha_{n+m} = a(x_1)\psi(x_1)x_1^m e^{-qx_1} - a(x_0)\psi(x_0)x_0^m e^{-qx_0}$$

for $m \in \mathbb{N}$, with the convention that $\phi_{-1} = 0$, x_0^n being interpreted as 1 when $x_0 = m = 0$, and $x_1^m e^{-qx_1}$ being interpreted as 0 if $x_1 = \infty$.

We next use the previous result in order to derive recursions to compute the mixed-Poisson probabilities with general rate for Generalized Pareto and Shifted Generalized Pareto distributions. For the Generalized Pareto distribution, $\mathrm{GP}(\kappa, \beta, \theta)$, we have

$$\frac{d \log a(x; \kappa, \beta, \theta)}{dx} = \frac{(\beta - 1)\theta - (\kappa + 1)x}{\theta x + x^2} = \frac{\eta(x)}{\psi(x)}.$$

Hence, $\phi(x) = \beta\theta + (1 - q\theta - \kappa)x - qx^2$ and the mixed-Poisson probabilities with rate q may be evaluated from the recursion

$$(n+2)(n+1)\alpha_{n+2} = q\theta(\beta + n)\alpha_n + (n+1)(n+1-\kappa-q\theta)\alpha_{n+1} \quad (5)$$

for $n = 0, 1, \ldots$.

From ([23], Theorem 1), we conclude that this recursion is strongly stable in the backward direction for $n \leq \kappa + q\theta - 1$ and in the forward direction for $n \geq \kappa + q\theta + 1$. Thus, to evaluate conveniently the mixed-Poisson probabilities from (5), we should previously evaluate the mixed-Poisson probabilities for n equal to the integer parts of $\kappa + q\theta$ and $\kappa + q\theta + 1$, as starting values for the recursions, and then use the recursion (5) in the backward (forward) direction to compute the mixed-Poisson probabilities α_n for $n \leq \kappa + q\theta - 1$ ($n > \kappa + q\theta + 1$). As noted in [23, 24], to evaluate the mixed-Poisson probabilities to start the recursions, it is convenient to express the mixed-Poisson probabilities as a function of the confluent hypergeometric function of the second kind

$$U(a, b, c) = \frac{1}{\Gamma(a)} \int_0^\infty e^{-cx} x^{a-1} (1+x)^{b-a-1} \, dx$$

in the form

$$\alpha_n = \frac{\Gamma(\kappa + \beta)\Gamma(n + \beta)(q\theta)^n}{\Gamma(\kappa)\Gamma(\beta)\Gamma(n+1)} U(n + \beta, n + 1 - \kappa, q\theta) \quad (6)$$

and evaluate $U(a, b, c) = c^{-a} y_0$ with y_0 obtained from the recursion

$$y_n = \frac{(n+1)(2n + 2a + 2 + c - b)y_{n+1} - (n+1)(n+2)y_{n+2}}{(n+a)(n+a+1-b)}$$

applying Miller's algorithm [25, 18].

Note that the Shifted Generalized Pareto distribution is the convolution of the distribution of a constant random variable with the Generalized Pareto distribution. Namely, if $X \sim \mathrm{SGP}(\kappa, \beta, \theta, \gamma)$, then $X \stackrel{d}{=} \gamma + Y$ with $Y \sim \mathrm{GP}(\kappa, \beta, \theta)$ and $\stackrel{d}{=}$ denoting equality in distribution. As mixed-Poisson distributions are closed under convolutions, the Shifted Generalized Pareto mixed-Poisson probabilities with rate q, α_n^X, can then be obtained from the Generalized Pareto mixed-Poisson probabilities with rate q, α_n^Y, by

$$\alpha_n^X = \sum_{m=0}^{n} e^{-q\gamma} \frac{(q\gamma)^m}{m!} \alpha_{n-m}^Y.$$

4. Numerical Results

In this section, we provide examples of application of our method that show that it is effective to solve $GP/M/s/c$ queueing systems and works well in situations where other methods do not. The algorithms used for the computations have been implemented using MATLAB and the results obtained by our method have been computed with an accuracy of $\varepsilon = 10^{-10}$.

In Figures 1 and 2 we present the steady-state distributions of the number of customers in $GP(\kappa, 0.4, \theta)/M/5/10$ systems and $GP(1.4, \beta, \theta)/M/5/10$ systems, with service rate $\mu = s^{-1}$ and traffic intensity $\rho = 1$, for four different values of the parameters κ and β, respectively. In turn, Figures 3 and 4 provide the steady-state blocking probability and steady-state mean waiting time in queue of non-blocked customers in the $GP(\kappa, \beta, \theta)/M/5/10$ system as a function of the traffic intensity.

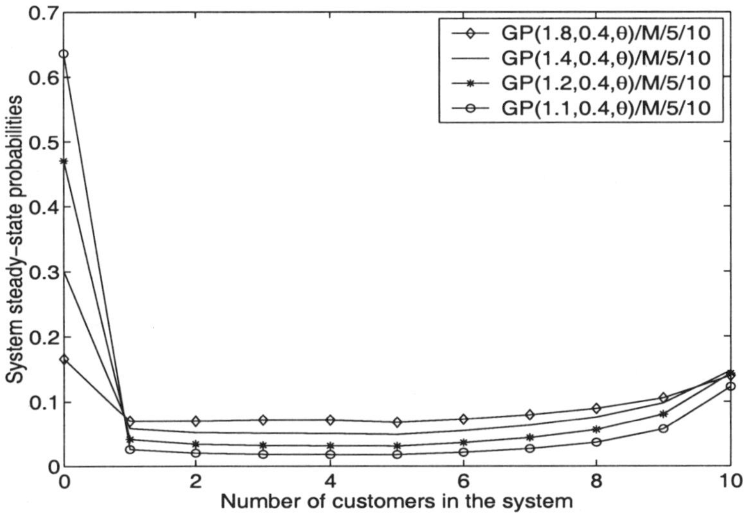

Figure 1. Steady-state distribution of the number of customers in $GP(\kappa, 0.4, \theta)/M/5/10$ systems with traffic intensity 1.

We observe that, although the $GP(\kappa, \beta, \theta)/M/s/c$ system may lead to the same kind of behavior that $GP(\kappa, \theta)/M/s/c$ systems, the performance of the queues with Generalized Pareto customer interarrival times may differ significantly from that of queues with two parameter Pareto customer interarrival times, specially for very small values of the parameter β. For fixed values of κ (β) and mean, the smaller is the value of β (κ) the greater

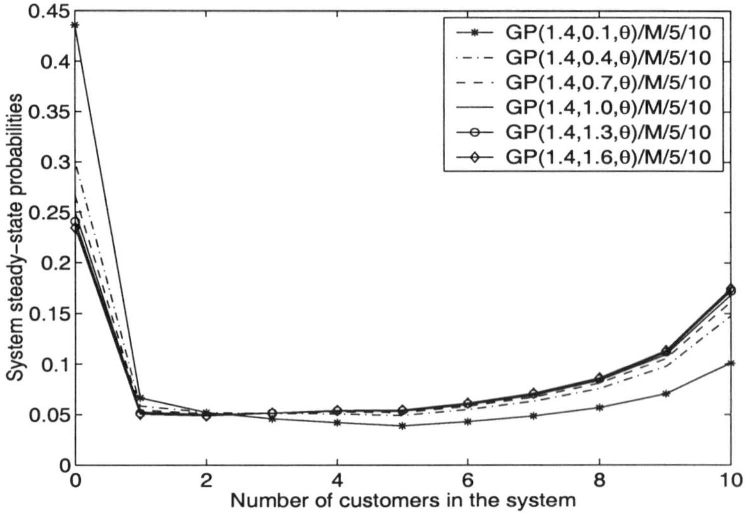

Figure 2. Steady-state distribution of the number of customers in $GP(1.4, \beta, \theta)/M/5/10$ systems with traffic intensity 1.

is the steady-state probability of having an empty system, as well as the steady-state blocking probability.

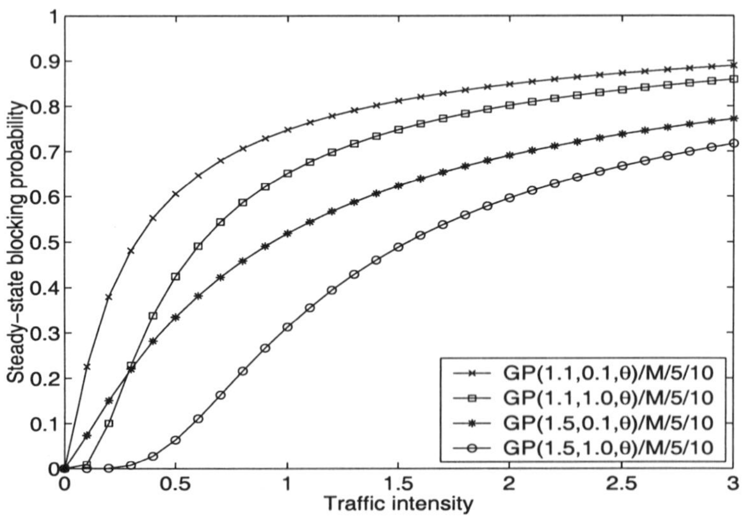

Figure 3. Steady-state blocking probabilities as a function of traffic intensity in $GP(\kappa, \beta, \theta)/M/5/10$ systems.

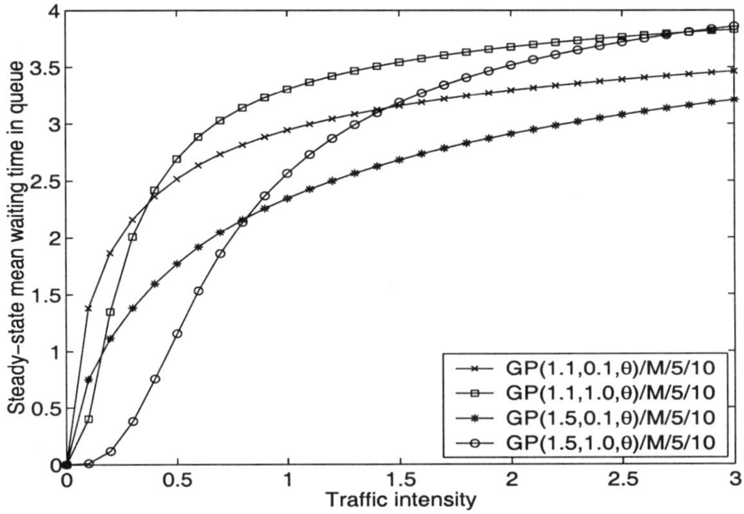

Figure 4. Steady-state mean waiting time in queue of non-blocked customers as a function of traffic intensity in $GP(\kappa, \beta, \theta)/M/5/10$ systems.

In addition, to show the advantages of our method *versus* others, we confront the results obtained for loss systems by our method with the approach proposed in [6], and explained after (2), after approximating the LT of the Pareto distribution with the methods presented in Subsection 2.3, and compare the results obtained by our method for $SP/M/1/c$ queueing systems with the ones obtained by the approximations proposed by Koh and Kim [14]. We have made extensive computations for systems with $P(\kappa, \theta)$ and $SP(\kappa, \theta)$ interarrival time distributions for a large range of parameters.

For queues with small/moderate capacities the results obtained by the approach explained after (2) are consistent with those provided by our method. This was an expected result as the LT may be approximated with any desired accuracy. However, for large system capacities, the former approach leads to incorrect results since the formula (2) is unstable as it involves sums of alternately positive and negative terms of different magnitudes. In contrast, our method is suited to deal with large capacity systems.

We observe that the truncation of the continued-fraction and the series expansions provides faster results than TAM to approximate the LT and that these methods combined with the steps explained after (2) are faster than ours. However, these expansions concern only to specific Pareto-type distributions, while both TAM and our method can be applied to general customer interarrival time distributions. We note that, nonetheless, our method provide fast results. For example, for queues with 10 servers and capacities 50, 100, 150, the algorithm running times were approximately 5, 11, and 46 seconds, respectively, on a 2.8 GHz computer.

When compared with Koh and Kim's method [14], ours has competitive running times, being slower only for small capacity systems – in which case the running times are only of a few seconds. This is not surprising since, as noted before, the complexity of Step 4 in Koh and Kim's method increases significantly with the queue capacity, whereas the complexity of our method increases more slowly with the queue capacity. However, for small queue capacities (the region where Koh and Kim's method is faster than ours) Koh and Kim's method does not provide accurate results, as can be seen in Figure 5. For (moderate to) large queue capacities, Koh and Kim's method leads to accurate results, as can be seen in Figure 6.

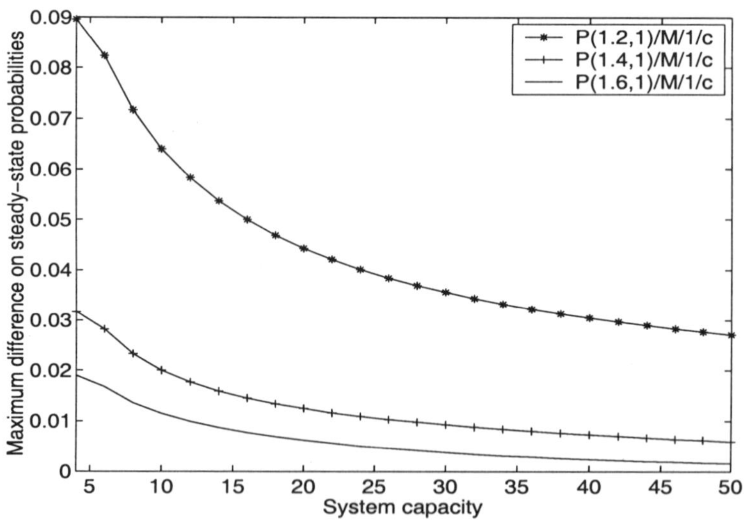

Figure 5. Maximum absolute difference between the steady-state probabilities obtained by our method (FP) and Koh and Kim's method (KK) as a function of the system capacity, in $P(\kappa, 1)/M/1/c$ systems with traffic intensity 0.8.

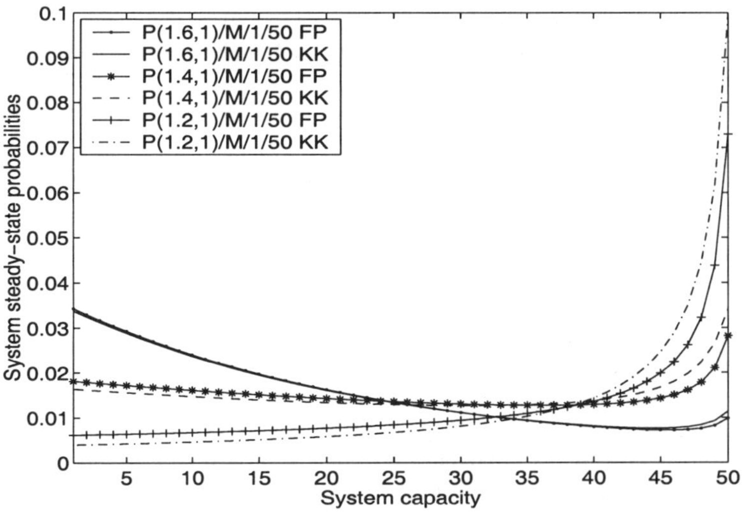

Figure 6. Steady-state distribution in $P(\kappa,1)/M/1/50$ systems, with traffic intensity 0.8, obtained by our method (FP) and by Koh and Kim's method (KK).

5. Conclusion

In this paper we have pointed out some of the difficulties arising in studying Pareto-Type$/M/\cdot/\cdot$ queues and gave an overview of existing methods to analyze these systems. In addition, we showed that the approach we have recently proposed in [3, 4], which combines the classical Markov chain embedding with uniformization and stochastic ordering techniques, is suited to study Generalized-Pareto$/M/\cdot/\cdot$ queues as it is possible to efficiently evaluate the associated Generalized-Pareto mixed-Poisson probabilities. The proposed approach produces results with any desired accuracy, opposed to other existing methods.

References

1. J. Abate and W. Whitt, Computing Laplace transforms for numerical inversion via continued fractions, *INFORMS Journal on Computing*, **11**, 394–405, (1999).
2. A. Feldmann and W. Whitt, Fitting mixtures of exponentials to long-tail distributions to analyze network performance models, *Performance Evaluation*, **31** (8), 963–976, (1998).
3. F. Ferreira and A. Pacheco, Analysis of $GI/M/s/c$ queues using uniformization, *Computers and Mathematics with Applications*, **51** (2), 291–304, (2006).
4. F. Ferreira and A. Pacheco, Analysis of $GI^X/M/s/c$ systems via uniformisa-

tion and stochastic ordering, In T. Czachórski and N. Pekergin, editors, *Proceedings of the Workshop in New Trends in Modelling, Quantitative Methods and Measurements, Zakopane, Poland, June 27-29, 2004*, pp. 97–133, (2004).
5. M. J. Fischer, A. M. Girard, D. M. Masi, and D. Gross, Simulating Internet-type queues with heavy-tailed service or interarrival times, In Bohdan Bodnar, editor, *Proceedings of the Applied Telecommunication Symposium, Advanced Simulation Technologies Conference, Seattle, Washington, April 22-26, 2001*, pp. 161–167, (2001).
6. M. J. Fischer, D. M. Masi, D. Gross, and J. Shortle, Using the correct heavy-tailed arrival distribution in modeling congestion systems, In *Proceedings of the Eleventh International Conference on Telecommunication Systems, Modeling and Analysis, Monterey, California, October 2-5, 2003*, (2003).
7. M. J. Fischer, D. M. Masi, D. Gross, and J. F. Shortle. Loss systems with heavy-tailed arrivals. *The Telecommunications Review*, pp. 95–99, (2004).
8. M. J. Fischer, D. M. Masi, D. Gross, J. Shortle, and P. H. Brill, Development of procedures to analyze queueing models with heavy-tailed interarrival and service times, In *Proceedings of the NSF Design, Service, and Manufacturing Grantees and Research Conference, January 3-6, 2005*, (2005).
9. J. Gordon, Pareto process as a model of self-similar packet traffic, In *Proceedings of the IEEE Global Telecommunications Conference (GLOBECOM '95)*, vol. 3, pp. 2232–2236, (1995).
10. C. M. Harris, P. H. Brill, and M. J. Fischer, Internet-type queues with power-tailed interarrival times and computational methods for their analysis, *INFORMS Journal on Computing*, **12**, 261–271, (2000).
11. C. M. Harris and W. G. Marchal, Distribution estimation using Laplace transforms, *INFORMS Journal on Computing*, **10**, 448–458, (1998).
12. R. Hogg and S. Klugman, editors, *Loss distributions*, John Wiley, New York, (1984).
13. J. Kim, Analysis of loss probability and convergence rate in queueing models when the capacity tends to infinity, Department of Mathematics, Korea University, Seoul, Korea. Available at: http://www.korea.ac.kr/ amcenter/dvi/insukwee/kim.pdf.pdf.
14. Y. Koh and K. Kim, Evaluation of steady-state probability of Pareto/$M/1/K$ experiencing tail-raising effect, *Lecture notes in Computer Science*, **2720**, 561–570, (2003).
15. V. G. Kulkarni, *Modeling and Analysis of Stochastic Systems*, Chapman and Hall, London, (1995).
16. M. Kwiatkowska, G. Norman, and A. Pacheco, Model checking CSL until formulae with random time bounds, *Lecture Notes in Computer Science*, **2399**, 152–168, (2002).
17. M. F. Neuts, *Matrix-Geometric Solutions in Stochastic Models: An Algorithmic Approach*, Dover Publications, New York, (1994).
18. W. H. Press, B. P. Flannery, S. A. Teukolsky, and W. T. Vetterling *Numerical Recipes in C*, Cambridge University Press, London, (1988).
19. M. Shaked and J. G. Shanthikumar, *Stochastic Orders and Their Applications*, Academic Press, San Diego, (1994).

20. J. F. Shortle, M. J. Fischer, D. Gross, and D. M. Masi, Using the Pareto distribution in queueing modeling,
Available at: http://www.mitretek.org/home.nsf/eCommerce/presentations
21. D. Stoyan, *Comparison Methods for Queues and Other Stochastic Models*, Wiley, Chichester, (1983).
22. L. Takács, *Introduction to the Theory of Queues*, Oxford University Press, New York, (1962).
23. S. Wang and H. Panjer, Critical starting points for stable evaluation of mixed-Poisson probabilities, *Mathematics and Economics*, **13**, 287–297, (1993).
24. G. E. Willmot, On recursive evaluation of mixed-Poisson probabilities and related quantities, *Scandinavian Actuarial Journal*, **2**, 114–133, (1993).
25. J. Wimp, Computation With Recurrence Relations, *Applicable Mathematics Series*, Pitman Advanced, Publishing Program, London, (1984).

AN OPTIMAL POLICY TO MINIMIZE DELAY COSTS DUE TO WAITING TIME IN A QUEUE

J. KOYANAGI and H. KAWAI

Tottori University, Koyama Minami 4-101, Tottori, 680-8552, Japan
E-mail: junji@sse.tottori-u.ac.jp, kawai@sse.tottori-u.ac.jp

abstracts We consider a discrete time queueing system. The service time and the interarrival time have geometric distributions. A decision maker(DM) arrives at the queue with two tasks, Task A and Task B. Task A must be processed by the server and DM must wait in the queue until it is finished if DM joins the queue. DM also needs a constant time to finish Task B. While DM is processing Task B, DM can interrupt Task B to join the queue and resume Task B after Task A is finished. There is a deadline for two tasks and the delay costs are incurred if two tasks are finished after the deadline. We consider a constant cost for breaking the deadline and a proportional cost for the delay from the deadline. We study the structure of the optimal policy to minimize the expected total delay costs.

1. Introduction

Queueing theory deals with various queueing systems and their analysis. In many queueing systems, it is assumed that customers arrive at the queue without decisions but there are some papers dealing with the customer who can choose whether to join the queue. Several such problems are introduced in Hassin and Haviv(2003)[1].

Among such decision problems, we introduce *smart* customer problem here. In Mandelbaum and Yechiali(1983)[2], they propose a queueing system where one customer called *smart* customer can decide whether to enter the queue or leave the system. As an additional action, the smart customer can defer the decision outside the queue observing the queue length and paying the waiting cost smaller than the waiting cost in the queue. The authors showed that the optimal policy has the following structure. When the queue length is short, smart customer should enter the system. When the queue length is middle, he should defer the decision and when it is long, he should leave.

In the above problem, the number of decisions is not restricted but in a real problem, it is usual the number of decisions is restricted. We have studied several models where the number of decisions is restricted. In that case, the decisions are affected the queue length and the number of the decisions. In Koyanagi and Kawai(2000)[3], we consider a decision maker (DM) with a task (TA) served in the queueing system who wants to minimize the waiting time in the queue. DM also has another task (TB) which needs several steps to be processed. At every step of TB, DM decides whether to interrupt TB and enter the queue. If DM chooses to enter the queue, he resumes TB after TA is finished. If DM finishes TB before entering the queue, DM must enter the queue and wait for the service.

One example of this problem is a repair problem of a machine with two functions. Consider a job which needs a hardcopy of thumbnail images of many photos. There is a printer which can be used as a scanner and we can use only this printer for the job. However, the print function is found to be broken, but the scan function can be used. If we bring the printer to the repair factory with a queue of other devices which need to be repaired, we must wait until the repair is completed and we cannot use the scan function while the printer is in the factory. Since print function is needed after all photos are scanned, we can postpone the repair and scan photos before the repair. Then the problem is when to bring the printer to the factory.

Such problem may be considered as a model of 'special order' in a factory which produces several kinds of products according to the orders by customers. Usually the products are produced by FIFO decipline and the products are delivered to the customers. However, if there is an accident while delivery and the products are broken or lost, the factory must produce the same products again. In that case, the factory deals with such order as 'special order' which has higher priority than other orders. Let us consider that several checks at the inspection machines are needed to complete the products. At each inspection machine, other products may wait to be checked in the line. At some of the inspection machines, 'special order' can cut in the line and at other inspection machines, the order cannot cut in the line because of the inspection machine structure or other reasons. If there exists only one mashine that the order cannot cut in, it is our problem when the special order should be put in the line while being checked in other machines.

We can consider various cost functions which depends on the time needed to process both TA and TB. We considered the expected time for TA and TB as the cost in the continuous time system[3]. In Koyanagi and Kawai(2001)[4] and Koyanagi and Kawai(2004)[5] we considered a discrete time queueing system and a deadline of the two tasks. As a cost function, we considered the probability not to finish TA and TB within the deadline[4] and we considered the tardiness cost which means the delay time from the deadline[5]. In this paper we consider a cost function which includes the costs in the two papers[4,5] and prove the monotone structure of the optimal policy.

2. Model

We consider one decision maker (DM) who has two tasks TA and TB. TA is processed in a discrete time queueing system. The arrival and the end of the service happens at every unit time with constant probability, the end of the service happens with probability q and the arrival does with probability p. The end of the service happens just before the arrival. We assume $p < q$ for the existence of the steady state of the queueing system. The other task TB needs b time units and DM observes the queueing system at every unit time for processing TB and decides whether to join the queue. Two tasks should be finished within $b + l$ time and if the two tasks are finished after $b + l$ time, a cost is incurred.

(1) Suppose that DM processes TB for $m(\leq b)$ time units.
(2) After m time units are spent in TB, DM observes queue length i and decides to join the queue as the $(i + 1)$st customer.
(3) DM cannot process TB while he is in the queue.
(4) If X time is needed for $i+1$ customers (including DM), DM processes the rest of TB which needs $b - m$ time after TA is processed in the queue.
(5) In the above situation, the cost becomes

$$\begin{cases} d + c'(X - l) & \text{if } X - l > 0, \\ 0 & \text{if } X - l \leq 0. \end{cases}$$

That is, d is the penalty cost for exceeding the deadline and cost c' per delay time is incurred. To minimize the cost, DM chooses an action between two actions at each time epoch: action A is to join the queue; action B is to process TB for one time unit. If DM chooses action B, DM makes a decision again after one time unit until TB is finished.

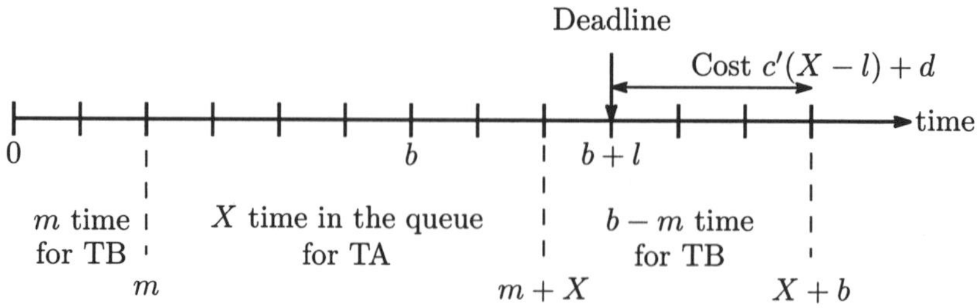

Figure 1. Case that the cost is incurred.

We define (i, m, l) as the system state where i is the queue length, m time units has been spent in TB, and l is the maximum time in the queue without the tardiness cost.

For the optimality equations, we define the following functions.

(1) $A(i, m, l)$ is the expected cost when DM chooses action A while in (i, m, l).
(2) $B(i, m, l)$ is the expected cost when DM chooses action B while in (i, m, l) and behaves optimally thereafter.
(3) $V(i, m, l)$ is the optimal expected cost for (i, m, l).

The queueing system is a discrete time system and at most one customer arrives or leaves in one time unit. Therefore, if action B is taken, the state transition from (i, m, l) is restricted to $(i-1, m+1, l)$, $(i > 0)$, $(i, m+1, l)$ or $(i+1, m+1, l)$. With these state transition, we have the following optimality equations.

$$A(i,m,l) = \sum_{k=0}^{i} \binom{l}{k} q^k (1-q)^{l-k} \{c'(i+1-k)/q + d\} \tag{1}$$

$$\begin{aligned} B(i,m,l) &= q(1-p)V(i-1,m+1,l) + (qp + (1-q)(1-p))V(i,m+1,l) \\ &\quad + (1-q)pV(i+1,m+1,l) \quad (i>0) \end{aligned} \tag{2}$$

$$V(i,m,l) = \min\{A(i,m,l), B(i,m,l)\} \quad (m<b), \quad V(i,b,l) = A(i,b,l) \tag{3}$$

The expression of $A(i, m, l)$ is obtained as follows.

(1) DM chooses to join the queue in (i, m, l), then DM becomes the $(i+1)$st customer in the queue.
(2) During l time, k customers are served; the distribution of k is binomial with success probability q.

(3) If $k \geq i+1$, no cost is incurred because TA is finished within l time units, otherwise $c'(i+1-k)/q+d$ is the expected cost.

For the simplicity, we define $c \equiv c'/q$, then

$$A(i,m,l) = \sum_{k=0}^{i} \binom{l}{k} q^k (1-q)^{l-k} \{c(i+1-k)+d\} \tag{4}$$

We define $\binom{l}{k} = 0$ for $k > l$, then this expression is valid even when $i > l$. If $i > l$, $A(i,m,l) = c(i+1) - cql + d$ by using binominal distribution. Note that $A(i,m,l)$ is independent of m.

3. Analysis

We analyze the structure of the optimal policy in this section.

Case $m = b$.

Only action A can be taken. Thus $V(i,b,l) = A(i,b,l)$

Case $m = b-1$.

If action B is taken, DM must take action A after one time unit and to deal with this case, we define $C(i,m,l)$ by

$$\begin{aligned} C(i,m,l) = q(1-p)A(i-1,m+1,l) + \{qp + (1-q)(1-p)\}A(i,m+1,l) \\ + (1-q)pA(i+1,m+1,l). \end{aligned} \tag{5}$$

That is, $C(i,m,l)$ is the cost if we take action B while in (i,m,l) and take action A after one time unit irrespective of the queue length. Note that $B(i,m,l) \leq C(i,m,l)$ $(m < b-1)$ and $B(i,b-1,l) = C(i,b-1,l)$ hold. We prove that the optimal action changes once in $(i,b-1,l)$ as i increases.

From Eq. (4),

$$\begin{aligned} A(i+1,m,l) = A(i,m,l) + \sum_{k=0}^{i+1} \binom{l}{k} q^k (1-q)^{l-k} c \\ + \binom{l}{i+1} q^{i+1} (1-q)^{l-i-1} d \end{aligned} \tag{6}$$

$$\begin{aligned} A(i-1,m,l) = A(i,m,l) - \sum_{k=0}^{i} \binom{l}{k} q^k (1-q)^{l-k} c \\ - \binom{l}{i} q^i (1-q)^{l-i} d \end{aligned} \tag{7}$$

Substituting these two equations to Eq. (5), we have

$$\begin{aligned} C(i,b-1,l) &= A(i,b,l) \\ &\quad - q(1-p)\left\{\sum_{k=0}^{i}\binom{l}{k}q^k(1-q)^{l-k}c + \binom{l}{i}q^i(1-q)^{l-i}d\right\} \\ &\quad + (1-q)p\left\{\sum_{k=0}^{i+1}\binom{l}{k}q^k(1-q)^{l-k}c + \binom{l}{i+1}q^{i+1}(1-q)^{l-i-1}d\right\} \\ &= A(i,b,l) + (p-q)\sum_{k=0}^{i}\binom{l}{k}q^k(1-q)^{l-k}c \\ &\quad + p\binom{l}{i+1}q^{i+1}(1-q)^{l-i}(c+d) \\ &\quad - (1-p)\binom{l}{i}q^{i+1}(1-q)^{l-i}d. \end{aligned}$$

Here, we define

$$\begin{aligned} S(i) \equiv (p-q)\sum_{k=0}^{i}\binom{l}{k}q^k(1-q)^{l-k}c + p\binom{l}{i+1}q^{i+1}(1-q)^{l-i}(c+d) \\ - (1-p)\binom{l}{i}q^{i+1}(1-q)^{l-i}d. \end{aligned} \tag{8}$$

Since $A(i,b,l) = A(i,b-1,l)$ by definition, we obtain $B(i,b-1,l) = C(i,b-1,l) = A(i,b-1,l) + S(i)$. Thus $B(i,b-1,l) \geq A(i,b-1,l)$ if and only if $S(i) \geq 0$.

Lemma 3.1. *$S(i) \leq 0$ if $i+1 \geq \dfrac{p(c+d)(l+1)}{pc+d}$.* □

Proof.
From Eq. (8), we obtain

$$\begin{aligned} S(i) &= (p-q)\sum_{k=0}^{i}\binom{l}{k}q^k(1-q)^{l-k}c \\ &\quad + \frac{q^{i+1}(1-q)^{l-i}l!}{(i+1)!(l-i)!}[p(c+d)(l+1) - (pc+d)(i+1)]. \end{aligned} \tag{9}$$

Since $p < q$, the first term is negative. Therefore, $S(i) \leq 0$ for $i+1 \geq \dfrac{p(c+d)(l+1)}{pc+d}$. This completes the proof. □

Next, we calculate $D(i) \equiv S(i+1) - S(i)$ to analyze $S(i)$ for $i+1 < p(c+d)(l+1)/(pc+d)$.

$$\begin{aligned}
D(i) &= c(p-q)\frac{q^{i+1}(1-q)^{l-i-1}l!}{(i+1)!(l-i-1)!} \\
&\quad + \frac{q^{i+2}(1-q)^{l-i-1}l!}{(i+2)!(l-i-1)!}\{p(c+d)(l+1)-(pc+d)(i+2)\} \\
&\quad - \frac{q^{i+1}(1-q)^{l-i}l!}{(i+1)!(l-i)!}\{p(c+d)(l+1)-(pc+d)(i+1)\} \\
&= \frac{q^{i+1}(1-q)^{l-i-1}l!}{(i+2)!(l-i)!}\Big[c(p-q)(i+2)(l-i) \\
&\quad + q(l-i)\{p(c+d)(l+1)-(pc+d)(i+2)\} \\
&\quad - (i+2)(1-q)\{p(c+d)(l+1)-(pc+d)(i+1)\}\Big]. \qquad (10)
\end{aligned}$$

The sign of $D(i)$ is determined by

$$\begin{aligned}
&c(p-q)(i+2)(l-i) + q(l-i)\{p(c+d)(l+1)-(pc+d)(i+2)\} \\
&- (i+2)(1-q)\{p(c+d)(l+1)-(pc+d)(i+1)\}.
\end{aligned}$$

Replace discrete variable i to continuous variable x and define

$$\begin{aligned}
f(x) &= c(p-q)(x+2)(l-x) + q(l-x)\{p(c+d)(l+1)-(pc+d)(x+2)\} \\
&\quad - (x+2)(1-q)\{p(c+d)(l+1)-(pc+d)(x+1)\}. \qquad (11)
\end{aligned}$$

$f(x)$ is a quadratic function of x and the coefficient of x^2 is $cq+d$ and $f(x) \leq 0$ when $x+1 = p(c+d)(l+1)/(pc+d)$. Then, the sign of $f(x)$ changes at most one for $1 \leq x \leq p(c+d)(l+1)/(pc+d) - 1$.

Therefore, the following lemma holds.

Lemma 3.2. *If $S(1) \geq 0$, then $S(i)$ becomes negative as i increases and never returns positive once $S(i)$ becomes negative.* □

Proof.

(1) If $f(1) > 0$ and $S(1) > 0$, $S(i)$ increases as i increases from 1, but as i increases, $f(i)$ becomes negative before i becomes larger than $p(c+d)(l+1)/(pc+d) - 1$. Then $S(i)$ begins to decrease while $i+1 \leq p(c+d)(l+1)/(pc+d)$ and $S(i)$ becomes negative for $i+1 > p(c+d)(l+1)/(pc+d)$ by Lemma 3.1.

(2) If $f(1) \leq 0$ and $S(1) > 0$, $S(i)$ decreases while $i+1 \leq p(c+d)(l+1)/(pc+d)$ and $S(i)$ is negative for $i+1 > p(c+d)(l+1)/(pc+d)$ by Lemma 3.1.

Thus the change of the sign of $S(i)$ occurs at most one. This completes the proof. □

$S(1) > 0$ indicates that the optimal action in $(1, b-1, l)$ is action A, then with Lemma 3.1 and Lemma 3.2, we have the following theorem.

Theorem 3.1. *If the optimal action is action A in $(1, b-1, l)$, the optimal action for $(i, b-1, l)$ changes from action A to action B once as i increases and the change occurs before $p(c+d)(l+1)/(pc+d)-1$.* □

Case $m < b - 1$.

It is shown that the optimal action changes once as i increases for $m = b-1$. Next we show that the same property exists for $m < b-1$ by induction.

First we show that $B(i, m, l)$ and $V(i, m, l)$ are increasing in m.

Lemma 3.3. *For $B(i, m, l)$ and $V(i, m, l)$, it holds that $B(i, m-1, l) \leq B(i, m, l)$ and $V(i, m-1, l) \leq V(i, m, l)$* □.

Proof. By definition, $V(i, b, l) = A(i, b, l)$ and $V(i, b-1, l) \leq A(i, b-1, l)$. It also holds that $A(i, m, l) = A(i, m-1, l)$ by (4). Thus

$$V(i, b-1, l) \leq A(i, b-1, l) = A(i, b, l) = V(i, b, l). \tag{12}$$

It is easy to show that if $V(i, m-1, l) \leq V(i, m, l)$, then $B(i, m-2, l) \leq B(i, m-1, l)$ and $V(i, m-2, l) \leq V(i, m-1, l)$. Thus, by induction Lemma 3.3 holds. This completes the proof. □

Lemma 3.3 shows that if $B(i, m, l) \leq A(i, m, l)$, then $B(i, k, l) \leq A(i, k, l)$ for $k \leq m$. Thus, as the time spent in TB increases, the optimal action changes from B to A if it changes, and never changes from action A to action B.

Let $I_m = \max\{i | A(i, m, l) \leq B(i, m, l)\}$.

Lemma 3.4. *If $A(i, m, l) \leq B(i, m, l)$ for $0 \leq i \leq I_m$, then $A(i, m-1, l) \leq B(i, m-1, l)$ holds for $0 \leq i \leq I_m - 1$.* □

Proof.
For $0 \leq i \leq I_m - 1$, $B(i, m-1, l) = C(i, m-1, l)$, because the queue length increases at most one. By definition, $C(i, m-1, l) = C(i, m, l)$ and $C(i, m, l) \geq B(i, m, l)$ hold. Thus $A(i, m-1, l) = A(i, m, l) \leq C(i, m, l) = C(i, m-1, l) = B(i, m-1, l)$ holds for $0 \leq i \leq I_m - 1$. This completes the proof. □

With the help of these lemmas we have the following theorem about the properties of the optimal policy.

Theorem 3.2. *If the optimal action is action A for* $(1, m-1, l)$, *then*

(1) *For* $0 \leq i \leq I_m$, *the optimal action is action A and* $I_m \leq p(c+d)(l+1)/(pc+d)-1$

(2) I_m *is increasing and increases at most one as* m *increases, i.e.,*

$$I_{m-1} \leq I_m \leq I_{m-1}+1.$$

By this theorem the optimal policy has switch curve structure and the threshold increases at most one as the time spent in Task B increases.

4. Conclusion

We analyzed the optimal policy in a queueing system. In our model, a decision maker(DM) has two tasks, one task (TA) is processed in the queueing system and DM must bring the task to the queueing system and wait for the task to be processed. The other task (TB) can be processed while DM is not in the queue. DM can choose whether to interrupt TB and bring TA to the queue at each time. If DM chooses to join the queue, DM processes the rest of TB after TA is finished. In this model, optimal action depends on the queue length and the time spent in TB. We proved that the optimal action changes at most one as the queue length increases and as the time spent in TB increases. We proved the result under the condition that the optimal action is to join the queue when there is only one customer. However, if the optimal action is not to join the queue even when there is only one customer, we think that the optimal policy is to join the queue only when there is no customer in the queue though we could not give the proof for this intuitive property.

References

1. R. Hassin and M. Haviv, *To queue or not to queue*, Kluwer Academic Pubishers, Boston (2003).
2. A. Mandelbaum and U. Yechiali, Optimal entering rules for a customer with wait option at an $M/G/1$ queue, *Management Science*, **29-2**, 174–187 (1983).
3. J. Koyanagi and H.Kawai, An optimal join policy to the queue in processing two kinds of jobs, *Proc. of the Int. Conf. Applied Stochastic System Modeling*, 140–147 (2000).
4. J. Koyanagi and H.Kawai, A maximization of the finishing probability of two jobs processed in a queue, *Proc. of the 32nd ISCIE International Symposium on Stochastic Systems Theory and Its Applications*, 171–176 (2001).

5. J. Koyanagi and H.Kawai, An optimal policy to minimize expected tardiness cost due to waiting time in the queue, *Proc. of the 2004 Asian International Workshop (AIWARM 2004)*, 285–292 (2004).

OPTIMAL CERTIFICATE UPDATE INTERVAL CONSIDERING COMMUNICATION COSTS IN PKI

S. NAKAMURA, M. ARAFUKA

Department of Human Life and Information,
Kinjo Gakuin University,
1723 Omori 2-chome, Moriyama-ku, Nagoya 463-8521, Japan
E-mail:snakam@kinjo-u.ac.jp, arafuka@kinjo-u.ac.jp

T. NAKAGAWA

Department of Marketing and Information Systems,
Aichi Institute of Technology, 1247 Yachigusa, Yagusa-cho, Toyota 470-0392, Japan
E-mail: toshi-nakagawa@aitech.ac.jp

CA (Certification Authority) in PKI (Public Key Infrastructure) issues the certification pubic keys to the user as the method of proving the owner of the public keys. When various circumstances may cause that a certificate becomes invalid prior to the expiration of the validity period, the PKI user confirms the certificate is effective in regularly acquiring CRL(Certification Revocation List) from the repository. For this, there is an inquiry method by online. However, supplying real-time information on revocation status during each of acquiring is computationally expensive. In addition, the transmission of large CRLs to potentially many clients can be prohibitively expensive. Then, we analytically discuss optimal issuing cycle times of CRL which minimize the various expected costs per unit of time.

1. Introduction

CA (Certification Authority) in PKI (Public Key Infrastructure) issues the certification pubic keys to the PKI user as the method of proving the owner of the public keys. When various circumstances may cause that a certificate becomes invalid prior to the expiration of the validity period, CA invalidates the certificate pubic keys, and notifies PKI users the revoked information as a CRL (Certificate Revocation List). As a notification method to PKI users, CA is stored CRL in the server that is called a repository and opened to the public to the PKI user.

The PKI user confirms the certificate is effective in regularly acquiring

CRL from the repository. However, since CRL is issued by CA at the decided cycle, the PKI user cannot obtain the revoked information until the next CRL is issued when the certificate lapses. Therefore, when the issuing cycle of CRL becomes long, the PKI user's acquiring time of revoked information becomes long. As a result, it is used as valid certificates though the certificate lapses, and so, the opportunity loss cost occurs[6]. Oppositely, when the issuing cycle of CRL is shortened, the PKI user's downloading and communication costs increase for the frequent issue CRLs.

Moreover, the operation cost of CA shows the tendency for the issuing frequency to increase, and to increase also by CA. Moreover, the issuing frequency of CRL increases, and PKI operation cost increases in CA. It is important to set an optimal cycle time of CRL in consideration of the security policy and the business system. Especially, we assume the probability distribution of the illegally used key and the various kinds of CRL operation costs for CRL issues.

We propose the model of two types CRL, obtain the expected costs of each types of CRL model and compare them. Further, we analytically discuss optimal issuing cycle times of CRL which minimize the expected costs per unit of time.

2. CRL Models

In order to validate a certificate among other things, a relying party must acquire a recently issue CRL to determine whether a certificate has been revoked. The confirmation method of certificate revocation is assumed to be a usual retrieval from the recent CRL issue. A relying party wishing to make use of the information in a certificate must first validate a certificate. Therefore, the CRL database based on the data issued from the CRL distribution point is constructed for a PKI user.

We obtain the expected costs of two models of Base CRL and Delta CRL, taking into consideration of various costs for different methods of CRL issues. Especially, we set an opportunity loss cost which a PKI user cannot acquire a new CRL information. For each model, the CA decides the issuing intervals of Base CRL which minimize the expected database construction costs[1].

The following notations are used:

M_0: Number of all certificates that have been revoked for i period ($i = 1, 2, \ldots, T$). M_0 is a number of CRLs.

T: Issuing period between CRLs ($T = 1, 2, \ldots$).

$F(i)$: Probability distribution with finite mean θ that the key is illegally used, where $F(0) \equiv 0$.

μ_i: Expected number of illegal use keys between $(i-1)$th and ith periods when the number of all certificates is M_0;
$\mu_i \equiv M_0[F(i) - F(i-1)] \quad (i = 1, 2, \ldots, T)$.

c_1: Issuing and communication (access to databases and maintenance are contained) costs per certificate for Base CRL.

c_2: Handling cost per issued Delta CRL file.

c_3: Confirmation cost of hand work of CRL.

It has been assumed in M_0 that when the certificate issued by CA lapses in the explanation date, CA puts the serial number in the lasping certificate and opens it with the server as CRL. Thus, if the lapse rate per day is constant and the expiration date passed is deleted, M_0 is constant when the CRL is issued at times $T, 2T, \ldots$ [3][4].

2.1. *Model 1*

We consider the time interval $(0, T]$ between Base CRL issues: Delta CRL is continuously issuing for T period after Base CRL issue (Figure 1). Delta CRL is a small CRL that provides information about certificates whose statuses have changed since the previous Base CRL[2]. That is, the number of revoked certificates in Delta CRL is the total of accumulated revoked certificates from the previous Base CRL issue. The full CRL database for a PKI user is constructed by Base CRL and the previous Delta CRL. Therefore, the number of handling files of CRL database updated at each issue is only files of the last one.

The expected cost of Model 1 is composed of the total of issuing cost of Base CRL, issue cost of Delta CRL, and handling cost of files for T period as follows: Delta CRL is continuously issuing for period after Base CRL issue. The number of handling files of CRL data base updated at each issue is only files of the last one. Then, the expected cost of Model 1 is

$$\begin{aligned} C_1(T) &= c_1 M_0 + c_1 \sum_{i=1}^{T-1} \sum_{j=1}^{i} \mu_j + c_2(T-1) \\ &= c_1 M_0 \{1 + \sum_{i=1}^{T-1} \sum_{j=1}^{i} [F(j) - F(j-1)]\} + c_2(T-1) \\ &= c_1 M_0 \{1 + \sum_{i=1}^{T-1} F(i)\} + c_2(T-1) \quad (T = 1, 2, \ldots). \end{aligned} \tag{1}$$

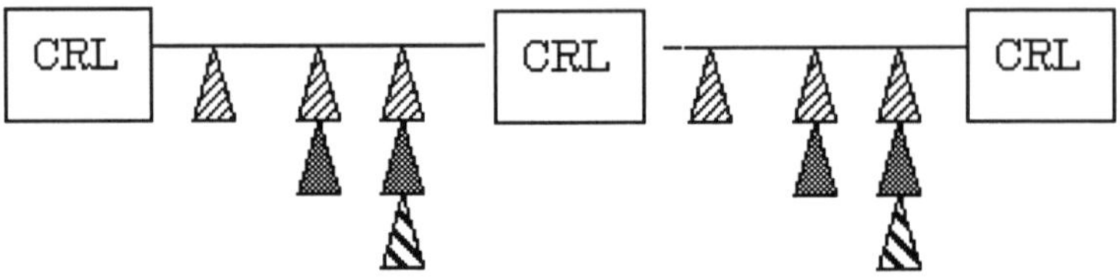

Figure 1. Operation process of Delta CRL.

2.2. *Model 2*

In Model 2, we consider that the data base of Base CRL is divided into two kinds of data bases: According to the date registered in CRL, it divides into two groups with an old date registered and with a new date registered in CRL. In general, there are a lot of accesses of the data registered in CRL recently. Oppositely, the access to data with old day of registration to CRL decreases. We assume the Base CRL division ratio of the group with a new date registered in CRL is p $(0 < p \leq 1)$. The PKI user issues the data base of the newly group date registered in CRL (we call Newly Base CRL) usually. When the access to the data base with an old registration to CRL (we call Old Base CRL) is generated, the PKI user takes the confirmation in the another way by hand works. The access probability to Old Base CRL is ε which is very small.

The cost c_3 is the hand work cost of PKI user that the hand work for Old Base CRL accessing. The expected cost of Model 2 is composed of the total issuing cost of Old Base CRL, issuing cost of Delta CRL, and handling cost of files for T period. Delta CRL is continuously issuing for period after Base CRL issue. If the amount of Base CRL is small, the communications cost of Delta CRL and data base updating costs are small. Then, the expected cost of Model 2 is

$$C_2(T) = pc_1M_0 + \varepsilon c_3(1-p)M_0 + pc_1 \sum_{i=1}^{T-1} \sum_{j=1}^{i} \mu_j + c_2(T-1)$$
$$= pc_1M_0\{1 + \sum_{i=1}^{T-1} F(i)\} + \varepsilon c_3(1-p)M_0 + c_2(T-1) \quad (T = 1, 2, \ldots). \tag{2}$$

The first term pc_1M_0 is the issuing and communication cost of Newly Base CRL. The second term $\varepsilon c_3(1-p)M_0$ is corresponding cost when there is an access for Old Base CRL, and the ratio of access for Old Base CRL is constant ε. The third term $pc_1 \sum_{i=1}^{T-1} \sum_{j=1}^{i} \mu_j$ is the total cost for the issuing of Delta CRL, the updating of Newly Base CRL data and the maintenance.

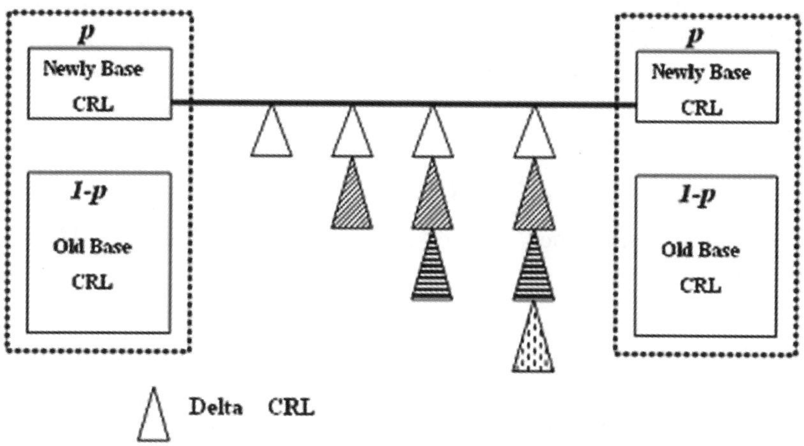

Figure 2. Operation process of Newly Base CRL and Delta CRL.

3. Comparisons of Expected Costs

We compare the expected costs $C_1(T)$ in (1) and $C_2(T)$ in (2). Firstly, it is noted that $C_1(T) = C_2(T)$ for $p = 1$. Forming $C_1(T) \geq C_2(T)$, we have the inequality

$$1 + \sum_{i=1}^{T-1} F(i) \geq \frac{\varepsilon c_3}{c_1} \quad (T = 1, 2, 3, \ldots). \tag{3}$$

We have two relations on $C_1(T)$ and $C_2(T)$ for $T = 1, 2, 3, \ldots$:

(i) If $\varepsilon c_3 \leq c_1$ then $C_1(T) \leq C_2(T)$.

(ii) If $\varepsilon c_3 > c_1$ then there exists a finite and unique minimum $\tilde{T}(1 < \tilde{T} < \infty)$ which satisfies (3), because $\sum_{i=1}^{\infty} F(i) = \infty$ for $F(i)$ is a probability distribution.

Therefore, we have

(a) If $T \leq \tilde{T}$ then $C_1(T) \leq C_2(T)$.

(b) If $T > \tilde{T}$ then $C_1(T) > C_2(T)$.

4. Optimal Polices

The costs for time T between Base CRL issues are independently and identically distributed because M_0 is costant between Base CRL issues. Thus, from the renewal reward process [5], the expexted cost per unit of time for an infinite time span is given by $C_i(T)/T \quad (i = 1, 2)$. We consider the optimal policies which minimize $C_i(T)/T$ as an objective function.

4.1. *Optimal time for Model 1*

We seek an optimal time T_1^* which minimizes $C_1(T)/T$, when $c_1 M_0 > c_2$, given in

$$\frac{C_1(T)}{T} = \frac{1}{T}\left\{c_1 M_0 \left[1 + \sum_{i=1}^{T-1} F(i)\right] + c_2(T-1)\right\} \quad (T = 1, 2, \ldots). \tag{4}$$

Suppose that $F(i)$ is strictly increasing from 0 to 1. Forming the inequality $C_1(T+1)/(T+1) - C_1(T)/T \geq 0$, we have

$$\sum_{i=0}^{T-1} [F(T) - F(i)] \geq 1 - \frac{c_2}{c_1 M_0} \quad (T = 1, 2, \ldots). \tag{5}$$

It can be evidently seen that the left-had side of (5) is strictly increasing $F(1)$ to $\theta = \sum_{i=0}^{\infty}[1 - F(i)]$. Thus, we have the following optimal issue period T_1^*:

(i) If $\theta > 1 - c_2/(c_1 M_0)$ then there exists a finite and unique $T_1^*(1 \le T_1^* < \infty)$ which satisfies (5), and the resulting cost is

$$c_1 M_0 F(T_1^* - 1) + c_2 \le \frac{C_1(T_1^*)}{T_1^*} \le c_1 M_0 F(T_1^*) + c_2. \tag{6}$$

(ii) If $\theta \le 1 - c_2/(c_1 M_0)$ then $T_1^* = \infty$, *i.e.*, the Base CRL issue should not be done, and the expected cost is

$$\lim_{T\to\infty} \frac{C_1(T)}{T} = c_1 M_0 + c_2. \tag{7}$$

It can be easily seen from (5) that T_1^* is an increasing function of c_2/c_1. Since θ represents the mean time to an illegal use key, it would be larger than 1, and hence, the case (ii) would not reflect a real world for $c_1 M_0 > c_2$.

In the particular case of $\mu \equiv M_0[F(i) - F(i-1)]$, equation (5) is

$$\frac{T(T+1)}{2} \ge \frac{1}{\mu}\left(M_0 - \frac{c_2}{c_1}\right). \tag{8}$$

4.2. *Optimal time for Model 2*

We seek an optimal time T_2^* which minimizes

$$\frac{C_2(T)}{T} = \frac{1}{T}\{pc_1 M_0[1 + \sum_{i=1}^{T-1} F(i)] + \varepsilon c_3(1-p)M_0 + c_2(T-1)\} \qquad (T = 1, 2, \ldots). \tag{9}$$

From the inequality $C_2(T+1)/(T+1) - C_2(T)/T$, we have

$$\sum_{i=0}^{T-1}[F(T) - F(i)] \ge 1 - \frac{c_2 - \varepsilon c_3(1-p)M_0}{pc_1 M_0} \quad (T = 1, 2, \ldots). \tag{10}$$

By the similar method to that of Model 1, we have

(i) If $\theta > 1 - \{[c_2 - \varepsilon c_3(1-p)M_0]/(pc_1 M_0)\}$ then there exists a finite and unique $T_2^*(1 \le T_2^* < \infty)$ which satisfies (10), and the resultig cost is

$$pc_1 M_0 F(T_2^* - 1) + c_2 < \frac{C_2(T_2^*)}{T_2^*} \le pc_1 M_0 F(T_2^*) + c_2. \tag{11}$$

(ii) If $\theta \leq 1 - \{[c_2 - c_3(1-p)M_0]/(pc_1M_0)\}$ then $T_2^* = \infty$), and

$$\lim_{T\to\infty} \frac{C_2(T)}{T} = pc_1M_0 + c_2. \tag{12}$$

It can be seen that T_2^* is an increasing function of c_2/c_1 and a decreasing function of $\varepsilon c_3/c_1$.

5. Numerical Examples

We compute the optimal time T_1^* and T_2^* when $F(i) = 1 - e^{-i\lambda}$,*i.e.*, $\mu_i = M_0[e^{-(i-1)\lambda} - e^{-i\lambda}]$. Then, the left-hand side of (5) and (10) is

$$\frac{1-(1+T)e^{-\lambda T} + Te^{-\lambda(1+T)}}{1-e^{-\lambda}}. \tag{13}$$

From the statistical information of the number of certificates stored in CRL, we assume $c_2/c_1 = 20$ and $M_0 = 10,000$, and we give the result in Table 1. This indicates that the optimal values of T_1^* are every month issuing for $\lambda = 0.002$. When $\lambda = 0.009$, the optimal value of T_1^* is 15 days. When λ increases, it is necessary to issue Base CRL frequently.

Table 2 shows the interval T_2^* and the cost $C_2(T_2^*)/c_1T_2^*$ for p and λ when $c_2/c_1 = 20$ and $\varepsilon c_3/c_1 = 0.1$. When $p = 0.5$, that is, Base CRL is divided in Newly Base CRL and Old Base CRL, and it divides into 1:1, the issuing interval of Base CRL of Models 1 and 2 is similar frequency to the value of all p. When $p = 0.1$, the issuing interval of Base CRL is $T_2^* = 12 \sim 10$ which are almost constant value for the change of λ.

6. Conclusions

We have proposed two stochastic models of Base CRL in PKI architecture and have obtained the expected costs of each model, introducing the costs of issuing CRL data and handling of Delta CRL files, and the opportunity loss cost. We have compared two expected costs and have derived analytically the optimal issue intervals which minimize them. Further, we have discussed in numerical examples which model is better.

Thus, by estimating the costs of issuing, handling and opportunity loss and the amount of revoked certificates from actual data, and by modifying some suppositions, we could practically determine an optimal issue interval of Base CRL.

Table 1: Optimal interval T_1^* and expected cost $C_1(T_1^*)/T_1^*$ for Model 1 when $M_0 = 10^4$ and $c_2/c_1 = 20$.

λ	T_1^*	$\frac{C_1(T_1^*)}{c_1 T_1^*}$
0.001	44	441
0.002	31	599
0.003	26	714
0.004	22	810
0.005	20	891
0.006	18	962
0.007	17	1025
0.008	16	1083
0.009	15	1136

Table 2: Optimal interval T_2^* and expected cost $C_2(T_2^*)/T_2^*$ when $M_0 = 10^4$, $c_2/c_1 = 20$ and $\varepsilon c_3/c_1 = 0.1$.

p	λ									
	0.001		0.003		0.005		0.007		0.009	
	T_2^*	$\frac{C_2(T_2^*)}{c_1 T_2^*}$	T_2^*	$\frac{C_2(T_2^*)}{c_1 T_2^*}$	T_2^*	$\frac{C_2(T_2^*)}{c_1 T_2^*}$	T_2^*	$\frac{C_2(T_2^*)}{c_1 T_2^*}$	T_2^*	$\frac{C_2(T_2^*)}{c_1 T_2^*}$
0.1	12	34	10	36	10	42	10	47	10	53
0.2	34	84	19	124	15	150	13	169	11	185
0.3	39	129	22	200	17	245	14	279	13	306
0.4	41	174	23	274	18	338	15	387	13	428
0.5	42	218	24	348	19	430	16	493	14	545
0.6	43	263	25	421	19	523	16	600	14	665
0.7	43	307	25	495	19	615	16	707	14	784
0.8	44	351	25	568	19	708	16	836	15	899
0.9	44	395	25	642	20	798	17	949	15	1017
1.0	44	441	26	714	20	891	17	1025	15	1136

References

1. Cooper, D.A., More Efficient Use of Delta-CRLs, *Proceedings of the 2000 IEEE Symposium on Security and Privacy*,pp.190-202,May (2000).

2. Chadwick, D.W., Young, A.J., , Merging and Extending the PGP and PEM Trust Models - The ICE-TEL Trust Model, *IEEE Networks Special Publication on Network and Internet Security*, May/June 1997, Vol.11, No.3, pp16-24
3. Housley.R., Ford.W., Polk,W., Solo.D., *Internet X.509 Public Key Infrastructure Certificate and CRL Profile*, The Internet Society (1999).
4. ITU/ISO *Recommendation X.509.-Information Technology-Open Systems Interconnection-The Directory:Public Key and Attribute Certificate Frameworks* (1997).
5. Ross,S.M., Stochastic Processes. Jon Wiley & Spns. New York (1992).
6. http://www.comodogroup.com/repository/docs/relying_party.html.

BURSTS AND GAPS OF MARKOV RENEWAL ARRIVAL PROCESSES

ANTÓNIO PACHECO
Department of Mathematics and CEMAT
Instituto Superior Técnico, Technical University of Lisbon
Av. Rovisco Pais, 1049-001 Lisboa, Portugal
E-mail: apacheco@math.ist.utl.pt

HELENA RIBEIRO
Department of Mathematics and CEMAT
Escola Superior de Tecnologia e Gestão, Polytechnic Institute of Leiria
Morro do Lena, Alto Vieiro, 2411-901 Leiria, Portugal
E-mail: mhcr@estg.ipleiria.pt

We investigate bursts and gaps of Markov renewal arrival processes (MRPs), which are able to model bursty traffic. A burst is a maximal interval of time during which all interarrival times are smaller or equal to some value ν. A gap begins at the time at which a burst ends and is the maximal interval of time during which all interarrival times are greater than ν. We show that bursts and gaps of MRPs may be interpreted as sojourn times in sets of states of related semi-Markov processes (SMPs). Firstly, we partition the state space of an SMP in two subsets and compute moments of the sojourn times in those sets, as well as of the cycle durations, where a cycle is constituted by a visit to one of the sets followed by a visit to the other one. This procedure is used to analyze bursts and gaps of the customer departure process from $M/G/1/K$ queues.

1. Introduction

We have been assisting to a growing interest in capturing and analyzing Internet traffic characteristics, stimulated by the fact that real time delivery of multimedia information over the internet requires high quality of service. However, we know that the traffic generated is typically bursty, having active and idle transmission periods; see, e.g., Crovella and Bestravos [1], Leland *et al.* [2]. With this in mind, we investigate bursts and gaps, respectively active and idle periods, as defined by Johnson *et al.* [3], of Markov renewal arrival processes (MRPs), which are dense in the family of time arrival processes on $[0, +\infty)$, and therefore able to model bursty traffic. In words: a burst is a maximal interval of time during which all interarrival times are smaller or equal to ν, for some value ν; and a gap begins at the time a burst ends and is a maximal interval during which all interarrival times are greater than ν.

For the analysis of bursts and gaps of MRPs, we show that they may be translated as sojourn times in sets of states of the associated semi-Markov process (SMP). Sojourn times in sets of states have been extensively studied in the literature and traditionally they have been characterized through their marginal distributions; see, e.g. Csenki [4] and Sericola [5], for Markov chains. However, Nunes and Pacheco [6] characterize the dependence structure of sequences of sojourn times in sets of states and reentrance times in the same sets in discrete time Markov chains (DTMCs).

We partition the state space of the SMP in two subsets, E_0 and E_1, such that a sojourn time in E_1 corresponds to a burst and a sojourn time in E_0 corresponds to a gap. A cycle is constituted by a visit to one of these sets followed by a visit to the other one. We compute the mean and the variance of the sojourn times in these sets of states and of the cycle durations, and the results obtained are illustrated with renewal processes. We intend to develop further the analysis of the paper to include effective statistical procedures to reach conclusions about the burstiness characteristics of Internet traffic. Thus, this article may be regarded as a first step in that direction.

This paper is structured in the following way. In Section 2 we show that bursts and gaps of a MRP are special cases of sojourn times in sets of states of the associated SMP. In sections 3 and 4 we introduce SMPs and the associated processes relative to (re)entrances into the sets E_0 and E_1 of a partition of the state space of the SMP. We derive vectors of moments associated to the duration of visits to the sets E_0 and E_1 and of cycle durations, respectively. In Section 5 we illustrate the results derived in the previous sections for the case of renewal arrival process and, finally, in Section 6 we apply the proposed methodology to analyze bursts and gaps of the customer departure process from $M/G/1/K$ queues.

We end the introduction with some notation to be used throughout the paper. We let I denote the identity matrix, $\mathbf{e}$ denote a column vector of ones, where the dimension is derived from the context, $\delta_{m,n}$ denote the delta Kronecker function, that is, $\delta_{m,n} = 1$ if $m = n$ and $\delta_{m,n} = 0$ otherwise, and $1_{\{a\}}$ denote the indicator function of proposition a. For every $i \in E$, $\mathbf{P}_i$ and $\mathbf{E}_i$ denote, respectively, the conditional probability and the conditional expected value given $X_0 = i$, e.g., $\mathbf{P}_i[\cdot] = \mathbf{P}[\cdot|X_0 = i]$. We denote matrices by capital letters, vectors by small boldface letters and the diagonal matrix with the elements of vector $\mathbf{a}$ in the diagonal by diag($\mathbf{a}$). Given a matrix A we denote its elements by A_{ij} or $[A]_{ij}$. If $A = [A_{ij}]_{i,j \in E}$, the partition of the state space in the subsets E_0 and E_1 induces a block-partitioning of A in the matrices $A_{\mathbf{00}}$, $A_{\mathbf{01}}$, $A_{\mathbf{10}}$ and $A_{\mathbf{11}}$, where, e.g., $A_{\mathbf{01}}$ denotes the submatrix formed by entries in rows belonging to E_0 and columns belong to E_1. Finally we write $A_{\mathbf{01}} = \boxed{0 \leftrightarrow 1}(A_{\mathbf{10}})$ to mean that knowing the matrix $A_{\mathbf{10}}$, the matrix $A_{\mathbf{01}}$ is obtained by changing $\mathbf{0}$ with $\mathbf{1}$, and vice versa, in $A_{\mathbf{10}}$.

2. Bursts and Gaps

In this section we interpret bursts and gaps produced by MRPs as sojourn times in sets of states of related SMPs. For that, let $S = \{S(t),\ t \geq 0\}$ denote the SMP constructed from the MRP (Y, T) with phase space H and kernel $R(t)$. That is, (Y, T) is an homogeneous Markov process with state space $H \times [0, +\infty)$.

In order to better explain how the sojourn times in sets of states can be applied to bursts and gaps we consider that at time T_{n-1}, $n \in \mathbb{N}$, the state of the SMP S is i, i.e., $Y_{n-1} = i$, and we pay attention to the time until the next transition, $T_n - T_{n-1}$.

If $Z_{n-1} = T_n - T_{n-1} \leq \nu$, we associate the pair (1, i) to state i; otherwise, if $Z_{n-1} > \nu$ we associate the pair (0, i) to state i. In this way, we append to each element of H the label 1 (0) if the sojourn time in the state is smaller or equal to ν (greater than ν). With this procedure we obtain a new set of states, $E = \{0, 1\} \times H$, with twice as many elements as H, and E can be partitioned in the sets $E_0 = \{0\} \times H$ and $E_1 = \{1\} \times H$.

We let $X_n = (1_{\{Z_n \leq \nu\}}, Y_n)$, for $n \in \mathbb{N}_0$, denote the state of the system at time T_n. Then, (X, T) is a MRP with phase space E and kernel $Q(t)$, where, for $(m, i), (n, j) \in E$ and $t \geq 0$, $Q_{(m,i)(n,j)}(t)$ is given by

$$\begin{cases} 0 & m = 0, t \leq \nu \\ (R_{ij}(t) - R_{ij}(\nu))\, \overline{R}_j(\nu)/\overline{R}_i(\nu) & (m,n) = (0,0), t > \nu \\ (R_{ij}(t) - R_{ij}(\nu))\, R_j(\nu)/\overline{R}_i(\nu) & (m,n) = (0,1), t > \nu \\ R_{ij}(t)\overline{R}_j(\nu)/R_i(\nu) & (m,n) = (1,0), t \leq \nu \\ R_{ij}(t)R_j(\nu)/R_i(\nu) & (m,n) = (1,1), t \leq \nu \\ Q_{(1,i)(n,j)}(\nu) & m = 1, t > \nu \end{cases} \tag{1}$$

where,

$$R_{ij}(t) = \mathbf{P}\left[Y_n = j, T_n - T_{n-1} \le t | Y_{n-1} = i\right],$$

$R_i(t) = \sum_{k \in H} R_{ik}(t)$ and $\overline{R}_i(t) = 1 - R_i(t)$. We exemplify the validity of the previous formula by computing $Q_{(1,i)(1,j)}(t)$, $t \le \nu$, given by

$$\begin{aligned} Q_{(1,i)(1,j)}(t) &= \mathbf{P}\left[X_n = (1,j), T_n - T_{n-1} \le t | X_{n-1} = (1,i)\right] \\ &= \frac{\mathbf{P}\left[X_n = (1,j), Y_{n-1} = i, T_n - T_{n-1} \le t, T_n - T_{n-1} \le \nu\right]}{\mathbf{P}\left[Y_{n-1} = i, T_n - T_{n-1} \le \nu\right]} \\ &= \frac{\mathbf{P}\left[Y_n = j, T_n - T_{n-1} \le t | Y_{n-1} = i\right] \mathbf{P}\left[X_n = (1,j) | Y_n = j\right]}{\mathbf{P}\left[T_n - T_{n-1} \le \nu | Y_{n-1} = i\right]} \\ &= R_{ij}(t) R_j(\nu) / R_i(\nu). \end{aligned}$$

Note that if we let $\{N(t),\, t \ge 0\}$ denote the counting process associated to the MRP $(X,\, T)$, i.e.,

$$N(t) = \sup\{n \ge 0 : T_n \le t\}$$

then the SMP constructed from the MRP $(X,\, T)$ is $\{J(t),\, t \ge 0\}$, with $J(t) = X_{N(t)}$, which has state space E and kernel $Q(t)$.

One visit to E_1 (E_0) corresponds to a maximal sequence of consecutive interarrival times in $(X,\, T)$ whose lengths are smaller or equal to ν (greater than ν) and which is completed when a visit to E_0 (E_1) starts. In this way, bursts and gaps originated from a MRP are special cases of sojourn times in sets of states of SPMs. Accordingly, in the next two sections we investigate sojourn times in sets of states of SMPs, namely sojourn times in E_1 and E_0. Note that we may interpret a burst as a sojourn time in E_1 and a gap a sojourn time in $E_0 = E \setminus E_1$.

3. Sojourn Times

In this section we compute expected values and variances of sojourn times in sets of states of a SMP. We first define the sequence of successive (re)entrance times in the sets E_0 and E_1, constituting a partition of the state space of the SMP, and next derive vectors of expected values and variances of sojourn times in these sets.

Let (X, T) denote a recurrent MRP with phase space $E = E_0 \cup E_1$, where $E_0 \cap E_1 = \emptyset$, and kernel $Q(t)$, and let $J = \{J(t),\, t \ge 0\}$ denote an SMP constructed from (X, T) as described above. The matrix $P = [P_{ij}]$ with $P_{ij} = \lim_{t \to \infty} Q_{ij}(t)$, is the transition probability matrix of the embedded DTMC X, and the partition of E in E_0 and E_1 induces the following partitions of the kernel $Q(t)$ along with the transition probability matrix P:

$$Q(t) = \begin{bmatrix} Q_{\mathbf{00}}(t) & Q_{\mathbf{01}}(t) \\ Q_{\mathbf{10}}(t) & Q_{\mathbf{11}}(t) \end{bmatrix} \quad \text{and} \quad P = \begin{bmatrix} P_{\mathbf{00}} & P_{\mathbf{01}} \\ P_{\mathbf{10}} & P_{\mathbf{11}} \end{bmatrix}. \tag{2}$$

Let $\{N_k^{(\cdot)},\ k \in \mathbb{N}\}$ denote the sequence of successive (re)entrance times of X into E_0 or E_1, i.e., $N_0^{(\cdot)} = 0$ and

$$N_k^{(\cdot)} = \inf\{n > N_{k-1}^{(\cdot)} : (X_{n-1} \in E_0 \wedge X_n \in E_1) \vee (X_{n-1} \in E_1 \wedge X_n \in E_0)\}$$

for $k \in \mathbb{N}$. In addition, we let $X^{(\cdot)}$ denote the embedded DTMC at the sequence of successive entrance times of X into E_0 or E_1, i.e., $X_n^{(\cdot)} = X_{N_n^{(\cdot)}}$, whose transition probability matrix is

$$P^{(\cdot)} = \begin{bmatrix} \mathbf{0} & P_{\mathbf{01}}^{(\cdot)} \\ P_{\mathbf{10}}^{(\cdot)} & \mathbf{0} \end{bmatrix} = \begin{bmatrix} \mathbf{0} & (I - P_{\mathbf{00}})^{-1} P_{\mathbf{01}} \\ (I - P_{\mathbf{11}})^{-1} P_{\mathbf{10}} & \mathbf{0} \end{bmatrix}. \tag{3}$$

Remark that $[T_{N_{k-1}^{(\cdot)}}, T_{N_k^{(\cdot)}})$ is the interval of time corresponding to the k-th visit to one of the sets E_0 or E_1, indistinctly, and the duration of that visit is $T_{N_k^{(\cdot)}} - T_{N_{k-1}^{(\cdot)}}$.

Let $\{N_k^{(0)},\ k \in \mathbb{N}_0\}$ and $\{N_k^{(1)},\ k \in \mathbb{N}_0\}$ denote the sequences of entrance times into E_0 and E_1, respectively, where, e.g., $N_0^{(0)} = \inf\{n \in \mathbb{N}_0 : X_n \in E_0\}$,

$$N_{k+1}^{(0)} = \inf\{n > N_k^{(0)} : X_{n-1} \in E_1 \wedge X_n \in E_0\},$$

for $k \in \mathbb{N}$, and let $X^{(0)}$ and $X^{(1)}$ denote the associated embedded DTMCs, i.e., $X_n^{(0)} = X_{N_n^{(0)}}$ and $X_n^{(1)} = X_{N_n^{(1)}}$, which have transition probability matrices $P_{\mathbf{01}}^{(\cdot)} P_{\mathbf{10}}^{(\cdot)}$ and $P_{\mathbf{10}}^{(\cdot)} P_{\mathbf{01}}^{(\cdot)}$, respectively.

In order to simplify the notation, in the following we do not write the index $k = 1$ in the first passage times to set of states; thus, e.g., we let $N^{(0)} = N_1^{(0)}$. We derive explicit expressions only for results concerning sojourn times in E_1, given that the results for the sequence of sojourn times in E_0 are similar to those of sojourn times in E_1.

Let $\{Z_k,\ k \in \mathbb{N}\}$ denote the sequence of random variables such that Z_k denotes the time between the $(k-1)$-th and k-th transitions of the SMP J, i.e., $Z_k = T_k - T_{k-1}$. For $r \in \mathbb{N}$, let $\mathbf{m}(r) = [m_i(r)]$ and $\mathbf{v} = [v_i]$ denote the vectors of moments of order r and variances of sojourn times in a set of states conditional to the initial state, respectively. In a more precise way,

$$m_i(r) = \mathbf{E}_i \left[\sum_{k=1}^{N^{(\cdot)}} Z_k^r \right] \quad \text{and} \quad v_i = \mathbf{Var}_i \left[\sum_{k=1}^{N^{(\cdot)}} Z_k \right]$$

for $i \in E$. The partition of E in E_0 and E_1 induces the partition of $\mathbf{m}(r)$ in the vectors $\mathbf{m_0}(r)$ and $\mathbf{m_1}(r)$, i.e., $\mathbf{m}(r) = \begin{bmatrix}\mathbf{m_0}(r)\\ \mathbf{m_1}(r)\end{bmatrix}$, where $m_i(r) = \mathbf{E}_i\left[\sum_{k=1}^{N^{(0)}} Z_k^r\right]$, for $i \in E_1$, and, to ease the notation, we let $\mathbf{m_1} = \mathbf{m_1}(1)$ denote the vector of expected values of sojourn times in E_1. In a similar way, $\mathbf{v}$ is partitioned in the vectors $\mathbf{v_0}$ and $\mathbf{v_1}$.

Before deriving vectors of expected values and variances it is convenient to define some matrices of conditional moments. For $r \in \mathbb{N}$, let $M(r)$ denote the matrix of moments of order r of Z_1 conditional to the initial state, i.e.,

$$M_{ij}(r) = \mathbf{E}_i\left[Z_1^r \delta_{X_1,j}\right] = P_{ij}\mathbf{E}\left[Z_1^r | X_0 = i, X_1 = j\right], \tag{4}$$

and we let $M = M(1)$. As in (2), the partition of the state space induces a block partition of the matrix $M(r)$ in the matrices $M_{\mathbf{00}}(r)$, $M_{\mathbf{01}}(r)$, $M_{\mathbf{10}}(r)$ and $M_{\mathbf{11}}(r)$. We finally introduce the matrices $U^{(\cdot)}(r)$ and $V^{(\cdot)}$, given by

$$U_{ij}^{(\cdot)}(r) = \mathbf{E}_i\left[\sum_{k=1}^{N^{(\cdot)}} Z_k^r \delta_{X_{N^{(\cdot)}},j}\right] \tag{5}$$

and

$$V_{ij}^{(\cdot)} = \mathbf{E}_i\left[\sum_{k=1}^{N^{(\cdot)}-1} \sum_{l=k+1}^{N^{(\cdot)}} Z_k Z_l \delta_{X_{N^{(\cdot)}},j}\right]. \tag{6}$$

These matrices can be block-partitioned in the form

$$U^{(\cdot)}(r) = \begin{bmatrix}\mathbf{0} & U_{\mathbf{01}}^{(\cdot)}(r)\\ U_{\mathbf{10}}^{(\cdot)}(r) & \mathbf{0}\end{bmatrix} \quad \text{and} \quad V^{(\cdot)} = \begin{bmatrix}\mathbf{0} & V_{\mathbf{01}}^{(\cdot)}\\ V_{\mathbf{10}}^{(\cdot)} & \mathbf{0}\end{bmatrix} \tag{7}$$

and let it be henceforth noted that, in consequence of the law of total probability,

$$\mathbf{m}(r) = U^{(\cdot)}(r)\mathbf{e} \quad \text{and} \quad \mathbf{m_1}(r) = U_{\mathbf{10}}^{(\cdot)}(r)\mathbf{e}. \tag{8}$$

The next two theorems express the matrices $U^{(\cdot)}(r)$ and $V^{(\cdot)}$ as a function of the matrices $M(r)$ and the probability transition matrix P.

Theorem 3.1. *For $r \in \mathbb{N}$,*

$$U_{\mathit{10}}^{(\cdot)}(r) = (I - P_{\mathit{11}})^{-1} M_{\mathit{11}}(r)\,(I - P_{\mathit{11}})^{-1} P_{\mathit{10}} + (I - P_{\mathit{11}})^{-1} M_{\mathit{10}}(r) \tag{9}$$

and $U_{\mathit{01}}^{(\cdot)}(r) = \boxed{0\leftrightarrow 1}(U_{\mathit{10}}^{(\cdot)}(r))$.

Proof: For $r, k, n \in \mathbb{N}$, with $k \le n$, and $i, j \in E$, let

$$[A_{\mathbf{10}}(r,k,n)]_{ij} = \mathbf{E}_i\left[Z_k^r \delta_{(N^{(0)},X_n),(n,j)}\right].$$

Then, for $k \le n-1$, $[A_{\mathbf{10}}(r,k,n)]_{ij}$ is equal to

$$\sum_{i_1,\cdots,i_{n-1}\in E_1} \mathbf{E}_i\left[Z_k^r \delta_{(N^{(0)},X_n),(n,j)} | X_1 = i_1, \cdots, X_{n-1} = i_{n-1}, X_n = j\right]$$
$$\times \mathbf{P}_i\left[X_1 = i_1, \cdots, X_{n-1} = i_{n-1}, X_n = j\right]$$
$$= \sum_{i_1,\cdots,i_{n-1}\in E_1} \mathbf{E}\left[Z_k^r | X_{k-1} = i_{k-1}, X_k = i_k\right] \times \prod_{l=1}^{n-1} [P_{\mathbf{11}}]_{i_{l-1},i_l} [P_{\mathbf{10}}]_{i_{n-1},j}$$

where $i_0 = i$. Thus, using (4), we obtain

$$[A_{\mathbf{10}}(r,k,n)]_{ij} = \begin{cases} \left[(P_{\mathbf{11}})^{k-1} M_{\mathbf{11}}(r) (P_{\mathbf{11}})^{n-k-1} P_{\mathbf{10}}\right]_{ij}, & k < n \\ \left[(P_{\mathbf{11}})^{n-1} M_{\mathbf{10}}(r)\right]_{ij}, & k = n \end{cases} \tag{10}$$

by proceeding similarly case $k = n$. As a result,

$$\left[U_{\mathbf{10}}^{(\cdot)}(r)\right]_{ij} = \mathbf{E}_i\left[\sum_{k=1}^{N^{(0)}} Z_k^r \delta_{X_{N^{(0)}},j}\right] = \mathbf{E}_i\left[\sum_{n=1}^{\infty}\sum_{k=1}^{n} Z_k^r \delta_{(N^{(0)},X_n),(n,j)}\right]$$
$$= \sum_{k=1}^{\infty}\sum_{n=k+1}^{\infty} [A_{\mathbf{10}}(r,k,n)]_{ij} + \sum_{n=1}^{\infty} [A_{\mathbf{10}}(r,n,n)]_{ij}$$

and, substituting (10) in the previous equation, we obtain (9). ■

Theorem 3.2. *The matrix $V_{\mathbf{10}}^{(\cdot)}$ is given by*

$$(I - P_{\mathbf{11}})^{-1} M_{11} (I - P_{\mathbf{11}})^{-1} M_{11} (I - P_{\mathbf{11}})^{-1} P_{\mathbf{10}}$$
$$+ (I - P_{\mathbf{11}})^{-1} M_{11} (I - P_{\mathbf{11}})^{-1} M_{10} \tag{11}$$

and $V_{\mathbf{01}}^{(\cdot)} = \boxed{0 \leftrightarrow 1}(V_{\mathbf{10}}^{(\cdot)})$.

Proof: For $k, l, n \in \mathbb{N}$, with $1 \le k < l \le n$, and $i, j \in E$, let

$$[B(k,l,n)]_{ij} = \mathbf{E}_i\left[Z_k Z_l \delta_{(N^{(0)},X_n),(n,j)}\right].$$

Similarly to (10), $[B_{\mathbf{10}}(k,l,n)]_{ij}$ is given by

$$\begin{cases} \left[(P_{\mathbf{11}})^{k-1} M_{11} (P_{\mathbf{11}})^{l-k-1} M_{11} (P_{\mathbf{11}})^{n-l-1} P_{\mathbf{10}}\right]_{ij}, & k < l < n \\ \left[(P_{\mathbf{11}})^{k-1} M_{11} (P_{\mathbf{11}})^{n-k-1} M_{10}\right]_{ij}, & k < l = n \end{cases} \tag{12}$$

and, in sequence,

$$\left[V_{\mathbf{10}}^{(\cdot)}\right]_{ij} = \sum_{n=2}^{\infty}\sum_{k=1}^{n-1}\sum_{l=k+1}^{n} \mathbf{E}_i\left[Z_k Z_l \delta_{\left(N^{(0)},X_n\right),(n,j)}\right]$$
$$= \sum_{k=1}^{\infty}\sum_{l=k+1}^{\infty}\sum_{n=l}^{\infty} \left[B_{\mathbf{10}}(k,l,n)\right]_{ij}.$$

Writing the previous equation for $k < l < n$ and $k < l = n$, and using (12) we obtain (11). ■

We end the section with the computation of the vectors of moments defined at the beginning of the section.

Corollary 3.1. *The vectors of expected values and variances of sojourn times in a set of states, conditional to the initial state, are given by*

$$\mathbf{m} = U^{(\cdot)}\mathbf{e} \quad and \quad \mathbf{v} = \left[U^{(\cdot)}(2) + 2V^{(\cdot)} - (diag(\mathbf{m}))^2\right]\mathbf{e} \tag{13}$$

where the matrices $U^{(\cdot)}(2)$ and $U^{(\cdot)} = U^{(\cdot)}(1)$ are given by (7) and (9), and $V^{(\cdot)}$ is given by (7) and (11). As a result, the vectors of expected values and variances of sojourn times in E_1, conditional to the initial state, are given by

$$\mathbf{m_1} = U_{\mathit{10}}^{(\cdot)}\mathbf{e} \quad and \quad \mathbf{v_1} = \left[U_{\mathit{10}}^{(\cdot)}(2) + 2V_{\mathit{10}}^{(\cdot)} - (diag(\mathbf{m_1}))^2\right]\mathbf{e}. \tag{14}$$

Proof: The stated formulas for $\mathbf{m}$ and $\mathbf{m_1}$ follow immediately from (8). On the other hand,

$$v_i = \mathbf{E}_i\left[\left(\sum_{k=1}^{N^{(\cdot)}} Z_k\right)^2\right] - \left(\mathbf{E}_i\left[\sum_{k=1}^{N^{(\cdot)}} Z_k\right]\right)^2$$
$$= \mathbf{E}_i\left[\sum_{k=1}^{N^{(\cdot)}} Z_k^2\right] + 2\mathbf{E}_i\left[\sum_{k=1}^{N^{(\cdot)}-1}\sum_{l=k+1}^{N^{(\cdot)}} Z_k Z_l\right] - m_i^2$$
$$= \sum_{j\in E}[U^{(\cdot)}(2) + 2V^{(\cdot)}]_{ij} - m_i^2$$

for $i \in E$, where the last equality follows from (5). Thus, in view of (8), we conclude the validity of the expression for $\mathbf{v}$ in (13). The expression for $\mathbf{v_1}$ in (14) follows from the one for $\mathbf{v}$ and the block-partitioning of the vectors $\mathbf{v}$ and $\mathbf{m}$ and the matrix $V^{(\cdot)}$ according to the sets E_0 and E_1. ■

4. Cycles

In this section we derive results for vectors of expected values and variances of cycle durations. Thus, we let $\mathbf{m}^c(r)$, $r \in \mathbb{N}$, and $\mathbf{v}^c$ denote the vector of moments of order r and vector of variances of cycle durations, conditional to the initial state, respectively. Before deriving the results of this section and following a similar procedure to the one used in the previous section, we define the sequence of successive times at which cycles are initiated along with matrices associated to cycle durations.

Let $\{N_k^{(\star)},\ k \in \mathbb{N}_0\}$ denote the sequence of times at which cycles are initiated, $X^{(\star)}$ denote the associated embedded DTMC, i.e., $X_k^{(\star)} = X_{N_k^{(\star)}}$, and $P^{(\star)}$ denote the transition probability matrix of $X^{(\star)}$. It follows that $P^{(\star)} = (P^{(\cdot)})^2$, so that the matrices $P^{(\star)}_{\mathbf{01}}$ and $P^{(\star)}_{\mathbf{01}}$ are null matrices, $P^{(\star)}_{\mathbf{00}} = P^{(\cdot)}_{\mathbf{01}} P^{(\cdot)}_{\mathbf{10}}$ and $P^{(\star)}_{\mathbf{11}} = P^{(\cdot)}_{\mathbf{10}} P^{(\cdot)}_{\mathbf{01}}$. We next define matrices associated with cycles:

$$C_{ij}^{(\star)}(r) = \mathbf{E}_i \left[\sum_{k=1}^{N^{(\star)}} Z_k^r \delta_{X_{N^{(\star)}},j} \right] \tag{15}$$

and

$$D_{ij}^{(\star)} = \mathbf{E}_i \left[\sum_{k=1}^{N^{(\cdot)}} \sum_{l=N^{(\cdot)}+1}^{N^{(\star)}} Z_k Z_l \delta_{X_{N^{(\star)}},j} \right] \tag{16}$$

for $r \in \mathbb{N}$. These matrices can be block-partitioned in the form

$$C^{(\star)}(r) = \begin{bmatrix} C^{(\star)}_{\mathbf{00}}(r) & \mathbf{0} \\ \mathbf{0} & C^{(\star)}_{\mathbf{11}}(r) \end{bmatrix} \quad \text{and} \quad D^{(\star)} = \begin{bmatrix} D^{(\star)}_{\mathbf{00}} & \mathbf{0} \\ \mathbf{0} & D^{(\star)}_{\mathbf{11}} \end{bmatrix}. \tag{17}$$

Let it be henceforth noted that, in view of the total probability law,

$$\mathbf{m}^c = C^{(\star)}\mathbf{e} \quad \text{and} \quad \mathbf{m}^c_{\mathbf{1}} = C^{(\star)}_{\mathbf{11}}\mathbf{e}. \tag{18}$$

We next state an auxiliary result which is very useful for the computation of the matrices $C^{(\star)}(r)$ and $D^{(\star)}$.

Lemma 4.1. *Let X and Y be nonnegative random variables and Z be a discrete random variable.*

If X and Y are conditionally independent given Z, then

$$\boldsymbol{E}[XY] = \sum_z \boldsymbol{E}[X\delta_{Z,z}]\, \boldsymbol{E}[Y|Z=z].$$

Proof: Under the stated conditions, the result is implied by the fact that

$$\begin{aligned}\mathbf{E}[XY] &= \sum_z \mathbf{E}[XY|Z=z]\,\mathbf{P}[Z=z] \\ &= \sum_z \mathbf{E}[X|Z=z]\,\mathbf{E}[Y|Z=z]\,\mathbf{P}[Z=z]\end{aligned}$$

where the last equality follows since X and Y are conditionality independents given Z. ■

Theorem 4.1. *For $r \in \mathbb{N}$,*

$$C^{(\star)}(r) = U^{(\cdot)}(r)P^{(\cdot)} + P^{(\cdot)}U^{(\cdot)}(r) \quad \textit{and} \quad D^{(\star)} = \left(U^{(\cdot)}\right)^2 \tag{19}$$

with $P^{(\cdot)}$ defined in (3) and $U^{(\cdot)}(r)$ given by (7) and (9). Moreover,

$$C^{(\star)}_{11}(r) = U^{(\cdot)}_{10}(r)P^{(\cdot)}_{01} + P^{(\cdot)}_{10}U^{(\cdot)}_{01}(r) \quad \textit{and} \quad D^{(\star)}_{11} = U^{(\cdot)}_{10}U^{(\cdot)}_{01} \tag{20}$$

and $C^{(\star)}_{00}(r) = \boxed{0\leftrightarrow 1}(C^{(\star)}_{11}(r))$ and $D^{(\star)}_{00} = U^{(\cdot)}_{01}U^{(\cdot)}_{10}$.

Proof: Each element of the matrix $C^{(\star)}(r)$ can be written as the sum of two conditional expected values, namely

$$C^{(\star)}_{ij}(r) = \mathbf{E}_i\left[\sum_{k=1}^{N^{(\cdot)}} Z^r_k \delta_{X_{N^{(\star)}},j}\right] + \mathbf{E}_i\left[\sum_{k=N^{(\cdot)}+1}^{N^{(\star)}} Z^r_k \delta_{X_{N^{(\star)}},j}\right]. \tag{21}$$

Lemma 4.1 implies that the first term on the right hand side of (21) is equal to

$$\begin{aligned}&\sum_{l\in E} \mathbf{E}_i\left[\sum_{k=1}^{N^{(\cdot)}} Z^r_k \delta_{X_{N^{(\cdot)}},l}\right] \mathbf{P}\left[X_{N^{(\star)}} = j | X_{N^{(\cdot)}} = l\right] \\ &\qquad = \sum_{l\in E} U^{(\cdot)}_{il}(r)\mathbf{P}_l\left[X_{N^{(\cdot)}=j}\right] = \left[U^{(\cdot)}(r)P^{(\cdot)}\right]_{ij}\end{aligned} \tag{22}$$

while the second term is equal to

$$\begin{aligned}&\sum_{l\in E} \mathbf{P}_i\left[X_{N^{(\cdot)}=l}\right] \mathbf{E}_i\left[\sum_{k=N^{(\cdot)}+1}^{N^{(\star)}} Z^r_k \delta_{X_{N^{(\star)}},j} | X_{N^{(\cdot)}} = l\right] \\ &\qquad = \sum_{l\in E} P^{(\cdot)}_{il}\mathbf{E}_l\left[\sum_{k=1}^{N^{(\cdot)}} Z^r_k \delta_{X_{N^{(\cdot)}},j}\right] = \left[P^{(\cdot)}U^{(\cdot)}(r)\right]_{ij}\end{aligned} \tag{23}$$

where the first equality of (23) follows from the strong Markov property. Substituting (22) and (23) in (21), the first equality of (19) follows. Moreover, the second equality of (19) follows directly from Lemma 4.1, taking into account that the random variables $\sum_{k=1}^{N^{(\cdot)}} Z_k$ and $\sum_{l=N^{(\cdot)}+1}^{N^{(\star)}} Z_l$ are conditionally independent given $X_{N^{(\cdot)}}$. The remaining equalities stated are immediate consequences of (19), taking into account (3) and (7). ■

As a result of the previous theorem, we have the following result concerning vectors of expected values and variances of cycle durations, conditional to the initial state.

Corollary 4.1. *The vectors of expected values and variances of cycle durations, conditional to the initial state, are given respectively by*

$$\mathbf{m}^c = [I + P^{(\cdot)}]U^{(\cdot)}e \tag{24}$$

$$\begin{aligned}\mathbf{v}^c =& [U^{(\cdot)}(2) + 2V^{(\cdot)}]e + P^{(\cdot)}[U^{(\cdot)}(2) + 2V^{(\cdot)}]e \\ &+ 2\left(U^{(\cdot)}\right)^2 e - [(diag(\mathbf{m}^c)]^2 e \end{aligned}\tag{25}$$

where $P^{(\cdot)}$ is given by (3), $U^{(\cdot)}(2)$ and $U^{(\cdot)} = U^{(\cdot)}(1)$ are given by (7) and (9) and $V^{(\cdot)}$ is given by (7) and (11). As a result, the vectors of expected values and variances of cycle durations beginning in E_1, given the initial state, are given by

$$\begin{aligned}\mathbf{m}_{\mathbf{1}}^c =& [U_{10}^{(\cdot)} + P_{10}^{(\cdot)}U_{01}^{(\cdot)}]e \\ \mathbf{v}_{\mathbf{1}}^c =& [U_{10}^{(\cdot)}(2) + 2V_{10}^{(\cdot)}]e + P_{10}^{(\cdot)}[U_{01}^{(\cdot)}(2) + 2V_{01}^{(\cdot)}]e \\ &+ 2U_{10}^{(\cdot)}U_{01}^{(\cdot)}e - [diag(\mathbf{m}_{\mathbf{1}}^c)]^2 e.\end{aligned}$$

Proof: From (18) and (19), it follows that

$$\mathbf{m}^c = C^{(\star)}\mathbf{e} = [U^{(\cdot)}P^{(\cdot)} + P^{(\cdot)}U^{(\cdot)}]\mathbf{e} = [I + P^{(\cdot)}]U^{(\cdot)}\mathbf{e},$$

taking into account that $P^{(\cdot)}\mathbf{e} = \mathbf{e}$. As a result, and in view of (7), $\mathbf{m}_{\mathbf{1}}^c = [U_{\mathbf{10}}^{(\cdot)} + P_{\mathbf{10}}^{(\cdot)}U_{\mathbf{01}}^{(\cdot)}]\mathbf{e}$. On the other hand, from the definition of v_i^c, $i \in E$, we have $v_i^c + (m_i^c)^2$ equals to,

$$
\begin{aligned}
&\mathbf{E}_i\left[\left(\sum_{k=1}^{N^{(\cdot)}} Z_k + \sum_{k=N^{(\cdot)}+1}^{N^{(*)}} Z_k\right)^2\right] \\
&=\mathbf{E}_i\left[\left(\sum_{k=1}^{N^{(\cdot)}} Z_k\right)^2\right] + \mathbf{E}_i\left[\left(\sum_{k=N^{(\cdot)}+1}^{N^{(*)}} Z_k\right)^2\right] + 2\mathbf{E}_i\left[\sum_{k=1}^{N^{(\cdot)}} \sum_{l=N^{[\cdot]}+1}^{N^{[*]}} Z_k Z_l\right] \\
&=\sum_{j\in E}\left\{\left[U^{(\cdot)}(2) + 2V^{(\cdot)}\right]_{ij} + \left[P^{(\cdot)}[U^{(\cdot)}(2) + 2V^{(\cdot)}]\right]_{ij} + 2D_{ij}^{(*)}\right\},
\end{aligned}
$$

where in the last equality: the first term follows from (5), the second term follows from (5) and the strong Markov property, and the third term follows from (16). Given that, in view of (19), $D^{(*)} = (U^{(\cdot)})^2$, the last equality corresponds in matrix notation to (25). The remaining equalities from (24)-(25) and the block-decomposition of the matrices and vectors involved. ■

5. Bursts and Gaps in Renewal Processes

In this section we apply the results previously obtained to analyze bursts and gaps for the particular case in which the arrival process is a renewal process. According to the steps followed in Section 2, we may view the state space H of the SMP associated to a renewal process has having only one element, $H = \{1\}$, and, as a result, the SMP J associated to the bursts and gaps of the renewal process has state space $E = \{(0,1), (1,1)\}$, whose elements may be denoted simply by 0 and 1, respectively.

Let Z be the random variable that denotes consecutive interarrival times of the renewal process, and let $Z_{\leq\nu}$ and $Z_{>\nu}$ be random variables that denote generic interarrival times that belong to bursts and gaps, respectively, so that

$$Z_{\leq\nu} \stackrel{\mathrm{d}}{=} Z|Z \leq v \quad \text{and} \quad Z_{>\nu} \stackrel{\mathrm{d}}{=} Z|Z > v$$

where $\stackrel{\mathrm{d}}{=}$ denotes equality in distribution. Note that the transition probabilities of the embedded DTMC associated to the SMP J only depend on the interarrival time distribution, and the matrices P and $M(r)$, defined respectively in (2) and (4) are given by

$$P = \begin{bmatrix} \overline{F}_Z(\nu) & F_Z(\nu) \\ \overline{F}_Z(\nu) & F_Z(\nu) \end{bmatrix} \tag{26}$$

and

$$M(r) = \begin{bmatrix} \overline{F}_Z(\nu)\mathbf{E}\left[Z^r_{>\nu}\right] & F_Z(\nu)\mathbf{E}\left[Z^r_{>\nu}\right] \\ \overline{F}_Z(\nu)\mathbf{E}\left[Z^r_{\leq\nu}\right] & F_Z(\nu)\mathbf{E}\left[Z^r_{\leq\nu}\right] \end{bmatrix}, \tag{27}$$

where F_Z ($\overline{F}_Z$) is the cumulative distribution (survival) function of Z. On the other hand, the matrices $U^{(\cdot)}(r)$ and $V^{(\cdot)}$, defined in (5) and (6), are 2×2 matrices and, by (9) and (11),

$$U^{(\cdot)}_{10}(r) = \mathbf{E}\left[Z^r_{\leq\nu}\right]/\overline{F}_Z(\nu) \qquad V_{10} = F_Z(\nu)\left[\mathbf{E}\left[Z_{\leq\nu}\right]\right]^2 / \left(\overline{F}_Z(\nu)\right)^2 \tag{28}$$

and $U^{(\cdot)}_{10} = U^{(\cdot)}_{10}(1)$ denotes the expected duration of a burst. In view of corollaries 3.1 and 4.1, we have the following result for bursts and gaps of renewal processes.

Corollary 5.1. *For a renewal process with generic interarrival time Z, the expected value and variance of a gap and a burst are, respectively,*

$$\boldsymbol{E}[gap] = \frac{\boldsymbol{E}[Z_{>\nu}]}{F_Z(\nu)} \qquad \boldsymbol{Var}[gap] = \frac{\boldsymbol{Var}[Z_{>\nu}]}{F_Z(\nu)} + \frac{\overline{F}_Z(\nu)}{(F_Z(\nu))^2}\left[\boldsymbol{E}[Z_{>\nu}]\right]^2$$

$$\boldsymbol{E}[burst] = \frac{\boldsymbol{E}[Z_{\leq\nu}]}{\overline{F}_Z(\nu)} \qquad \boldsymbol{Var}[burst] = \frac{\boldsymbol{Var}[Z_{\leq\nu}]}{\overline{F}_Z(\nu)} + \frac{F_Z(\nu)}{\left(\overline{F}_Z(\nu)\right)^2}\left[\boldsymbol{E}[Z_{\leq\nu}]\right]^2.$$

Moreover, the expected value and variance of a cycle are, respectively $\boldsymbol{E}[gap] + \boldsymbol{E}[burst]$ *and* $\boldsymbol{Var}[gap] + \boldsymbol{Var}[burst]$.

6. Customer Departures from $M/G/1/K$ Systems

In this section we use the derived results to analyze bursts and gaps associated to the customer departure process in $M/G/1/K$ systems, i.e., single-server queueing systems with finite capacity K, such that: costumers arrive to the system according to a Poisson process and their service times are independent and identically distributed random variables.

We let λ denote the customer arrival rate, $G(\cdot)$ denote the service time distribution function, and μ^{-1} denote the corresponding mean, so that

$$\mu^{-1} = \int_0^\infty (1 - G(t))\mathrm{d}t.$$

We let $\bar{S}(t)$ denote the number of costumers in the system at time t and T_n denote the time of the nth customer departure from the system. In addition, we let $Y_n = \bar{S}(T_n^+)$ denote the number of customers that are left in the system at the departure of the nth customer. It then follows that $(Y,T) = \{(Y_n, T_n), n \geq 0\}$ is a MRP with state space $H \times [0, +\infty)$, with $H = \{0, 1, \cdots, K-1\}$, and we let $\{N(t), t \geq 0\}$ denote the counting process associated to the MRP (Y, T), i.e.,

$$N(t) = \sup\{n \geq 0 : T_n \leq t\}$$

denotes the number of customer departures until time t. Then, $\{S(t), t \geq 0\}$, with $S(t) = Y_{N(t)}$, is a SMP with state space H and kernel $R(t)$ such that [for the corresponding analysis for $M/G/1$ systems see, e.g., (Kulkarni [7], Example 9.4)]

$$R_{ij}(t) = \begin{cases} \int_0^t (1 - e^{-\lambda(t-u)}) e^{-\lambda u} \frac{(\lambda u)^j}{j!} G(\mathrm{d}u) & i = 0, j \leq K-2 \\ \int_0^t (1 - e^{-\lambda(t-u)}) & \\ \quad \times \sum_{l=K-1}^{\infty} e^{-\lambda u} \frac{(\lambda u)^l}{l!} G(\mathrm{d}u) & i = 0, j = K-1 \\ \int_0^t e^{-\lambda u} \frac{(\lambda u)^{j-i+1}}{(j-i+1)!} G(\mathrm{d}u) & 0 \leq i-1 \leq j \leq K-2 \\ \int_0^t e^{-\lambda u} \sum_{l=K-i}^{\infty} \frac{(\lambda u)^l}{l!} G(\mathrm{d}u) & i > 0, j = K-1 \\ 0 & \text{otherwise} \end{cases}. \quad (29)$$

If, as in Section 2, we let $X_n = (1_{\{T_n - T_{n-1} \leq \nu\}}, Y_n)$, for $n \in \mathbb{N}_+$, then (X, T) is a MRP with phase space $E = E_0 \cup E_1$ where $E_0 = \{0\} \times H$ and $E_1 = \{1\} \times H$. Moreover, the SMP $J = \{J(t), t \geq 0\}$, with $J(t) = X_{N(t)}$, constructed from the MRP (X,T) has transition kernel given by (1), as a function of the kernel of $\{S(t), t \geq 0\}$ given in (29).

To compute moments of bursts, gaps, and cycles associated to the departure of the customers from the system we need to compute the matrix of moments, $M(r)$, as defined in (4), and the transition probability matrix, P, defined in (2), with $P = M(0)$, for which it is useful to introduce some additional notation.

First, we let A_m denote the mth moment of the distribution (function) $A(\cdot)$ on the positive reals, i.e.,

$$A_m = \int_0^\infty t^m A(\mathrm{d}t)$$

and, for positive ν, we let

$$A_m(\nu) = \int_0^\nu t^m A(\mathrm{d}t) \quad \text{and} \quad \bar{A}_m(\nu) = \int_{(\nu,\infty)} t^m A(\mathrm{d}t) = A_m - A_m(\nu).$$

In addition, we let

$$\alpha_m = \int_0^{+\infty} e^{-\lambda t} \frac{(\lambda t)^m}{m!} G(\mathrm{d}t)$$

denote the mixed-Poisson probability with mixing distribution $G(\cdot)$ and (positive) rate λ, for $m \in \mathbb{N}$. We note that these probabilities may be computed in linear time for a large class of service time distributions by means of simple recursions schemes – see, e.g., Kwiatkowska *et al.* [8] and German [9].

In order to simplify the writing, in the following we let, for positive ν and nonnegative integers m and r,

$$\alpha_m(\nu) = \int_0^\nu e^{-\lambda s} \frac{(\lambda s)^m}{m!} G(\mathrm{d}s)$$

$$\bar{\alpha}_m(\nu) = \int_{(\nu,+\infty)} e^{-\lambda s} \frac{(\lambda s)^m}{m!} G(\mathrm{d}s)$$

so that $\bar{\alpha}_m(\nu) = \alpha_m - \alpha_m(\nu)$,

$$\beta_{(m,r)}(\nu) = \int_0^\nu u^r e^{-\lambda u} \frac{(\lambda u)^m}{m!} G(\mathrm{d}u) = \frac{(m+r)!}{\lambda^r m!} \alpha_{m+r}(\nu)$$

$$\bar{\beta}_{(m,r)}(\nu) = \int_{(\nu,\infty)} u^r e^{-\lambda u} \frac{(\lambda u)^m}{m!} G(\mathrm{d}u) = \frac{(m+r)!}{\lambda^r m!} \bar{\alpha}_{m+r}(\nu)$$

and

$$\gamma_{(m,r)}(\nu) = \int_0^\nu u^r \lambda e^{-\lambda u} \int_0^u \frac{(\lambda s)^m}{m!} G(\mathrm{d}s)\, \mathrm{d}u$$

$$\bar{\gamma}_{(m,r)}(\nu) = \int_{(\nu,\infty)} u^r \lambda e^{-\lambda u} \int_0^u \frac{(\lambda s)^m}{m!} G(\mathrm{d}s)\, \mathrm{d}u.$$

Moreover, we let

$$H(t) = \int_0^t \left[1 - e^{-\lambda(t-u)}\right] G(\mathrm{d}u)$$

for $t \geq 0$, i.e., $H(\cdot)$ is the distribution function of the convolution of the exponential distribution with rate λ with the service time distribution $G(\cdot)$, and let $\bar{H}(t) = 1 - H(t)$.

Note that, for $i = 0, 1$ and $l, m = 0, 1, \ldots, K-1$,

$$M_{(i,l)(0,m)}(r) = \begin{cases} \frac{\bar{H}(\nu)}{H(\nu)} M_{(i,l)(1,m)}(r) & m = 0 \\ \frac{\bar{G}(\nu)}{G(\nu)} M_{(i,l)(1,m)}(r) & m > 0 \end{cases}. \tag{30}$$

Thus, in view of (1), (4), (29), and (30), to compute the matrix of moments, $M(r)$, it suffices to obtain

$$\{M_{(i,l)(1,m)}(r),\ i = 0, 1,\ 0 \leq l \leq K-1,\ \max(0, l-1) \leq m \leq K-1\}$$

and, in particular, $M_{(i,l)(j,m)}(r) = 0$ if $m < l-1$, for $i, j = 0, 1$. Thus, $M(r)$ is completely characterized trough (30) along with the following equalities that follow easily using (1), (4), and (29):

$$M_{(0,0)(1,0)}(r) = \frac{H(\nu)}{\bar{H}(\nu)}\bar{\gamma}_{(0,r)}(\nu)$$

$$M_{(0,0)(1,m)}(r) = \frac{G(\nu)}{\bar{H}(\nu)}\bar{\gamma}_{(m,r)}(\nu), \quad 0 < m < K-1$$

$$M_{(0,0)(1,K-1)}(r) = \frac{G(\nu)}{\bar{H}(\nu)}\left[\bar{H}_r(\nu) - \sum_{j=0}^{K-2} \bar{\gamma}_{(j,r)}(\nu)\right]$$

$$M_{(1,0)(1,0)}(r) = \gamma_{(0,r)}(\nu)$$

$$M_{(1,0)(1,m)}(r) = \frac{G(\nu)}{H(\nu)}\gamma_{(m,r)}(\nu), \quad 0 < m < K-1$$

$$M_{(1,0)(1,K-1)}(r) = \frac{G(\nu)}{H(\nu)}\left[H_r(\nu) - \sum_{j=0}^{K-2} \gamma_{(j,r)}(\nu)\right]$$

$$M_{(0,1)(1,0)}(r) = \frac{H(\nu)}{\bar{G}(\nu)}\bar{\beta}_{(0,r)}(\nu), \quad l = 1$$

$$M_{(0,l)(1,m)}(r) = \frac{G(\nu)}{\bar{G}(\nu)}\bar{\beta}_{(m-l+1,r)}(\nu), \quad l \geq 1 \wedge \max(l-1,1) \leq m < K-1$$

$$M_{(0,l)(1,K-1)}(r) = \frac{G(\nu)}{\bar{G}(\nu)}\left[\bar{G}_r(\nu) - \sum_{j=0}^{K-l-1} \bar{\beta}_{(j,r)}(\nu)\right], \quad l \geq 1$$

$$M_{(1,1)(1,0)}(r) = \frac{H(\nu)}{G(\nu)}\beta_{(0,r)}(\nu), \quad l = 1$$

$$M_{(1,l)(1,m)}(r) = \beta_{(m-l+1,r)}(\nu), \quad l \geq 1 \wedge \max(l-1,1) \leq m < K-1$$

$$M_{(1,l)(1,K-1)}(r) = G_r(\nu) - \sum_{j=0}^{K-l-1} \beta_{(j,r)}(\nu), \quad l \geq 1.$$

Note that, as a consequence of the definition,

$$\gamma_{(m,r)}(\nu) = \frac{r!}{\lambda^r}\left[\sum_{j=0}^{r}\binom{m+j}{j}\alpha_{m+j}(\nu) - \frac{\lambda^m G_m(\nu)}{m!}\sum_{j=0}^{r} e^{-\lambda\nu}\frac{(\lambda\nu)^j}{j!}\right]$$

$$\bar{\gamma}_{(m,r)}(\nu) = \frac{r!}{\lambda^r}\left[\sum_{j=0}^{r}\binom{m+j}{j}\bar{\alpha}_{m+j}(\nu) + \frac{\lambda^m G_m(\nu)}{m!}\sum_{j=0}^{r} e^{-\lambda\nu}\frac{(\lambda\nu)^j}{j!}\right].$$

Moreover, if we let

$$G^\star(\nu) = \int_0^\nu e^{\lambda u}\, G(\mathrm{d}u)$$

then

$$H_r(\nu) = \sum_{j=0}^{r} \frac{r!}{\lambda^j\,(r-j)!}\, G_{r-j}(\nu) - \frac{r!}{\lambda^r}\, G^\star(\nu,\lambda) \sum_{j=0}^{r} e^{-\lambda\nu} \frac{(\lambda\nu)^j}{j!}$$

$$\bar{H}_r(\nu) = H_r - H_r(\nu) = \sum_{j=0}^{r} \frac{r!}{\lambda^j\,(r-j)!}\, G_{r-j} - H_r(\nu)$$

where in the last equality we have used the fact that $H(\cdot)$ is the convolution of $G(\cdot)$ with the exponential distribution with rate λ along with the binomial formula

$$\mathbf{E}[(Z+W)^r] = \sum_{j=0}^{r} \binom{r}{j}\, \mathbf{E}[Z^j]\, \mathbf{E}[W^{r-j}]$$

for independent random variables Z and W.

Combining the derived results, we conclude that to compute the matrices $\{M(j)\}_{0\le j\le r}$ we need to compute $(\alpha_m(\lambda), \alpha_m(\nu,\lambda))_{0\le m\le K+r-2}$, $(G_m(\nu))_{0\le m\le K-2}$, and $(G_m, G^\star_m(\nu))_{0\le m\le r}$. In view of Theorems 3.1 and 3.2 we express the matrices $U^{(\cdot)}(j)$ and $V^{(\cdot)}$, defined in (5)-(7), as a function of the matrices $M(j)$ and the probability transition matrix $P = M(0)$. From this, by resorting to the corollaries 3.1 and 4.1, we are able to compute the expected values and variances of bursts, gaps, and cycles associated to the customer departure process in $M/G/1/K$ systems. The authors are implementing the computation of these performance measures, whose results will be published elsewhere.

Acknowledgments

This research was supported in part by *Programa de Formação Avançada de Docentes do Ensino Superior Medida 5/Acção 5.3* (PRODEP III) and the *Programa Operacional "Ciência, Tecnologia, Inovação"* (POCTI) of the *Fundação para a Ciência e a Tecnologia* (FCT), cofinanced by the European Community fund FEDER, and the projects POSC/EIA/60061/2004 and Euro-NGI.

References

1. M. Crovella and A. Bestavros. Self-similarity in world wide web traffic: evidence and possible causes. *IEEE/ACM Transactions on Networking*, **5**(12), 835–846, (1997).
2. W. E. Leland, M. S. Taqqu, W. Willinger, and D. V. Wilson. On the self-similar nature of Ethernet traffic. *IEEE/ACM Transactions on Networking*, **2**(1), 1–15, (1994).
3. M. A. Johnson, D. Liu, and S. Narayana. Burstiness descriptores for Markov renewal processes and Markovian arrival processes. *Stochastic Models*, **13**(3), 619–646, (1997).
4. A. Csenki. The joint distribuition of soujourn times in finite Markov processes. *Advances in Applied Probabability*, **24**, 141–160, (1992).
5. B. Sericola. Occupation times in Markov processes. *Stochastic Models* **16**(5), 479-510, (2000).
6. C. Nunes and A. Pacheco. Sojourn and passage times in Markov chains. In *Matrix-Analytic Methods: Theory and Applications* (Edited by G. Latouche and P. Taylor), pp. 311–332, World Scientific, River Edge, New Jersey, (2002).
7. V. G. Kulkarni, *Modeling and Analysis of Stochastic Systems.* London: Chapman and Hall, 1995.
8. M. Kwiatkowska, G. Norman, and A. Pacheco, "Model checking CSL until formulae with random time bounds," *Lecture Notes in Computer Science*, vol. 2399, pp. 152–168, 2002.
9. R. German, *Performance Analysis of Communication Systems: Modeling with Non-Markovian Stochastic Petri Nets.* Chichester : Wiley, 2000.
10. G. Casella and R. L. Berger *Statistical Inference.* Belmont : Duxbury Press, 1990.

PART E

Management Science

NASH EQUILIBRIUM FOR THREE RETAILERS IN AN INVENTORY MODEL WITH A FIXED DEMAND RATE

HITOSHI HOHJO†

Department of Mathematics and Information Sciences, Osaka Prefecture University, 1-1 Gakuen-cho, Naka-ku, Sakai, Osaka 599-8531, Japan
hojo@mi.s.osakafu-u.ac.jp

YOSHINOBU TERAOKA

Department of Mathematics and Information Sciences, Osaka Prefecture University, 1-1 Gakuen-cho, Naka-ku, Sakai, Osaka 599-8531, Japan
teraoka@mi.s.osakafu-u.ac.jp

In this paper we consider a competitive inventory control model with a fixed demand rate and initial sudden demands. The model is described as follows: There are three retailers which handle the same kind of products. They simultaneously open their stores and begin to sell it. They make their orders only at the beginning of period. Demands for a retailer occur with a fixed rate and the remaining retailers get initial sudden demands. If shortages happen in a store, unsatisfied demands are reallocated to a neighbor store with strong purchasing power. Moving to buy products causes a time lag in our model. In such a situation, each retailer is planning to minimize the sum of costs related with ordering, holding inventory, shortages and sales. The purpose of each retailer is to decide his order quantity minimizing his total cost. By mathematical formulation, we attempt investigating the optimal strategies for three players. We are interested in a Nash equilibrium analysis giving the optimal strategy in a deterministic demand.

1. Introduction

Some studies on an equilibrium point for an inventory model with many retailers have been discussed in recent years. Hohjo[3] and Hohjo, Teraoka[4-6] developed an inventory control model with reallocation from a viewpoint of a Nash equilibrium point. Their models were analyzed for two retailers under various assumptions with respect to demand and distribution of customers. Lippman and McCardle[8] examined the relation between equilibrium inventory levels and the splitting rule. They also provided conditions under which there is

† This work is supported by a Grant-in-Aid for Young Scientists (B) from the Ministry of Education, Culture, Sports, Science and Technology of Japan under Grant. No.16710120.

a unique equilibrium point for a competitive newsboy problem. Parlar[9] proved the existence of a Nash equilibrium point for an inventory model with two substitutable products having random demands. They treated the competitive inventory problems with the general number of retailers, their abstract result prevented application to realistic problems.

In this paper we consider a competitive inventory control model for three retailers so as to be able to correspond to realistic problems, and we explore a concrete strategy. We are interested in a Nash equilibrium analysis giving the optimal strategies in a deterministic demand. This paper is organized as follows. In Section 2, we describe the competitive inventory problem under consideration and define the notations and assumptions. We calculate total cost functions which are indispensable to our analysis in Section 3. Section 4 obtains an equilibrium point which we wish as a result. We give the concluding remarks in Section 5.

2. Model Description

In this paper we consider a single-period competitive inventory control model with a fixed demand rate and initial sudden demands. The model is described as follows: There are three retailers which deal with the same kind of products. We define that Player 1 is a retailer such that the sum of distances from him to each of others is shortest in three retailers, and Player 2 is a retailer near from Player 1. We also call the remainder Player 3. They simultaneously open their stores and begin sales. Their stock levels start zero. Their orders for products are placed only once at the beginning of the planning period and are arrived without lead-time. They can not replenish products during the planning period. Let z_i denote the order quantity for Player $i, i=1,2,3$. Customers to Player 1 sequentially arrive with a fixed demand rate. And some customers simultaneously arrive at Players 2 and 3 at the beginning of the planning period. For Player $i, i=1,2,3$, let b_i be the quantity of cumulative demands planned when each player does not occur shortage. When customers' demands are satisfied by the first visiting retailer, there is no new demand except for these customers. If customers unsatisfied their demands exist and they move to another player to satisfy their demands, players treat reallocated customers as additional demands. Customers don't know the information with respect to the quantities ordered by players and the stock levels of players at any time. Customers visit the first player and try to buy products. If they fail in purchase, they visit the nearest player and try to buy products. If customers can not satisfy their demands at the second visiting player, they go to the remainder. And if the last visiting player doesn't have stock, customers give up buying products at that time. We assume that customers move at a constant speed.

In this paper the unsatisfied demands cause the time lag by moving, because there are distances among players. Let λ_{ij} denote time which customers consume to move from Player i $(i = 1,2,3)$ to Player j $(j = 1,2,3, i \neq j)$. Then it holds that

$$\lambda_{ij} = \lambda_{ji} \quad \text{for } i, j = 1,2,3, \ i \neq j, \qquad \lambda_{ii} = 0 \quad \text{for } i = 1,2,3,$$

$$\text{and} \quad \lambda_{12} \leq \lambda_{13} \leq \lambda_{23}. \tag{1}$$

Let t denote the planning period common to all players. We assume the planning period is longer than time which first visiting customers wander distances among three players, that is, it holds that $\lambda_{12} + \lambda_{23} < t$.

The following defines the cost parameters relevant to our model. Let c_i be the unit ordering cost charged in the initial orders. The products are sold at the selling price r_i per unit. The holding cost is h_i for each unit on-hand stock carried over. On the other hand, the penalty cost of p_i per unit is incurred for shortage.

Each of the three players wants to decide his order quantity minimizing his total cost given consideration to ordering, stock holding, shortage in inventory and sales.

3. Cost Functions

3.1. *Inventory transition*

In this section, we investigate many cases of inventory transition which we encounter in our model. We try to calculate total cost functions for each of retailers in these cases. To simplify the descriptions of conditions, we give the following notations, which often appear in cases.

C1: $0 \leq z_1 < (\lambda_{13}/t)b_1$ C2: $z_1 \geq (\lambda_{13}/t)b_1$ C3: $z_1 \geq (\lambda_{12}/t)b_1$
C4: $0 \leq z_2 < b_2$ C5: $z_2 \geq b_2$ C6: $0 \leq z_3 < b_3$
C7: $z_3 \geq b_3$ C8: $2\lambda_{12} + \lambda_{23} \leq t$ C9: $2\lambda_{12} + \lambda_{23} > t$
C10: $\lambda_{12} + \lambda_{13} + \lambda_{23} \leq t$ C11: $\lambda_{12} + \lambda_{13} + \lambda_{23} > t$
C12: $(1 - (\lambda_{12} + \lambda_{23})/t)b_1 \leq z_1 < (1 - \lambda_{12}/t)b_1$
C13: $z_1 + z_2 + z_3 \geq (1 - (\lambda_{12} + \lambda_{23})/t)b_1 + b_2 + b_3$
C14: $z_1 + z_2 + z_3 \geq (1 - \lambda_{12}/t)b_1 + b_2 + b_3$

C15: $0 \le z_1 + z_2 + z_3 < (\lambda_{13}/t)b_1 + b_2 + b_3$

C16: $0 \le z_1 + z_2 + z_3 < (1-(\lambda_{12}+\lambda_{23})/t)b_1 + b_2 + b_3$

This model has 50 cases which we can write by using these notations as follows:

(I) $z_1 \ge b_1$, C5, C7;

(II-1) $z_1 + z_2 \ge b_1 + b_2$, C4, C7;

(II-2) $z_1 + z_3 \ge b_1 + b_3$, C5, C6;

(II-3) $(1-\lambda_{12}/t)b_1 \le z_1 < b_1$, C5, C7;

(II-4) $z_1 + z_2 \ge (1-\lambda_{12}/t)b_1 + b_2$, C7, C12;

(III) $z_1 + z_2 + z_3 \ge b_1 + b_2 + b_3$, C4, C6;

(IV-1) $(1-\lambda_{12}/t)b_1 + b_2 \le z_1 + z_2 < b_1 + b_2$, C4, C7;

(IV-2) $\max\{\lambda_{12}/t, 1-(\lambda_{12}+\lambda_{23})/t\}\cdot b_1 + b_2 \le z_1 + z_2 < (1-\lambda_{12}/t)b_1 + b_2$, C4, C7;

(IV-3) $(\lambda_{12}/t)b_1 + b_2 \le z_1 + z_2 < (1-(\lambda_{12}+\lambda_{23})/t)b_1 + b_2$, C4, C8, C13;

(IV-4) $0 \le z_1 + z_2 < (\lambda_{12}/t)b_1 + b_2$, $z_1 + z_2 + z_3 \ge (\lambda_{12}/t)b_1 + b_2 + b_3$, C3, C4, C9;

(IV-5) [$0 \le z_1 + z_2 < (\lambda_{12}/t)b_1 + b_2$, C3] instead of $(\lambda_{12}/t)b_1 + b_2 \le z_1 + z_2 < (1-(\lambda_{12}+\lambda_{23})/t)b_1 + b_2$ in Case (IV-3);

(IV-6) $(1-(\lambda_{12}+\lambda_{23})/t)b_1 + b_2 \le z_1 + z_2 < (1-\lambda_{12}/t)b_1 + b_2$, C5, C7, C12;

(IV-7) $0 \le z_1 < (1-\lambda_{12}/t)b_1$, $z_1 + z_2 < (1-\lambda_{12}/t)b_1 + b_2$, C5, C13;

(IV-8) $z_2 + z_3 \ge b_2 + b_3$, C4, C12;

(IV-9) $0 \le z_1 < (1-(\lambda_{12}+\lambda_{23})/t)b_1$, C4, C13;

(V-1) $(1-\lambda_{12}/t)b_1 + b_3 \le z_1 + z_3 < b_1 + b_3$, C5, C6;

(V-2) $(\lambda_{13}/t)b_1 + b_3 \le z_1 + z_3 < (1-\lambda_{12}/t)b_1 + b_3$, C6, C14;

(V-3) $z_1 + z_3 < (\lambda_{13}/t)b_1 + b_3$, C2, C6, C14;

(V-4) C1, C6, C14;

(VI-1) $(1-(\lambda_{12}+\lambda_{23})/t)b_1 \le z_1 < (\lambda_{12}/t)b_1$, C4, C6, C9;

(VI-2) $0 \le z_1 < \min\{\lambda_{12}/t, 1-(\lambda_{12}+\lambda_{23})/t\}\cdot b_1$, C4, C6;

(VI-3) $0 \le z_1 + z_2 < (\lambda_{12}/t)b_1 + b_2$, C3, C4, C6, C9;

(VI-4) C8 instead of C9 in Case (VI-3);

(VI-5) $\max\{\lambda_{12}/t, 1-(\lambda_{12}+\lambda_{23})/t\}\cdot b_1 + b_2 \le z_1 + z_2 < (\lambda_{13}/t)b_1 + b_2$, C4, C6, C11;

(VI-6) $(\lambda_{12}/t)b_1 + b_2 \le z_1 + z_2 < \min\{\lambda_{13}/t, 1-(\lambda_{12}+\lambda_{23})/t\}\cdot b_1 + b_2$, C4, C6, C8;

(VI-7) $z_1 + z_2 \ge (\lambda_{13}/t)b_1 + b_2$, C4, C6, C11, C15;

(VI-8) C10 instead of C11 in Case (VI-7);

(VI-9) $(1-\lambda_{12}/t)b_1 + b_2 + b_3 \le z_1 + z_2 + z_3 < b_1 + b_2 + b_3$, C4, C6;

(VI-10) $\max\{\lambda_{13}/t, 1-(\lambda_{12}+\lambda_{23})/t\}\cdot b_1 + b_2 + b_3 \le z_1 + z_2 + z_3 < (1-\lambda_{12}/t)b_1 + b_2 + b_3$, C4, C6;

(VI-11) $(\lambda_{13}/t)b_1 + b_2 + b_3 \le z_1 + z_2 + z_3 < (1-(\lambda_{12}+\lambda_{23})/t)b_1 + b_2 + b_3$, C4, C6, C10;

(VII-1) C1 instead of C4 in Case (VI-10);
(VII-2) C1 instead of C4 in Case (VI-11);
(VII-3) $z_1 + z_2 \geq (\lambda_{13}/t)b_1 + b_2$, C1, C6, C11, C15;
(VII-4) C10 instead of C11 in Case (VII-3);
(VII-5) $(1-(\lambda_{12}+\lambda_{23})/t)b_1 + b_2 \leq z_1 + z_2 < (\lambda_{13}/t)b_1 + b_2$, C1, C5, C6, C11;
(VII-6) $0 \leq z_1 + z_2 < \min\{\lambda_{13}/t, 1-(\lambda_{12}+\lambda_{23})/t\} \cdot b_1 + b_2$, C1, C5, C6;
(VII-7) [$z_1 + z_3 < (\lambda_{13}/t)b_1 + b_3$, C2] instead of C4 in Case (VI-10);
(VII-8) [$z_1 + z_3 < (\lambda_{13}/t)b_1 + b_3$, C2] instead of C1in Case (VII-2) ;
(VII-9) $z_1 + z_3 < (\lambda_{13}/t)b_1 + b_3$, C1, C5, C6, C11, C15;
(VII-10) C10 instead of C11 in Case (VII-9);
(VII-11) $(\lambda_{13}/t)b_1 + b_3 \leq z_1 + z_3 < b_1 + b_3$, C5, C6,
$(1-(\lambda_{12}+\lambda_{23})/t)b_1 + b_2 + b_3 \leq z_1 + z_2 + z_3 < (1-\lambda_{12}/t)b_1 + b_2 + b_3$;
(VII-12) $(\lambda_{13}/t)b_1 + b_3 \leq z_1 + z_3 < b_1 + b_3$, C5, C6,C16;
(VIII-1) $(1-(\lambda_{12}+\lambda_{23})/t)b_1 \leq z_1 < (\lambda_{12}/t)b_1$, $0 \leq z_2 + z_3 < b_2 + b_3$, C7, C9;
(VIII-2) [$0 \leq z_2 + z_3 < b_2 + b_3$, C7] instead of C6 Case (VI-2);
(VIII-3) $0 \leq z_1 < (\lambda_{12}/t)b_1$, $z_2 + z_3 \geq b_2 + b_3$, C4, C16;
(VIII-4) $0 \leq z_1 + z_2 < (\lambda_{12}/t)b_1 + b_2$, $0 \leq z_1 + z_2 + z_3 < (\lambda_{12}/t)b_1 + b_2 + b_3$, C3, C4, C7, C9;
(VIII-5) C8 instead of C9 in Case (VIII-4);
(VIII-6) $0 \leq z_1 + z_2 < (\lambda_{12}/t)b_1 + b_2$, C3, C4, C8,
$(\lambda_{12}/t)b_1 + b_2 + b_3 \leq z_1 + z_2 + z_3 < (1-(\lambda_{12}+\lambda_{23})/t)b_1 + b_2 + b_3$;
(VIII-7) $(\lambda_{12}/t)b_1 + b_2 \leq z_1 + z_2 < b_1 + b_2$, C4, C7, C16;
(IX) $0 \leq z_1 < (1-(\lambda_{12}+\lambda_{23})/t)b_1$, $0 \leq z_1 + z_2 < (1-\lambda_{12}/t)b_1 + b_2$, C5, C7, C16.

Case (I) represents the situation in which all players are not allowed to be sold out. Case (II) is the situation in which one of three players can not satisfy his demands and the others keep positive inventory levels until the end of the planning period. Case (III) gives the situation in which Players 2 and 3 do not have enough stocks and Player 1 makes good shortages from them. In Case (IV) Players 1 and 2 become to be out of stock. And Players 1 and 3 also behave so in Case (V). In the other cases, all players drop into shortage by the end of the planning period. In particular we distinguish cases by whether Player 2 and/or Player 3 can satisfy initial sudden demands. In Case (VI) both players can not satisfy their demands. Player 2 and Player 3, respectively, can satisfy initial sudden demands in Cases (VII) and (VIII). Then another player can not do it. Case (IX) gives the situation in which both players can satisfy sudden initial demands.

3.2. *Reduction of calculation*

As seeing from the definition of players' locations, Player 1 is the special player that inevitably receives demands reallocated from a player with shortage. At this point he is different from other players. Although many cases can be considered in this model, we can intuitively obtain the optimal solution in most cases. If the quantity of initial sudden demands b_i is more than the order quantity z_i for either Player 2 or Player 3, the player becomes to be sold out at all time. To cut his total cost down, he takes place an initial order so as not to be sold out at the beginning of the period. This argument leads us to reduce calculations in cases under consideration in a deterministic demand model. And, since the excess stock yields a high total cost, a player having the excess stock adjusts his inventory level so as to be equal to zero at the end of the planning period. Furthermore, this argument and continuity of cost functions give us the reduction of considerable cases. For example, Case (II-4) results in the situation of Case (IV-6) by using the optimum of Player 2. And the optimum of Player 3 guides Case (IV-7) to Case (IX). As a result, we should investigate three important cases, (II-3), (IV-6) and (IX), in a deterministic model.

We go through calculating process to the total cost functions only for these cases described above on this report.

Let $C^i(\mathbf{z},\mathbf{b})$ represent the total cost function for Player i, $i=1,2,3$, where $\mathbf{z}=(z_1,z_2,z_3)$ and $\mathbf{b}=(b_1,b_2,b_3)$. We give the inventory quantity $Q_i(T)$ at time T and calculate the total cost $C^i(\mathbf{z},\mathbf{b})$ for Player i.

3.3. *Case (II-3)*

We first explain the transition of stock levels in Case (II-3). Players 2 and 3 satisfy their initial sudden demands, respectively, and keep positive stock levels. Player 1's stock decreases gradually and reaches a negative level at the end of the planning period. Though Player 1 makes shortages, there are no demands reallocated from Player 1 to Player 2.

For Player 1 the inventory quantity $Q_1(T)$ on hand is represented by

$$Q_1(T)=z_1-(T/t)b_1, \quad 0\le T\le t. \tag{2}$$

Then the average inventory quantity I_1^1 and the average shortage quantity in inventory I_2^1 are calculated by

$$I_1^1=(1/t)\int_0^{z_1 t/b_1}\{z_1-b_1T/t\}dT \tag{3}$$

and

$$I_2^1 = (1/t)\int_{z_1 t/b_1}^{t} \{b_1 T/t - z_1\} dT. \tag{4}$$

Using these values, the total cost of Player 1, denoted by $C^1(\mathbf{z},\mathbf{b})$, in this case is given by

$$C^1(\mathbf{z},\mathbf{b}) = c_1 z_1 + h_1 I_1^1 + p_1 I_2^1 - r_1 z_1. \tag{5}$$

For Player i, $i = 2,3$, their inventory quantities are represented by

$$Q_i(T) = z_i - b_i, \quad 0 \le T \le t. \tag{6}$$

The average inventory quantity is calculated by

$$I_1^i = z_i - b_i. \tag{7}$$

The average shortage quantity in inventory is equal to zero. Hence, the total cost $C^i(\mathbf{z},\mathbf{b})$ of Player i, $i = 2,3$ is given by

$$C^i(\mathbf{z},\mathbf{b}) = c_i z_i + h_i I_1^i - r_i b_i. \tag{8}$$

3.4. *Case (IV-6)*

Next we explain the inventory transition in Case (IV-6). Player 1's stock level decreases gradually and reaches negative by the end of the planning period. Players 2 and 3 satisfy their initial sudden demands. The demands reallocated from Player 1 give influence to Player 2's stock level, which shifts into a negative value by the end of the planning period. Though Player 2 can not satisfy all demands to him, there are no demands reallocated to Player 3.

Player 1's stock level has same transition as in Case (II-3). Hence, the total cost $C^1(\mathbf{z},\mathbf{b})$ is given by Eq.(5).

For Player 2 the inventory quantity $Q_2(T)$ is represented by

$$Q_2(T) = \begin{cases} z_2 - b_2, & 0 \le T < z_1 t/b_1 + \lambda_{12} \\ z_1 + z_2 - b_1(T - \lambda_{12})/t - b_2, & z_1 t/b_1 + \lambda_{12} \le T \le t. \end{cases} \tag{9}$$

Then the average inventory quantity I_1^2 and the average shortage quantity in inventory I_2^2 are calculated by

$$\begin{aligned} I_1^2 &= (1/t)\int_0^{z_1 t/b_1 + \lambda_{12}} \{z_2 - b_2\} dT \\ &+ (1/t)\int_{z_1 t/b_1 + \lambda_{12}}^{(z_1 + z_2 - b_2)t/b_1 + \lambda_{12}} \{z_1 + z_2 - b_1(T - \lambda_{12})/t - b_2\} dT \end{aligned} \tag{10}$$

and

$$I_2^2 = (1/t)\int_{(z_1+z_2-b_2)t/b_1+\lambda_{12}}^{t} \{b_1(T-\lambda_{12})/t + b_2 - z_1 - z_2\}dT. \tag{11}$$

Using these values, the total cost $C^2(\mathbf{z},\mathbf{b})$ of Player 2 is given by

$$C^2(\mathbf{z},\mathbf{b}) = c_2 z_2 + h_2 I_1^2 + p_2 I_2^2 - r_2 z_2. \tag{12}$$

Player 3's stock level traces same transition as in Case (II-3). The total cost $C^3(\mathbf{z},\mathbf{b})$ is given by Eq.(8).

3.5. *Case (IX)*

At last, we explain the inventory transition in Case (IX). The stock levels of Players 1 and 2 trace same transitions as in Case (IV-6). Therefore their total costs give Eqs. (5) and (12), respectively. Player 3 satisfies his initial sudden demands. The demands reallocated from Player 1 to Player 2 give influence to Player 3's stock level again. Player 3's level shifts into a negative value by the end of the planning period.

For Player 3 the inventory quantity $Q_3(T)$ is represented by

$$Q_3(T) = \begin{cases} z_3 - b_3, & 0 \le T < (z_1+z_2-b_2)t/b_1 + \lambda_{12} + \lambda_{23} \\ \sum_{i=1}^{3} z_i - b_1(T-\lambda_{12}-\lambda_{23})/t - b_2 - b_3, & \\ & (z_1+z_2-b_2)t/b_1 + \lambda_{12} + \lambda_{23} \le T \le t. \end{cases} \tag{13}$$

Then the average inventory quantity I_1^3 and the average shortage quantity in inventory I_2^3 are calculated by

$$I_1^3 = (1/t)\int_0^{(z_1+z_2-b_2)t/b_1+\lambda_{12}+\lambda_{23}} \{z_3 - b_3\}dT + (1/t)\int_{(z_1+z_2-b_2)t/b_1+\lambda_{12}+\lambda_{23}}^{(z_1+z_2+z_3-b_2-b_3)t/b_1+\lambda_{12}+\lambda_{23}} \left\{\sum_{i=1}^{3} z_i - b_1(T-\lambda_{12}-\lambda_{23})/t - b_2 - b_3\right\}dT \tag{14}$$

and

$$I_2^3 = (1/t)\int_{(z_1+z_2+z_3-b_2-b_3)t/b_1+\lambda_{12}+\lambda_{23}}^{t} \left\{b_1(T-\lambda_{12}-\lambda_{23})/t + b_2 + b_3 - \sum_{i=1}^{3} z_i\right\}dT. \tag{15}$$

Using these values, the total cost $C^3(\mathbf{z},\mathbf{b})$ of Player 3 is given by

$$C^3(\mathbf{z},\mathbf{b}) = c_3 z_3 + h_3 I_1^3 + p_3 I_2^3 - r_3 z_3. \tag{16}$$

4. Equilibrium

In previous section we found the total cost functions for some cases. These functions lead us to obtain an equilibrium point of players' strategies which we wish.

Let k_i be defined as $k_i = (r_i - c_i + p_i)/(h_i + p_i), \quad i = 1,2,3$. Observing the relation of partial derivatives of total cost functions, we can get candidates of the optimal solution. The results are as follows:

Case (II-3):

$$z_1 = \begin{cases} (1-\lambda_{12}/t)b_1, & k_1 < 1-\lambda_{12}/t \\ k_1 b_1, & 1-\lambda_{12}/t \le k_1 < 1 \\ b_1, & k_1 \ge 1, \end{cases}$$

$$z_2 = b_2,$$

$$z_3 = b_3;$$

Case (IV-6):

$$z_1 = \begin{cases} (1-(\lambda_{12}+\lambda_{23})/t)b_1, & k_1 < 1-(\lambda_{12}+\lambda_{23})/t \\ k_1 b_1, & 1-(\lambda_{12}+\lambda_{23})/t \le k_1 < 1-\lambda_{12}/t \\ (1-\lambda_{12}/t)b_1, & k_1 \ge 1-\lambda_{12}/t, \end{cases}$$

$$z_2 = \begin{cases} b_2, & k_2 < z_1/b_1 + \lambda_{12}/t \\ (k_2 - \lambda_{12}/t)b_1 + b_2 - z_1, & z_1/b_1 + \lambda_{12}/t \le k_2 < 1 \\ (1-\lambda_{12}/t)b_1 + b_2 - z_1, & k_2 \ge 1, \end{cases}$$

$$z_3 = b_3;$$

Case (IX):

$$z_1 = \begin{cases} (1-(\lambda_{12}+\lambda_{23})/t)b_1, & k_1 < 1-(\lambda_{12}+\lambda_{23})/t \\ k_1 b_1, & k_1 \ge 1-(\lambda_{12}+\lambda_{23})/t \end{cases}$$

$$z_2 = \begin{cases} b_2, & k_2 < z_1/b_1 + \lambda_{12}/t \\ (k_2 - \lambda_{12}/t)b_1 + b_2 - z_1, & z_1/b_1 + \lambda_{12}/t \le k_2 < 1 \\ (1 - \lambda_{12}/t)b_1 + b_2 - z_1, & k_2 \ge 1, \end{cases}$$

$$z_3 = \begin{cases} b_3, & k_3 < (z_1 + z_2 - b_2)/b_1 + (\lambda_{12} + \lambda_{23})/t \\ (k_3 - (\lambda_{12} + \lambda_{23})/t)b_1 + b_2 + b_3 - z_1 - z_2, & \\ \qquad (z_1 + z_2 - b_2)/b_1 + (\lambda_{12} + \lambda_{23})/t \le k_3 < 1 & \\ (1 - (\lambda_{12} + \lambda_{23})/t)b_1 + b_2 + b_3 - z_1 - z_2, & k_3 \ge 1. \end{cases}$$

By using these candidates and the continuity of the total cost functions, we conclude Theorem 4.1.

Theorem 4.1. *Let* $k_i = (r_i - c_i + p_i)/(h_i + p_i)$, $i = 1,2,3$. *We then obtain following results as an equilibrium point* (z_1^*, z_2^*, z_3^*) *under various conditions with respect to cost parameters.*

1. *If it holds that* $k_1 \ge 1$, *the equilibrium point is* (b_1, b_2, b_3).
2. *If it holds one of three conditions:*
   (i) $1 - \lambda_{12}/t \le k_1 < 1$;
   (ii) $1 - (\lambda_{12} + \lambda_{23})/t \le k_1 < 1 - \lambda_{12}/t,\ k_2 < k_1 + \lambda_{12}/t$;
   (iii) $k_1 < 1 - (\lambda_{12} + \lambda_{23})/t,\ k_2 < k_1 + \lambda_{12}/t,\ k_3 < k_1 + (\lambda_{12} + \lambda_{23})/t$,
   the equilibrium point is $(k_1 b_1, b_2, b_3)$.
3. *If it holds one of two conditions:*
   (i) $1 - (\lambda_{12} + \lambda_{23})/t \le k_1 < 1 - \lambda_{12}/t,\ k_1 + \lambda_{12}/t \le k_2 < 1$;
   (ii) $k_1 < 1 - (\lambda_{12} + \lambda_{23})/t,\ k_1 + \lambda_{12}/t \le k_2 < 1,\ k_3 < k_2 + \lambda_{23}/t$,
   the equilibrium point is $(k_1 b_1, (k_2 - \lambda_{12}/t - k_1)b_1 + b_2, b_3)$.
4. *If it holds one of two conditions:*
   (i) $1 - (\lambda_{12} + \lambda_{23})/t \le k_1 < 1 - \lambda_{12}/t,\ k_2 \ge 1$;
   (ii) $k_1 < 1 - (\lambda_{12} + \lambda_{23})/t,\ k_2 \ge 1,\ k_3 < 1 - \lambda_{23}/t$,
   the equilibrium point is $(k_1 b_1, (1 - \lambda_{12}/t - k_1)b_1 + b_2, b_3)$.
5. *If* $k_1 < 1 - (\lambda_{12} + \lambda_{23})/t,\ k_2 < k_1 + \lambda_{12}/t,\ k_1 + (\lambda_{12} + \lambda_{23})/t \le k_3 < 1$, *the equilibrium point is* $(k_1 b_1, b_2, (k_3 - (\lambda_{12} + \lambda_{23})/t - k_1)b_1 + b_3)$.
6. *If* $k_1 < 1 - (\lambda_{12} + \lambda_{23})/t,\ k_2 < k_1 + \lambda_{12}/t,\ k_3 \ge 1$, *the equilibrium point is* $(k_1 b_1, b_2, (1 - (\lambda_{12} + \lambda_{23})/t - k_1)b_1 + b_3)$.
7. *If* $k_1 < 1 - (\lambda_{12} + \lambda_{23})/t,\ k_1 + \lambda_{12}/t \le k_2 < 1,\ k_2 + \lambda_{23}/t \le k_3 < 1$, *the equilibrium point is* $(k_1 b_1, (k_2 - \lambda_{12}/t - k_1)b_1 + b_2, (k_3 - \lambda_{23}/t - k_2)b_1 + b_3)$.

8. *If* $k_1 < 1-(\lambda_{12}+\lambda_{23})/t,\ k_1+\lambda_{12}/t \le k_2 < 1,\ k_3 \ge 1$, *the equilibrium point is* $(k_1b_1, (k_2-\lambda_{12}/t-k_1)b_1+b_2, (1-\lambda_{23}/t-k_2)b_1+b_3)$.
9. *If* $k_1 < 1-(\lambda_{12}+\lambda_{23})/t,\ k_2 \ge 1,\ 1-\lambda_{23}/t \le k_3 < 1$, *the equilibrium point is* $(k_1b_1, (1-\lambda_{12}/t-k_1)b_1+b_2, (k_3-\lambda_{23}/t-1+k_1)b_1+b_3)$.
10. *If* $k_1 < 1-(\lambda_{12}+\lambda_{23})/t,\ k_2 \ge 1,\ k_3 \ge 1$, *the equilibrium point is* $(k_1b_1, (1-\lambda_{12}/t-k_1)b_1+b_2, (k_1-\lambda_{23}/t)b_1+b_3)$.

5. Concluding Remarks

In this paper, we developed a competitive inventory control model with a fixed demand rate for one retailer and initial sudden demands for the others. We showed that Player 1 need not to make arrangements for his order quantity whichever strategies Players 2 and 3 take. We treated a model in which players are non-cooperative. It is enough to consider three situations in a deterministic model. The simplicity of analysis is caused by a characteristic of sudden demands. However, when at lease two players are cooperative, the equilibrium point should be decided in cooperation with each other if there is not treachery in their actions. Of course, a stochastic model with uncertain demands needs total cost functions in all situations and the perseverance is required to analyze it.

References

1. J. Bryant, *Int. Econ. Rev.*, **21**, 619 (1980).
2. D.P. Heyman and M.J. Sobel, *Stochastic Models, Handbooks in Operations Research and Management Science* **2**, Elsevier Science Publishers, North-Holland (1990).
3. H. Hohjo, *Comput. & Math. with Appli.*, **41,** 523 (2001).
4. H. Hohjo and Y. Teraoka, *J. of the Oper. Res. Soc. of Jpn.*, **43**, 355 (2000).
5. H. Hohjo and Y. Teraoka, *Scientiae Mathematicae Japonicae*, **55**, 361 (2002).
6. H. Hohjo and Y. Teraoka, *Math. and Comput. Modelling*, **38**, 1191 (2003).
7. M.Kodama, *The Basis of Production and Inventory Control Systems (in Japanese)*, Kyushu University Press, Japan, (1996).
8. S.A. Lippman and K.F. McCardle, *Oper. Res.*, **45**, 54 (1997).
9. M. Parlar, *Nav. Res. Logist.*, **35**, 397 (1988).

A SEQUENTIAL EXPENDITURE PROBLEM FOR PUBLIC SECTOR BASED ON THE OUTCOME*

TŌRU NAKAI

Faculty of Economics, Kyushu University,
Hakozaki 6-19-1, Fukuoka, Japan
E-mail: nakai@en.kyushu-u.ac.jp

It is usual to grasp the activity of the public sector as a cycle of inputs, outputs and outcomes. The inputs are the resources or expenditures, the outputs are the products or services achieved, and the outcome is the criterion to measure the results. We will consider an optimal expenditure policy by formulating it as a sequential decision problem with Markovian transition with the outcome as a state. Especially, the outcome relates to the cumulative amount of resources, but the strict relationship is unknown. The problem is how much to expend to the public services to improve the outcomes. We also consider this problem in which the state changes according to a Markovian transition based on the total positivity of order two. The total positivity of order two is a fundamental property to investigate the sequential decision problem, and it also plays an important role in the Bayesian learning procedure for a partially observable Markov process.

1. Introduction

It is important to activate a management system of the public sector and to make this system efficient. It is usual to grasp the activity of the public sector as a management cycle of inputs, outputs and outcomes. The inputs are the resources or expenditures, and the outputs are the products and services achieved as a result. An outcome is the criterion to measure the performance and the results for the goal or the target. The relation among inputs and outputs is comparatively easy to evaluate when the performances (or the results) can be numerically measured, but, in fields where a qualitative evaluation is required, it is not easy to evaluate the performances and the results.

*This research was partially supported by the Grant-in-Aid for Scientific Research of the Japan Society for the Promotion of Science.

In this paper, we will consider an optimal expenditure policy in public sectors and formulate this problem as a sequential decision problem with Markovian transition based on the total positivity of order two, where the outcome is considered as a state. The problem is how much to expend to the public services to improve the outcomes. Especially, the outcome relates to the cumulative amount of resources, but the strict relationship is not known in general.

Consider an activity of public sectors like a fire service or police service, and also consider to expend within the range of the budget for each stage or period. For these public services, there exists some relationships between the number of equipments or staffs and the grade of a fulfillment level or a satisfaction rating. On the other hand, the satisfaction rating may change as the public situation changes. For this purpose, the satisfaction rating or a grade of a fulfillment level as an outcome is treated as a state. This state is not only reflected by public situations, but also changed by expending some additional amount within the range of the budget. We will start to treat a problem where the state of the process is observable.

The evaluation about the management cycle in the public sector and the relationship among inputs, outputs and outcomes will be summarized in Sec. 2 according to Nakai[11]. In Sec. 3, we will start to treat this problem where the state of the process is observable, and the state is changed only by expending some additional amount within the range of the budget.

The total positivity of order two is a fundamental property to investigate sequential decision problems, and it also plays an important role in the Bayesian learning procedure for a partially observable Markov process. In this paper, we will consider an optimal expenditure policy in public sectors such as the police services and the fire services. This problem is formulated as a sequential decision problem with Markovian transition based on the total positivity of order two, where the state is considered as a fulfillment level or a satisfaction rating for these services. These state changes according to a Markov chain and it is also reflected by expending some additional amount within the range of the budget. The problem is how much to expend to the public services to improve these level or rating.

In Sec. 4, we will consider a case where the state changes according to the Markov process under several assumptions. Sec. 5 concerns the total positivity of order two to investigate the learning procedure for a partially observable Markov process. In this problem, it cannot be known which one of these state it is now, but there is some information regarding what a present state is. All information about unobservable state are summarized by probability distributions on the state space, and we employ the Bayes' theorem as a learning procedure. By using a property called a totally positive of order two, or simply TP_2, some relationships among prior and posterior informations, an optimal policy and the expected values obtainable by employing this policy are considered. The properties of the total positivity are also investigated by Karlin and McGregor[4], Karlin[3] and Karlin and Rinott[5] and by others regarding the stochastic processes. Finally, we will treat this sequential expenditure problem with uncertainty for this partially observable Markov process in Sec. 6.

2. Evaluation of the Management Cycle of Public Sectors

In a private company, profit as a numerically expressed indicator can determine whether performance is good or bad. However, when the performance of the public sector is evaluated by a criterion similar to that of a private company, those in which profits cannot be taken are not sufficiently evaluated. Therefore, we are faced with the problem of how to evaluate the performance of the public sector.

Though there are various concepts of the action cycle or the management cycle of the public sector, it is usual to grasp the concept of activity of the public sector as a management cycle of

$$\text{inputs} \longrightarrow \text{outputs} \longrightarrow \text{outcomes}$$

as Hedley[2]. These inputs, outputs and outcomes can be classified as follows:

- The inputs are amounts of the resources expended.
- The outputs are amounts of the products or public services as a result.
- The outcomes are goals or targets used to measure the performance and the benefit.

In this management cycle, the product and service are produced (as the outputs) based on the resource (as the inputs) expended. It is considered that the criterion or the expectation (as the target values) is achieved as outcomes of the produced things. Therefore, if these are brought together, it becomes as follows.

- The relation between the inputs and the outputs is proportional to the relation of the resources expended and the products obtained as a result. Therefore, it is possible to conceive of the relation as the ratio of inputs *vs.* outputs or cost *vs.* benefit.

- The relation between the outputs and the outcomes is proportional to the relation of the products obtained and the criterion (the targets or the goals). The judgement or evaluation depends on whether these products reach these criterions or the targets (goals).

Therefore, though it is comparatively easy to conceptualize the relation between the inputs and the outputs, concerning the relation between the outputs and the outcomes, it is difficult to evaluate whether it qualifies for the targets or the goals.

It is important to activate the management system and to make this system efficient by employing the feedback system and the evaluation process. It is usual to consider the concepts of 'economy', 'efficiency' and 'effectiveness' as the criteria for such an evaluation system.

Especially, efficiency and effectiveness are important in evaluating the performance, but the content is not sufficiently defined. However, efficiency is to evaluate the relation between the inputs and the outputs, and effectiveness is to evaluate the relation between the outputs and the outcomes. In addition, efficiency has on the one hand the desirability of suppressing the inputs for the desired outputs, but on the other hand it is oriented toward enlarging the outputs within the limit of the given inputs. Therefore, in economy, efficiency, and effectiveness, it is understood that economy corresponds to suppressing inputs, and enlarging outputs corresponds to efficiency.

It is necessary to note that 'effectiveness' can be evaluated only by a relation between the outputs and the outcomes. Since it is difficult to conceptualize the causal relations between the inputs and the outcomes in the management cycle, the inputs are not related directly to the outcomes; the output is placed between these inputs and outcomes. However, it is also true that the relation between the inputs and the outcomes is incomprehensible.

In many cases, a comparatively concrete numerical value can be obtained regarding the inputs and the outputs. On the other hand, measuring the outcomes and expressing them numerically for evaluating is accompanied by a plethora of trouble. The following things are thought to be the reason. (1) It is difficult to measure the criterion or the expectation (goals) which corresponds to the outcomes concretely. (2) The causal relation between the outcomes and the inputs is not clear. (3) The relation between the budget to manage them and the service obtained as a result does not manifest itself clearly. (4) When evaluating the outcomes, it is possible to recognize that the policy and the program is related to the achievements and the results, but the degree of the contribution of the policy and the program to the achievements is not clear.

3. Sequential Expenditure Problem: A Deterministic Case

Consider an activity of public sectors like a fire service or police service, and also consider to expend some amount within the range of the budget for each fiscal year or period. For these public services, there exist some relationships between the number of equipments or staffs and the grade of a fulfillment level or a satisfaction rating as an outcome. On the other hand, even if the number of equipments or staffs grow larger, the satisfaction rating may decreases because the demand increases as the public situation or the circumstances changes. For this purpose, the satisfaction rating is considered as the state, and this state can be changed by expending some additional amounts within the range of the budget. It is assumed that the satisfaction rating for these public services is represented by a value s in $[0, 1]$, which is a state space of this problem. In this case, if $s = 1$, then this service is complied with the people, and it is not sufficiently complied with their demands as s becomes smaller. It is also possible to analyse the case where the state space is $(-\infty, \infty)$.

When the cumulative amount z of resources is expended to satisfy the request for these services by allocating equipments or staffs, it is assumed that the satisfaction rating $S(z)$ is realized, *i.e.* $S(z)$ is a function representing a relation between satisfaction rating and the cumulative amount z of the resources without considering other factors. On the other hand, when the satisfaction rating is s and $s = S(z)$, then this rating can be realized by expending an amount of z, and this z can be considered as an estimated value of the currently available services or resources.

Assume that the function $S(z)$ is an increasing and a concave function of z. Putting $z(s) = \inf\{z|S(z) \geq s, z \geq 0\}$ yields $z(s)$ to be a cumulative amount of the resources to realize the satisfaction rating s, *i.e.* an amount $z(s)$ is necessary to realize this grade in this time. When the state changes according to uncontrollable matters as the public situation or the circumstances, it represents a gap between the outcome and the currently available resources by using this function. Example 1 shows a simple case for these $S(z)$ and $z(s)$. Let $c(s, t)$ be a cost function to increase a satisfaction rating to t from the current rating s $(t \geq s)$.

In this section, we initially consider a case where the state can be changed by only expending some additional amounts within the range of the budget, and the state is not influenced by other uncontrollable matters. When there are n stages to go and the range of the budge at the present period is K, the problem is to increase the satisfaction rating by expending the equipments, the staffs and so on within this range. When the satisfaction rating is s, let $v_n(s)$ be an expected value obtainable by employing the optimal policy, then the optimality equation is

$$v_n(s) = \max_{0 \le x \le K} \{-c(s, s + d_s(x)) + v_{n-1}(s + d_s(x))\} \tag{1}$$

with $v_0(s) = u(s)$, where $d_s(x) = S(x + z(s)) - s$. As for the initial condition, the terminal reward $u(s)$ is assumed to be an increasing and concave function of t.

A function $d_s(x)$ represents an increment of the satisfaction rating by expending an additional amount x when the satisfaction rating is s. $z(s)$ is an amount to realize the current satisfaction rating s, and, at the next time, the satisfaction rating becomes $S(x+z(s))$ by expending an additional amount x to $z(s)$. In this section, since we treat a case where the satisfaction rating depends only on the cumulative amount of expenditure, it is possible to consider that $S(z(s)) = s$. In this case, the cost $-c(s, s+d_s(x))$ depends a new satisfaction rating as a result by expending an amount x when the satisfaction rating is s. If $-c(s, s + d_s(x)) = x$, this cost function can be considered to equal to an amount of expenditure. For the function $d_s(x)$, it is easy to show the following properties.

Lemma 1. *If $s < t$, then $d_s(x) \ge d_t(x)$.*

Lemma 2. *If $s < t$, then $s + d_s(x) \le t + d_t(x)$ for all $x \ge 0$.*

Example 1. When the cumulative amount of expenditure is z, let a function $S(z)$ representing a satisfaction rating be

$$S(z) = 1 - e^{-z} \quad (z \ge 0).$$

On the other hand, when the satisfaction rating is s, the cumulative amount $z(s)$ of resources is estimated as

$$z(s) = -\log(1 - s) \quad (0 \le s \le 1),$$

and $\dfrac{dS(z)}{dz} = e^{-z}$, $\dfrac{dz(s)}{ds} = -\dfrac{1}{(1-s)}$. If the satisfaction rating varies from s to t, the additional amount $-\log \dfrac{1-t}{1-s}$ is necessary to expend, and, therefore, it is possible to estimate the cost for this decision as

$$c(s,t) \equiv y - z = -\log\frac{1-t}{1-s}.$$

When the satisfaction rating is s, the cumulative amount of the resources can be estimated as $z(s) = -\log(1-s)$, and if the additional amount x is expended within the range of the budget, then the cumulative amount of the resources becomes $-\log(1-s)+x$ and the satisfaction rating will be $s(-\log(1-s)+x) = 1-(1-s)e^{-x}$. For this case, the satisfaction rating varies from s to $s+d_s(x) = 1-e^{-x}(1-s) = e^{-x}s+1-e^{-x} \geq 0$ and $c(s,t) = x$, and, therefore, $d_s(x) = e^{-x}s+1-e^{-x}-s = (1-e^{-x})(1-s) \geq 0$ is an increment of the satisfaction rating.

Assume that the cost function $c(s,t)$ is an increasing and convex function of t and a decreasing function of s, and it is also assumed that $c(s,t) \to 0$ as $t \to s$. By the induction method on n, it is easy to show the following properties.

Lemma 3. *$v_n(s)$ is a decreasing function of s, i.e. if $s \leq t$, then $v_n(s) \leq v_n(t)$.*

Lemma 4. *$v_n(s)$ is a non-decreasing function of n, i.e. $v_n(s) \leq v_{n+1}(s)$ for all $n \geq 1$.*

3.1. *Cost Function* $c(x)$

Consider this problem when the cost function depends only on x, which is denoted as $c(x)$. In subsequent sections, we formulate this problem with this $c(x)$, and, therefore, the optimality equation is

$$v_n(s) = \max_{0 \leq x \leq K}\{-c(x) + v_{n-1}(s + d_s(x))\}.$$

Similarly to the previous case, it is assumed that this $c(x)$ is an increasing and concave function of x, and suppose the following assumption.

Assumption 1. *$d_s(x)$ is concave with respect to s.*

For the last example, $s(z(s+u)+x) - s(z(s)+x) = ue^{-x} = s(z(t+u)+x) - s(z(t)+x)$, and, therefore, Example 1 satisfies Assumption 1. Here we obtain the following properties.

Lemma 5. *$v_n(s)$ is a concave function of s.*

Lemma 6. *Suppose that there are n stages to go and the satisfaction rating (state) is s, and let the optimal amount of the expenditure be $x_n^*(s)$, then $x_n^*(s) \leq x_n^*(t)$ for all $s \leq t$.*

Remark 1. As Eq. (1), since a cost $c(s, s + d_s(x))$ is represented by s and x, it is possible to denote a cost function as $c(s, x)$. If $c(t, x) - c(s, x)$ is a decreasing function of x for all $s < t$, then this property implies Lemma 6, *i.e.* an inequality $c(t, x) - c(t, x^*) \leq c(s, x) - c(s, x^*)$ for all $0 \leq x \leq x^*$ implies Lemma 6.

Lemma 7. *Suppose that there are n stages to go and the satisfaction rating (state) is s, and let the optimal amount of the expenditure be $x_n^*(s)$, then $x_{n-1}^*(s) \geq x_n^*(s)$.*

Lemma 8. *If $s < t$ then $v_{n-1}(t) - v_{n-1}(s) \geq v_n(t) - v_n(s)$ for all $n \geq 1$.*

4. Sequential Expenditure Problem: A Stochastic Case

In Sec. 3, the state can be changed by an additional expenditure only, but, in subsequent sections, we will treat this problem for a case where the state changes according to a Markov process, *i.e.* this Markov process represents the influence of uncontrollable matters. This is a case where, even if the number of equipments or staffs grow larger, the satisfaction rating may decreases by circumstances *et al.*, *i.e.* the satisfaction rating can be changed by additional expenditures within the range of the budget and also depends on uncontrollable matters.

In this problem, a state changes according to a Markov process. Let the state space be $[0, 1]$ as before, and suppose the transition probability of this Markov process as $(p_s(t))_{s,t \in [0,1]}$. Initially, we introduce a property of total positive of order two by Definition 1.

Definition 1. For a set function $\boldsymbol{P} = (p_s(t))_{s,t \in [0,1]}$, if $\begin{vmatrix} p_s(u) & p_s(v) \\ p_t(u) & p_t(v) \end{vmatrix} \geq 0$ for any s, t, u and v, where $s \leq t$ and $u \leq v$ $(s, t, u, v \in [0, 1])$, then this $\boldsymbol{P}$ is said to be total positive of order two, or simply TP_2.

Concerning this transition probability of a Markov process treated here, an assumption is introduced.

Assumption 2. *The transition probability $(p_s(t))_{s,t \in [0,1]}$ is TP_2.*

Suppose that there are n stages to go and the range of the budget at the present period is K. When the satisfaction rating (state) is s, let the expected value obtainable by employing the optimal policy be $V_n(s)$, then the optimality equation is

$$V_n(s) = \max_{x \geq 0} \left\{ -c(x) + \int_0^1 p_{s(x)}(t) V_{n-1}(t) dt \right\} \tag{2}$$

with $V_0(s) = u(s)$, where $s(x)$ represents a new state after expending an additional amount x when the current state is s. For a case treated in the last section, $s(x) = s + d_s(x)$. In subsequent sections, $s(x)$ is assumed to be an increasing and concave function of x and an increasing function of s.

Lemma 9. *Suppose a set function* $\boldsymbol{P} = (p_s(t))_{s,t \in [0,1]}$ *for the transition probability and an increasing function* $s(x)$ *of* x. *If* $x \leq y$ *and* $u \leq v$ $(u, v \in [0,1])$, *then* $\begin{vmatrix} p_{s(x)}(u) & p_{s(x)}(v) \\ p_{s(y)}(u) & p_{s(y)}(v) \end{vmatrix} \geq 0$ *for any* s *and* t $(s, t \in [0,1])$.

Proof. Since $x \leq y$, $s(x) \leq s(y)$, and this implies this property by Definition 1.□

Next we introduce an order among random variables by using the total positivity of order two as Definition 2, where the random variables are defined on a complete separable metric space with a total order $\geq$. It is easy to show that this order is a partial order.

Definition 2. Suppose that two random variables X and Y has the respective probability density functions $f(x)$ and $g(x)$. If $f(y)g(x) \leq f(x)g(y)$ for all x and y where $x \geq y$, then X is said to be greater than Y by means of the likelihood ratio, or simply $X \succeq Y$.

Let $\mathcal{F}_{SSD} = \{u \mid u(x) \text{ is increasing and concave in } x\}$, then it is also possible to define a partial order by using this set as Definition 3.

Definition 3. Suppose two random variables X and Y. If $E[u(X)] \geq E[u(Y)]$ for all $u(x)$ in $\mathcal{F}_{SSD}$, then $X \geq_{SSD} Y$.

Lemma 10 represents a relation between the orders defined in Definitions 2 and 3. When the transition probability is TP_2, Lemma 11 is known in Kijima and Ohnishi[6], and this lemma yields Lemma 12.

Lemma 10. *Suppose two random variables* X *and* Y. *If* $X \succeq Y$, *then* $X \geq_{SSD} Y$.

Lemma 11. *If* $u(t)$ *is an increasing and concave function of* t, *then* $\int_0^1 p_s(t)u(t)dt$ *is also an increasing function of* s.

Lemma 12. *If $x < y$, then $\int_0^1 p_{s(x)}(t)u(t)dt \le \int_0^1 p_{s(y)}(t)u(t)dt$ for all $u(x)$ in $\mathcal{F}_{SSD}$.*

If $V_{n-1}(t)$ is an increasing function of t, then $\int_0^1 p_s(t)V_{n-1}(t)dt$ is also an increasing function of s from Lemma 11. This fact and Lemma 12 imply the following properties by the induction principle on n.

Lemma 13. *$V_n(s)$ is a non-decreasing function of n, i.e. $V_n(s) \ge V_{n-1}(s)$.*

Lemma 14. *$V_n(s)$ is a non-decreasing function of s, i.e. if $s < s'$, then $V_n(s) \ge V_n(s')$.*

Example 2 satisfies the assumptions used in this paper.

Example 2. Let $(p_s(t))_{s,t\in[0,1]}$ be a transition probability of Markov process with state space $[0,1]$, and

$$p_s(t) = \begin{cases} 0 & 0 \le t < \underline{s-\alpha} \\ \dfrac{1}{\alpha(s)} & \underline{s-\alpha} \le t \le \overline{s+\alpha} \\ 0 & \overline{s+\alpha} < t \le 1 \end{cases},$$

where $\alpha(s) = (2\alpha + (s-\alpha)\wedge 0) - ((s+\alpha-1)\vee 0)$ in which $a \vee b = \max\{a,b\}$, $a \wedge b = \min\{a,b\}$, $\underline{a} = 0 \vee a$ and $\overline{a} = a \wedge 1$. The following calculations imply this $(p_s(t))_{s,t\in[0,1]}$ to be TP_2, i.e. $p_s(u)\,p_t(v) \ge p_t(u)\,p_s(v)$ or $\begin{vmatrix} p_s(u) & p_s(v) \\ p_t(u) & p_t(v) \end{vmatrix} \ge 0$ for all s,t,u,v where $s \le t$ and $u \le v$, $(s,t,u,v \in [0,1])$.

When $0 \le s-\alpha < t+\alpha \le 1$ and $|t-s| < 2\alpha$, since

$$\begin{aligned} &p_s(u)\,p_t(v) - p_t(u)\,p_s(v) \\ &\quad = \begin{cases} \dfrac{1}{2\alpha}\dfrac{1}{2\alpha} & if \begin{cases} s-\alpha \le u < t-\alpha \le v < t+\alpha \\ t-\alpha \le u < s+\alpha \le v < t+\alpha \end{cases} \\ 0 & \text{otherwise} \end{cases}, \end{aligned}$$

$p_s(u)\,p_t(v) - p_t(u)\,p_s(v) \ge 0$ for all s,t,u,v where $s \le t$ and $u \le v$. When $0 \le s-\alpha < t+\alpha \le 1$ and $|t-s| \ge 2\alpha$, similar calculations imply

$$\begin{aligned} &p_s(u)\,p_t(v) - p_t(u)\,p_s(v) \\ &\quad = \begin{cases} \dfrac{1}{2\alpha}\dfrac{1}{2\alpha} & if\; s-\alpha \le u < s+\alpha < t-\alpha \le v < t+\alpha \\ 0 & \text{otherwise} \end{cases} \end{aligned}$$

and $p_s(u)\,p_t(v) - p_t(u)\,p_s(v) \geq 0$ for all s, t, u, v where $s \leq t$ and $u \leq v$.

It is the same for a general case. If $|t - s| < 2\alpha$, then

$$
\begin{aligned}
&p_s(u)\,p_t(v) - p_t(u)\,p_s(v) \\
&\quad = \begin{cases} \dfrac{1}{\alpha(s)}\dfrac{1}{\alpha(t)} & if \begin{cases} \underline{s-\alpha} \leq u < \underline{t-\alpha} \leq v < \overline{t+\alpha} \\ \underline{t-\alpha} \leq u < \overline{s+\alpha} \leq v < \overline{t+\alpha} \end{cases} \\ 0 & \text{otherwise} \end{cases}
\end{aligned}
$$

and if $|t - s| \geq 2\alpha$ then

$$
\begin{aligned}
&p_s(u)\,p_t(v) - p_t(u)\,p_s(v) \\
&\quad = \begin{cases} \dfrac{1}{\alpha(s)}\dfrac{1}{\alpha(t)} & if\, \underline{s-\alpha} \leq u < \overline{s+\alpha} < \underline{t-\alpha} \leq v < \overline{t+\alpha} \\ 0 & \text{otherwise} \end{cases},
\end{aligned}
$$

and, therefore, $(p_s(t))_{s,t\in[0,1]}$ is TP_2.

Now, we introduce a following assumption, and the above example satisfies Assumption 3.

Assumption 3. *The transition probability $(p_s(t))_{s,t\in[0,1]}$ satisfies the property that $\int_0^1 p_s(t)u(t)dt$ is concave with respect to s for an increasing and concave function $u(t)$ of t.*

It is possible to show Lemma 15 and Proposition 1 under this assumption.

Lemma 15. *$V_n(s)$ is a concave function of s.*

Proposition 1. *Suppose that there are n stages to go and the satisfaction rating (state) is s, and let the optimal amount of the expenditure be $x_n^*(s)$, then $x_n^*(s) \leq x_n^*(t)$ for all $x \leq y$.*

Proof: By the assumption concerning $s(x)$, if $s < s'$, then, $s(x) \geq s'(x)$ and $s(x) - s \geq s'(x) - s'$ for any $x \geq 0$. Put $x_n^*(s) = x^*$, then

$$
\begin{aligned}
v_n(s) &= \max_{0\leq x\leq K}\left\{-c(x) + \int_0^1 p_{s(x)}(t)V_{n-1}(t)dt\right\} \\
&= -c(x^*) + \int_0^1 p_{s(x^*)}(t)V_{n-1}(t)dt, \qquad (3)
\end{aligned}
$$

$$
v_n(s') = \max_{0\leq x\leq K}\left\{-c(x) + \int_0^1 p_{s'(x)}(t)V_{n-1}(t)dt\right\}. \qquad (4)
$$

For any $0 \leq x \leq x^*$, if it is possible to show the inequality that

$$-c(x) + \int_0^1 p_{s'(x)}(t)V_{n-1}(t)dt \leq -c(x^*) + \int_0^1 p_{s'(x^*)}(t)V_{n-1}(t)dt,$$

then comparing Eqs. (3) and (4) yields this proposition. By Eq. (3), since

$$-c(x) + \int_0^1 p_{s(x)}(t)V_{n-1}(t)dt \leq -c(x^*) + \int_0^1 p_{s(x^*)}(t)V_{n-1}(t)dt$$

for any $x \geq 0$,

$$\begin{aligned}
-c(x) + c(x^*) &\leq \int_0^1 p_{s(x^*)}(t)V_{n-1}(t)dt - \int_0^1 p_{s(x)}(t)V_{n-1}(t)dt \\
&\leq \int_0^1 p_{s'(x^*)}(t)V_{n-1}(t)dt - \int_0^1 p_{s'(x^*)+s(x)-s(x^*)}(t)V_{n-1}(t)dt \\
&\leq \int_0^1 p_{s'(x^*)}(t)V_{n-1}(t)dt - \int_0^1 p_{s'(x)}(t)V_{n-1}(t)dt. \qquad (5)
\end{aligned}$$

Assumption 3 implies that $\int_0^1 p_{s(x)}(t)V_{n-1}(t)dt$ is concave function of s, and $s'(x^*) + s(x) - s(x^*) \geq s'(x)$ since $s(x)$ is a concave function of s and x. The inequality (5) is derived from these facts, and, therefore, the proof of this property is obtained. □

The following assumption is used to show Proposition 2.

Assumption 4. *The transition probability $(p_s(t))_{s,t\in[0,1]}$ satisfies that $\int_0^1 p_s(t)u(t)dt - u(t)$ is decreasing with respect to s for an increasing and concave function $u(t)$ of t.*

If this assumption is satisfied for the transition probability $(p_s(t))_{s,t\in[0,1]}$, then

$$\int_0^1 p_{s'}(t)u(t)dt - u(s') \leq \int_0^1 p_s(t)u(t)dt - u(s)$$

or $\int_0^1 p_{s'}(t)u(t)dt - \int_0^1 p_s(t)u(t)dt \leq u(s') - u(s)$. This inequality implies Lemmas 16 and 17, and Proposition 2 is derived from these lemmas.

Lemma 16. *If $s < s'$, then*

$$0 \leq \int_0^1 p_{s'}(t)(V_n(t) - V_{n-1}(t))dt \leq \int_0^1 p_s(t)(V_n(t) - V_{n-1}(t))dt \qquad (6)$$

for any $n \geq 1$.

Equation (6) is the same to

$$\int_0^1 p_{s'}(t)V_{n-1}(t)dt - \int_0^1 p_s(t)V_{n-1}(t)dt \geq \int_0^1 p_{s'}(t)V_n(t)dt - \int_0^1 p_s(t)V_n(t)dt.$$

Lemma 17. *If $s < s'$, then*

$$V_n(s') - V_n(s) \leq \int_0^1 p_{s'}(t)V_{n-1}(t)dt - \int_0^1 p_s(t)V_{n-1}(t)dt$$

for any $n \geq 1$.

Proposition 2. *Suppose that there are n stages to go and the satisfaction rating (state) is s, and let the optimal amount of the expenditure be $x_n^*(s)$, then $x_{n-1}^*(s) \geq x_n^*(s)$ for all $n \geq 1$.*

Proof: Let $x_n^*(s) = x^*$, then

$$\begin{aligned} v_n(s) &= \max_{0 \leq x \leq K} \left\{ -c(x) + \int_0^1 p_{s(x)}(t)V_{n-1}(t)dt \right\} \\ &= -c(x^*) + \int_0^1 p_{s(x^*)}(t)V_{n-1}(t)dt \end{aligned} \tag{7}$$

$$v_{n-1}(s) = \max_{0 \leq x \leq K} \left\{ -c(x) + \int_0^1 p_{s(x)}(t)V_{n-2}(t)dt \right\}. \tag{8}$$

If the inequality that

$$-c(x) + \int_0^1 p_{s(x)}(t)V_{n-2}(t)dt \leq -c(x^*) + \int_0^1 p_{s(x^*)}(t)V_{n-2}(t)dt \tag{9}$$

is valid for any $0 \leq x \leq x^*$, then comparing Eqs. (7) and (8) yields this proposition. Equation (8) implies

$$-c(x) + \int_0^1 p_{s(x)}(t)V_{n-1}(t)dt \leq -c(x^*) + \int_0^1 p_{s(x^*)}(t)V_{n-1}(t)dt.$$

On the other hand, Lemma 16 implies

$$\begin{aligned} &\int_0^1 p_{s(x^*)}(t)V_{n-1}(t)dt - \int_0^1 p_{s(x)}(t)V_{n-1}(t)dt \\ &\qquad \leq \int_0^1 p_{s(x^*)}(t)V_{n-2}(t)dt - \int_0^1 p_{s(x)}(t)V_{n-2}(t)dt \end{aligned}$$

and

$$\begin{aligned} c(x^*) - c(x) &\leq \int_0^1 p_{s(x^*)}(t)V_{n-1}(t)dt - \int_0^1 p_{s(x)}(t)V_{n-1}(t)dt \\ &\leq \int_0^1 p_{s(x^*)}(t)V_{n-2}(t)dt - \int_0^1 p_{s(x)}(t)V_{n-2}(t)dt. \end{aligned}$$

Inequality (9) is derived from these inequalities, and, therefore, the proof of this lemma is obtained.□

5. Partially Observable Markov Process and the Total Positivity

The total positivity of order two is a fundamental property to investigate sequential decision problems, and it also plays an important role in the Bayesian learning procedure for a partially observable Markov process. Consider a Markov process with state space $[0, 1]$ and transition probability $(p_s(t))_{s,t\in[0,1]}$, then $\boldsymbol{p}_s = (p_s(t))_{t\in[0,1]}$ is a probability distribution on $[0, 1]$ for any $s \in [0, 1]$. In subsequent sections, the state of this process can not be observed, *i.e.* this sequential decision problem will be treated on a partially observable Markov process. Associated to each state s corresponding to an outcome ($s \in [0, 1]$), there exists a non-negative random variable Y_s as an information process, *i.e.* an information system or an observation process exist to obtain information about unobservable state.

For this purpose, we introduce an assumption concerning the random variables Y_s ($s \in [0, 1]$) as Assumption 5, since we employ the Bayes' theorem as a learning procedure. For each state s, the random variables Y_s are absolutely continuous with density $f_s(y)$ ($s \in [0, 1]$). It is possible to generalize it as in Nakai[10], and also to apply this partially observable Markov process to sequential decision problems as Nakai[7,8,9] and so on.

Assumption 5. *For the random variables* $\{Y_s\}_{s\in[0,1]}$, *if* $s \le s'$, *then* $Y_{s'} \succeq Y_s$ $(s, s' \in [0, 1])$, *i.e.* Y_s *is increasing with respect to* s *by means of the likelihood ratio.*

In Assumption 5, $Y_s \succeq Y_{s'}$ implies that; if $x < y$, then $f_s(y)f_{s'}(x) \le f_s(x)f_{s'}(y)$ for s and s' where $s \le s'$ $(s, s' \in [0, 1])$. From this fact, the random variable Y_s takes on smaller values as s becomes smaller, and the state 0 represents the lowest class, $\cdots$, and state 1 is the highest class. The assumption concerning the transition probability implies that the probability of moving from the current state to 'better' states increases with improvement in the current state. From this fact, as a number s associated with each state becomes larger, the probability to make a transition into the higher class increases.

Information about unobservable state is assumed to be a probability distribution $\boldsymbol{\mu}$ on the state space $[0, 1]$. Let $\mathcal{S}$ be a set of all information about unobservable state, then

$$\mathcal{S} = \left\{ \boldsymbol{\mu} = (\mu(s))_{s\in[0,1]} \,\middle|\, \int_0^1 \mu(s)ds = 1, \mu(s) \geq 0\,(s \in [0,1]) \right\}.$$

Among informations in $\mathcal{S}$, we introduce an order by using the total positivity of order two as Definition 2, *i.e.* for two probability distributions $\boldsymbol{\mu}, \boldsymbol{\nu}$ on $[0,1]$, if $\mu(s')\nu(s) \leq \mu(s)\nu(s')$ for any s, s' $(s \leq s',\ s, s' \in [0,1])$ and $\mu(s')\,\nu(s) < \mu(s)\,\nu(s')$ at least one pair of s and s', then $\boldsymbol{\mu}$ is said to be greater than $\boldsymbol{\nu}$, or simply $\boldsymbol{\mu} \succ \boldsymbol{\nu}$. This order is a partial order and also said to be total positive of order two, or simply TP_2. By this definition, if $\boldsymbol{\mu} \succeq \boldsymbol{\nu}$ $(\boldsymbol{\mu}, \boldsymbol{\nu} \in \mathcal{S})$, then, as s becomes large, the ratio $\dfrac{\mu(s)}{\nu(s)}$ of the densities increases whenever $\nu(s) \neq 0$. On the other hand, if we put $\boldsymbol{p}_s = (p_s(u))$ and $\boldsymbol{p}_{s'} = (p_{s'}(u))$, then $\boldsymbol{p}_{s'} \succeq \boldsymbol{p}_s$ for all s, s' $(s \leq s',\ s, s' \in [0,1])$ since $\boldsymbol{P}$ satisfies Assumption 2. It is possible to generalize this order relation to investigate a partially observable Markov process, and the details are shown in Nakai[10] with the applications to sequential decision problems. Concerning this order relation, Lemma 18 is also obtained under Assumptions 2 and 5.

Lemma 18. *If* $\boldsymbol{\mu} \succeq \boldsymbol{\nu}$ *in* $\mathcal{S}$, *then* $\displaystyle\int_0^\infty h(x)dF_{\boldsymbol{\mu}}(x) \geq \int_0^\infty h(x)dF_{\boldsymbol{\nu}}(x)$ *for a non-decreasing non-negative function* $h(x)$ *of* x.

In this lemma, $F_{\boldsymbol{\mu}}(x) = \displaystyle\int_0^1 \mu(s)F_s(x)$ is a weighted distribution function as in De Vylder[1]. Regarding the unobservable state of the process, the random variables $\{Y_s\}_{s\in[0,1]}$ corresponds to an information process, since we improve information about unobservable state by observing these random variables. When prior information is $\boldsymbol{\mu}$, we first observe these random variables $\{Y_s\}_{s\in[0,1]}$ and improve information about it by employing the Bayes' theorem. After that, we see time moving forward by one unit and thus this process will make a transition to a new state. It is also possible to formulate and analyze this model by other order. If an observation is y, we improve information as $\boldsymbol{\mu}(y) = (\mu(y,s))_{s\in[0,1]} \in \mathcal{S}$ by employing the Byes' theorem, and, after changing to a new state according to $\boldsymbol{P}$, information at the next stage becomes $\overline{\boldsymbol{\mu}(y)} = (\overline{\mu(y,s)})_{s\in[0,1]} \in \mathcal{S}$.

For a set function $\boldsymbol{h}(x) = (h(x,s))_{s\in[0,1]}$, we introduce a monotonic property as Definition 4.

Definition 4. For any $x \in \Re_+$, suppose a set valued non-negative function $\boldsymbol{h}(x) = (h(x,s))_{s\in[0,1]}$. When $x < y$, $\boldsymbol{h}(y) \succeq \boldsymbol{h}(x)$ $(\boldsymbol{h}(x) \succeq \boldsymbol{h}(y))$, if and only if $h(x, s')\, h(y, s) \le h(x, s)\, h(y, s')$ $(h(x, s')\, h(y, s) \ge h(x, s)\, h(y, s'))$ for any s' and s ($s \le s'$ and $s, s' \in [0, 1]$). This function $\boldsymbol{h}(x)$ is said to be a increasing (decreasing) function of x.

Since the density functions $\{f_s(y) \mid s \in [0, 1]\}$ of $\{Y_s\}_{s \in [0,1]}$ satisfy Assumption 5, put $\boldsymbol{f}(y) = (f_s(y))_{s \in [0,1]}$ then $\boldsymbol{f}(x) \succeq \boldsymbol{f}(y)$, *i.e.* if $x > y$, then $f_s(y) f_{s'}(x) \le f_s(x) f_{s'}(y)$ for any s and s' ($s \le s'$ and $s, s' \in [0, 1]$). From this fact, $\boldsymbol{f}(x)$ is an increasing function of x.

Regarding a relationship between prior information $\boldsymbol{\mu}$ and posterior information $\overline{\boldsymbol{\mu}(x)}$, the essential properties (Lemma 19) is obtained under Assumptions 2 and 5, which is obtained in Nakai[10,11] and others.

Lemma 19. *If $\boldsymbol{\mu} \succ \boldsymbol{\nu}$, then $\boldsymbol{\mu}(y) \succ \boldsymbol{\nu}(y)$ and $\overline{\boldsymbol{\mu}(y)} \succ \overline{\boldsymbol{\nu}(y)}$ for all y. For any $\boldsymbol{\mu}$, $\boldsymbol{\mu}(y)$ and $\overline{\boldsymbol{\mu}(y)}$ is an increasing function of y.*

Lemma 19 implies that an order relation among prior information $\boldsymbol{\mu}$ is preserved in $\boldsymbol{\mu}(y)$ and posterior information $\overline{\boldsymbol{\mu}(y)}$. Furthermore, for same prior information $\boldsymbol{\mu}$, as an amount of an observation y increases, posterior information $\overline{\boldsymbol{\mu}(y)}$ becomes better by means of the likelihood ratio.

6. Sequential Expenditure Problem: An Incomplete Information Case

In this section, the state (outcome) changes according to uncontrollable matters and this state is not known. Information about current state is obtained through an information process. This problem is formulated by using a partially observable Markov process treated in Sec. 5.

As a preparation, first consider a sequential expenditure problem without learning, *i.e.* the state changes according to a partially observable Markov process but there exists no information about unobservable state of the process. Suppose that there are n stages to go, and let the range of the budget at the present period be K. When information about unobservable state (satisfaction rating) is known as a probability distribution $\boldsymbol{\mu} = (\mu(s))_{s \in [0,1]}$ on the state space, let $\overline{V}_n(\boldsymbol{\mu})$ be an expected value obtainable by employing the optimal policy, then optimality equation is determined by

$$\overline{V}_n(\boldsymbol{\mu}) = \max_{0 \le x \le K} \left\{ -c(x) + \overline{V}_{n-1}(\widetilde{\boldsymbol{\mu}(x)}) \right\} \tag{10}$$

with $\overline{V}_0(\boldsymbol{\mu}) = \int_0^1 u(t)d\mu(t)$. In this equation, we use a notation

$$\widetilde{\boldsymbol{\mu}(x)} = \int_0^1 \mu(s)p_{s(x)}(t)ds \tag{11}$$

for a probability distribution on the state after expending an amount x when prior information about unobservable state is $\boldsymbol{\mu}$. Under assumptions treated in the previous sections, it is possible to show the following properties, since the transition probability $(p_{s(x))}(t))_{0\le s\le 1}$ is TP_2 for any x by Lemma 9.

Lemma 20. *If $x > y$, then $\widetilde{\boldsymbol{\mu}(x)} \succeq \widetilde{\boldsymbol{\mu}(y)}$, and if $\boldsymbol{\mu} \succeq \boldsymbol{\nu}$, then $\widetilde{\boldsymbol{\mu}(x)} \succeq \widetilde{\boldsymbol{\nu}(x)}$ for any $x(\ge 0)$.*

By a method similar to one used in Sec. 4, the following properties are obtained for a sequential expenditure problem on a partially observable Markov process of Sec. 5 without learning. A monotonic properties for posterior information stated in Lemma 18 shows that the expected value obtainable by employing the optimal policy has also a monotonic property under the assumptions in Sec. 5.

Lemma 21. *If $\boldsymbol{\mu} \succeq \boldsymbol{\nu}$, then $\overline{V}_n(\boldsymbol{\mu}) \ge \overline{V}_n(\boldsymbol{\nu})$ for any s.*

Lemma 22. *For any $n \ge 1$, then $\overline{V}_n(\boldsymbol{\mu}) \ge \overline{V}_{n-1}(\boldsymbol{\mu})$ for any $\boldsymbol{\mu}$.*

Finally, for a sequential expenditure problem on a partially observable Markov process, information about unobservable state is obtained through information process and we employ the Bayes' theorem as a learning procedure. This is a problem in which the state changes according to a partially observable Markov process treated in Sec. 5, and associated to each state s ($s \in [0,1]$), there exists a random variable Y_s as an observation process to obtain information about it. Under Assumption 2, after obtaining information by observing these Y's, we employ the Bayes' theorem as a learning procedure. When information about unobservable state is known as $\boldsymbol{\mu}$, let $\widetilde{V}_n(\boldsymbol{\mu})$ be an expected value obtainable by employing the optimal policy, then the principle of optimality implies a recursive equation as

$$\begin{aligned} \widetilde{V}_n(\boldsymbol{\mu}) &= \int_0^\infty \widetilde{V}_n(\boldsymbol{\mu}|y)d\boldsymbol{\mu}(y) \\ \widetilde{V}_n(\boldsymbol{\mu}|y) &= \max_{x\ge 0}\left\{-c(x) + \widetilde{V}_{n-1}(\widetilde{\boldsymbol{\mu}(y)(x)})\right\} \end{aligned} \tag{12}$$

with $\widetilde{V}_0(\boldsymbol{\mu}) = \int_0^1 u(t)d\mu(t)$. In Eq. (12), $\boldsymbol{\mu}(y)$ is posterior information about new state after improving information by using y from the observation process, *i.e.* when prior information is $\boldsymbol{\mu}$, we first observe a realized value y from information process and improve information about it as $\boldsymbol{\mu}(y)$ by using the Bayes' theorem. After expending an additional amount x, we see time moving forward by one unit according to a transition probability $(p_{s(x)}(t))_{0\leq s\leq 1}$ whenever the unobservable state is s. Thus this process will make a transition to a new state, and information about this new state becomes $\widetilde{\boldsymbol{\mu}(y)(x)}$ as Eq. (11), which is a probability distribution on the state space after moving forward by one unit. After that, the expected value obtainable by employing the optimal policy is $\widetilde{V}_{n-1}(\widetilde{\boldsymbol{\mu}(y)(x)})$. By a method similar to one used in Lemmas 21 and 22, the following properties are obtained under the assumptions treated in Sec. 4.

Proposition 3. *$\widetilde{V}_n(\boldsymbol{\mu})$ is a non-decreasing function of n.*

Proposition 4. *$\widetilde{V}_n(\boldsymbol{\mu})$ is a non-decreasing function of $\boldsymbol{\mu}$.*

Proposition 3 is derived from an induction principle on n. On the other hand, if $\boldsymbol{\mu} \succ \boldsymbol{\nu}$, then $\boldsymbol{\mu}(y) \succ \boldsymbol{\nu}(y)$ for any observation y by Lemma 19 and $\widetilde{\boldsymbol{\mu}(x)} \succeq \widetilde{\boldsymbol{\nu}(x)}$ for any amount x to expend by Lemma 20. These monotonic properties concerning posterior information imply that; if $\boldsymbol{\mu} \succ \boldsymbol{\nu}$, then $\widetilde{\boldsymbol{\mu}(y)(x)} \succeq \widetilde{\boldsymbol{\nu}(y)(x)}$ for any additional amount x of expenditure and any observed value y, and, therefore, Proposition 4 is derived from these things by employing an induction principle on n.

7. Conclusion

When we grasp the activity of the public sector as a management cycle of inputs, outputs and outcomes, it is usual to consider the concepts of 'economy', 'efficiency' and 'effectiveness' as the criteria of an evaluation for such a system. Especially, efficiency and effectiveness are important in evaluating the performance and the result of the administration activity. In this paper, we considered an optimal expenditure problem as a sequential decision problem with Markovian transition based on the TP_2, and an outcome is considered as a state. These states relate to the cumulative amount of resources but is influenced by not relating to the resources directly. The problem is how much to spend to the public services to improve the outcomes within the range of the budget. Under some assumptions about transition rule, there exist some monotonic properties concerning the optimal policy and the expected value obtainable by employing the optimal policy.

References

1. F. De Vylder, Duality Theorem for Bounds in Integrals with Applications to Stop Loss Premiums, *Scandinavian Actuarial Journal*, 129–147 (1983).
2. T. P. Hedley, Measuring Public Sector Effectiveness Using Private Sector Method, *Public Productivity & Management Review*, **21** (3), 251–258 (1998).
3. S. Karlin, *Total Positivity*, Stanford University Press, Stanford, California (1968).
4. S. Karlin and J. L. McGregor, Classical Diffusion Process and Total Positivity, *Journal of Mathematical Analysis and Applications*, **1**, 163–183 (1960).
5. S. Karlin and Y. Rinott, Total Positivity Properties of Absolute Value Multinomial Variables with Applications to Confidence Interval Estimates and Related Probabilistic Inequalities, *The Annals of Statistics*, **9**, 1035–1049 (1981).
6. M. Kijima and M. Ohnishi, Stochastic Orders and Their Applications in Financial Optimization, *Mathematical Methods of Operations Research*, **50**, 351–372 (1999).
7. T. Nakai, A Sequential Stochastic Assignment Problem in a Partially Observable Markov process, *Mathematics of Operations Research*, **11**, 230–240 (1986).
8. T. Nakai, An Optimal Selection Problem on a Partially Observable Markov process, In *Stochastic Modelling in Innovative Manufacturing*, Lecture Notes in Economics and Mathematical Systems **445**, (Eds. A. H. Christer, S. Osaki and L. C. Thomas), pp. 140–154, Springer-Verlag, Berlin (1996).
9. T. Nakai, An Optimal Assignment Problem for Multiple Objects per Period – Case of a Partially Observable Markov process, *Bulletin of Informatics and Cybernetics*, **31**, 23–34 (1999).
10. T. Nakai, A Generalization of Multivariate Total Positivity of Order Two with an Application to Bayesian Learning Procedure, *Journal of Information & Optimization Sciences*, **23**, 163–176 (2002).
11. T. Nakai, Properties of a Job Search Problem on a Partially Observable Markov Chain in a Dynamic Economy, *Computers & Mathematics with Applications*, **51**, 189–198 (2005).
12. T. Nakai, Economy, Efficiency and Effectiveness, *Policy Analysis in the Era of Globalization and Localization* (Eds. Research Project Group for Policy Evaluation in Kyushu University), Kyushu University Press, 165–193 (2006).

CALCULATING THE PROBABILITIES OF WINNING THE ASIAN QUALIFIERS FOR 2006 FIFA WORLD CUP

MASABUMI SUZAKI and SHUNJI OSAKI

Faculty of Mathematical Sciences and Information Engineering,
Nanzan University, 27 Seirei-cho, Seto-shi, Aichi 489-0863, Japan
E-mail: {m05mm030, shunji}@nanzan-u.ac.jp

NOBUYOSHI HIROTSU

Department of Sports Information, Japan Institute of Sports Sciences,
3-15-1 Nishigaoka, Kita-ku, Tokyo 115-0056, Japan
E-mail: hirotsu.nobuyoshi@jiss.naash.go.jp

This paper addresses a problem of how to calculate the probabilities of winning for the sports contest under the assumption that the strength among contestants can be described by a linearly ordered set. The general formulae for calculating the probabilities of winning are given. Applying the general formulae, we give the probabilities of winning the Asian Qualifiers for 2006 FIFA World Cup.

1. Introduction

Operations Research in Sports is one of the hottest issues, and many contributions have been and will be devoted (see, e.g., References [1], [2]). In this paper we describe a method for calculating the probabilities of winning for the sports contest. In general, for the calculation of these probabilities, two approaches can be considered. One is that the probabilities can be preassigned for the contest. The other is that the strength among the contest teams is deterministic, i.e., the order of the strength is a linearly ordered set. We cite two example for explanation.

The first example is the famous "Triangular Contest" ("Tomoe Sen" in Japanese) for choosing one of three contestants as follows: We denote three contestants as A, B, and C. In the first match A plays against B. Then, the winner (A or B) of the first match plays against C in the second match, and so on. The winner of each match has the right to join the next match. If a contestant wins two consecutive matches, the contestant becomes the champion. This procedure has been used for the playoff of the traditional

Japanese "Sumo Wrestling", in the case where three sumoists have the same highest scores for 15 days contest. It is well known (see, e.g., Tamaki [3]) that under the assumption of equal probabilities (each match among A, B and C is a tie), the contestants A or B can become the champion with the probability $\frac{5}{14}$, and the remaining contestant C with $\frac{4}{14}$. That is, the contestants A and B have an advantage of C with the difference of probabilities $\frac{1}{14}$. Therefore, "Triangular Contest" can be considered "unfair" under the above assumption.

The second example appeared in the famous tale "Alice's Adventure in Wonderland", by Lewis Carroll, 1865. Of course, Lewis Carroll was a pen name. His real name was C. L. Dodgson, who was a mathematician. Knuth [4] cited a famous "Wimbledon Tournament" that is a typical "knockout tournament" (see also Mosteller [5]). Let us explain the knockout tournament (cited in Knuth [4]) in Fig. 1. Among 32 players, the strength for ranks can be labeled by 01, 02, $\cdots$, 32, where rank "xx" denotes the strength as a linearly ordered set such that $01 > 02 > \cdots > 32$. That is, the player 01 is the first-best, 02 the second-best, $\cdots$, and 32 the 32nd-best. It is clear in Fig. 1 that the first-best can become the champion in Round 5. However, the second-best is defeated by the first-best in Round 3 even though the second-best is stronger than the remaining players except the first-best. Knuth [4] surveyed a procedure of finding the minimum number of determining the first- and second-best players in such a tournament.

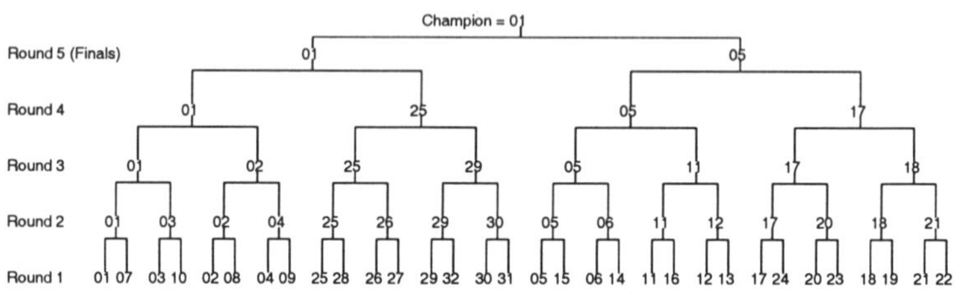

Figure 1. A knockout tournament with 32 players

In this paper we are interested in a problem of calculating the probabilities of winning based on the second approach (the strength among contestants can be described by a linearly ordered set). In particular, we cite 2006 FIFA World Cup: Asian Qualifiers (in June, 2005, four countries became the Asian Qualifiers, four countries including Japan became the Asian Qualifiers). In Section 2 we give the formulae for calculating the probabilities of winning based on a generalized assumption. In Section 3 we explain the Asian Qualifiers for 2006 FIFA World Cup and calculate the probabilities for the Asian Qualifiers. In Section 4, we conclude with some comments.

2. General Formulae for the Combinations of League Contestants

We assume in general that there are n teams and l leagues, where $m = n/l$ (an integer) is the number of teams within the same league. That is,

$$n = ml \tag{1}$$

Of course, we assume that the strength among teams can be described by a linearly ordered set labeled from 1 to n. We can understand that the strongest team can become the champion from the assumption. All the combinations for the potential matches are given by

$$\frac{n!}{(m!)^l \cdot l!} \tag{2}$$

We can label all the teams 1, 2, $\cdots$, n for the first round. After finishing the first round, only l teams can survive (l champion teams for each league). Then we can relabel l teams $1'$, $2'$, $\cdots$, l' after the first round.

The combination of winning the second round assuming that π-best team ($\pi = 1, 2, \cdots, n$) for the first round and π'-best team for the second round is given by

$$C(n, m, l, \pi, \pi', 1) = \sum_A \frac{(\pi - 1)! \cdot (n - \pi)!}{\prod_{k=1}^{l} S_k! \cdot \prod_{k=1}^{l} W_k! \cdot \prod_{i=0}^{m} T_i!} \tag{3}$$

where A denotes all the combinations of satistying

$$S_1 \geq S_2 \geq \cdots \geq S_l$$
$$0 \leq S_k \leq m, k = 1, \cdots, l$$
$$W_1 \leq W_2 \leq \cdots \leq W_l$$
$$0 \leq W_k \leq m, k = 1, \cdots, l$$
$$S_k + W_k = m, k = 1, \cdots, l$$
$$\sum_{k=1}^{l} S_k = \pi - 1, \sum_{k=1}^{l} W_k = n - \pi$$
$$T_m = l - \pi', \sum_{i=0}^{m} T_i = l - 1$$

Note that S_k denotes the number of teams stronger than π-best, and W_k the number of teams weaker than π-best. Note also that T_i is the number of the type for each league, where i denotes the number of teams weaker than π-best, except the league including π-best team.

2.1. *A concrete example from the asian qualifiers*

The candidates of the Asian Qualifiers are 32 teams (countries) and 8 leagues for the first round. 8 champion teams from 8 leagues can proceed the second round. For the second round, there are 2 leagues with each 4 teams. After the league matches for 2 leagues, the first- and second-best teams can become the Asian Qualifiers.

Let us consider a special case of 9-best team proceeding the second round in the Asian Qualifiers with 4-best team relabeled in the second round. That is, $\pi = 9$ and $\pi' = 4$. All the potential cases are shown in Fig. 2, where there are 3 cases in Fig. 2, respectively, where "S" and "W" denote the stronger and weaker than 9-best team in Fig. 2. For instance, in Fig. 2, three champion teams selected from left-side three columns are stronger than 9-best team (that is, $\pi' = 4$ in the second round). In this case, we have $S_1 = 4$, $S_2 = 3$, $S_3 = 1$, $S_4 = 0$, $S_5 = 0$, $S_6 = 0$, $S_7 = 0$, $S_8 = 0$, $W_1 = 0$, $W_2 = 1$, $W_3 = 3$, $W_4 = 3$, $W_5 = 4$, $W_6 = 4$, $W_7 = 4$, $W_8 = 4$, $T_0 = 1$, $T_1 = 1$, $T_2 = 0$, $T_3 = 1$, $T_4 = 4$ (see the details in Fig. 2). In a similar fashion, we have all S_k's, W_k's and T_i's ($k = 1, 2, 3, 4, 5, 6, 7, 8$, $i = 0, 1, 2, 3, 4$) for Fig. 2. From these facts, we have

$$\begin{aligned} C(32,4,8,9,4',1) &= \frac{8!23!}{4!3!3!3!4!4!4!4!4!} + \frac{8!23!}{4!2!2!2!2!3!4!4!4!4!2!4!} \\ &\quad + \frac{8!23!}{3!3!2!2!3!4!4!4!4!2!4!} \\ &= 25,251,881,955,300,000 + 28,408,367,199,712,500 \\ &\quad +75,755,645,865,900,000 \\ &= 129,415,895,020,912,500 \end{aligned}$$

As shown in Fig. 2, we can assign the league champion teams as the left-side column from π-best team in the first round (and π'-best in the second round) and the right-side columns from π-best team. Then we can rewrite Eq. (3) into

$$C(n,m,l,\pi,\pi',1) = \sum_A \frac{(\pi-1)! \cdot (n-\pi)!}{\prod_{k=1}^{\pi'-1} S_k! \cdot \prod_{k=1}^{\pi'-1} W_k! \cdot (m-1)! \cdot (m!)^{l-\pi'} \cdot \prod_{i=0}^{m-1} T_i! \cdot (l-\pi')!} \quad (4)$$

which is easier than the cumbersome Eq. (3)

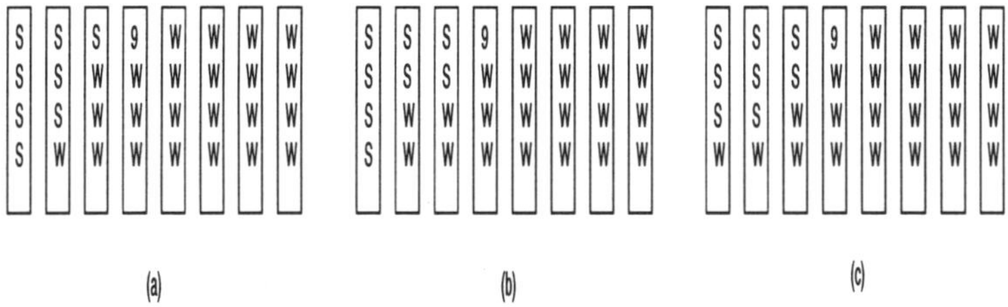

Figure 2. 3 cases of the combinations

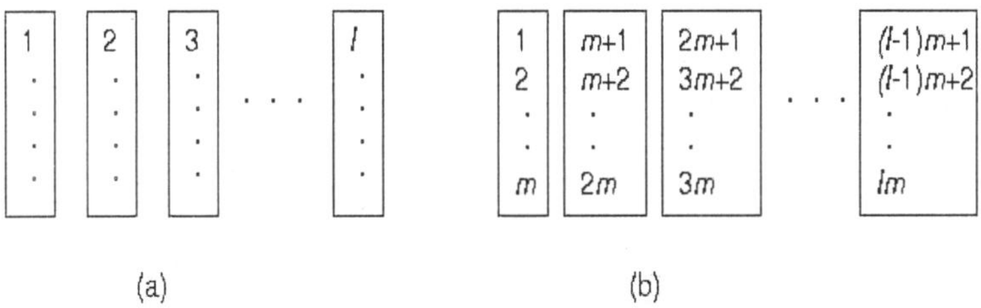

Figure 3. The explanation of the potential rank, where (a) is the strongest case and (b) the weakest case

2.2. *The weakest team surviving the final selection*

In the first round the league champion teams (l teams) can proceed the second round. If there are 2 leagues in the second round, the weakest team that can become the final qualifiers must be $m(l-3)+1$ (if two stronger teams can become the final qualifiers). In the Asian Qualifiers, we assume that $m = 4$, $l = 8$, then $m(l-3)+1 = 21$, that is, the 21st-best team among 32 teams can become the Asian Qualifiers. Of course, even though the probability of winning is quite low, the 21st-best team has the chance (see the probability in Table 9 below).

2.3. *The rank of the strength in the second round*

As shown in the proceeding subsection, we have shown the weakest team surviving the final selection. In the other hand, the stronger teams in the second round qualify from the rank of the strength in the first round. For instance, $1'$-best team can become only from 1-best team. $2'$-best team can become from 2-best to $(m+1)$-best teams, $3'$-best team can become from 3-best to $(2m+1)$-best teams. In general, k'-best team can become from k-best to $((k-1)m+1)$-best teams. See Fig. 3.

2.4. *The probabilities of winning*

The probability of winning the champion for a league is given by the ratio of Eq. (4) divided by Eq. (2)

$$P(n,m,l,\pi,\pi',1) =$$

$$\sum_A \frac{\frac{(\pi-1)!\cdot(n-\pi)!}{\prod_{k=1}^{\pi'-1} S_k!\cdot\prod_{k=1}^{\pi'-1} W_k!\cdot(m-1)!\cdot(m!)^{l-\pi'}\cdot\prod_{i=0}^{m-1} T_i!\cdot(l-\pi')!}}{\frac{n!}{(m!)^l\cdot l!}} =$$

$$\sum_A \frac{(m!)^l\cdot l!\cdot(\pi-1)!\cdot(n-\pi)!}{n!\cdot\prod_{k=1}^{\pi'-1} S_k!\cdot\prod_{k=1}^{\pi'-1} W_k!\cdot(m-1)!\cdot(m!)^{l-\pi'}\cdot\prod_{i=0}^{m-1} T_i!\cdot(l-\pi')!} =$$

$$\sum_A \frac{\prod_{k=1}^{\pi'-1}\binom{m}{S_k}\cdot m\cdot l!\cdot(\pi-1)!\cdot(n-\pi)!}{n!\cdot\prod_{i=0}^{m-1} T_i!\cdot(l-\pi')!} =$$

$$\sum_A \frac{\prod_{k=1}^{\pi'-1}\binom{m}{S_k}\cdot m\cdot P(l,\pi')\cdot(\pi-1)!}{P(n,\pi)\cdot\prod_{i=0}^{m-1} T_i!} \tag{5}$$

where A denotes all the combinations of satistying

$$S_1 \geq S_2 \geq \cdots \geq S_{\pi'-1}$$
$$0 \leq S_k \leq m, k = 1,\cdots,\pi'-1$$
$$W_1 \leq W_2 \leq \cdots \leq W_{\pi'-1}$$
$$0 \leq W_k \leq m, k = 1,\cdots,\pi'-1$$
$$S_k + W_k = m, k = 1,\cdots,\pi'-1$$
$$\sum_{k=1}^{\pi'-1} S_k = \pi - 1, \sum_{k=1}^{\pi'-1} W_k = n - \pi - (l - \pi' + 1)m + 1$$
$$\sum_{i=0}^{m} T_i = l - 1$$

3. The Asian Qualifiers

There are 32 candidate teams (countries) in the Asian Qualifiers. In the first round 8 teams can become the champions for each league, in which each league has 4 teams. In the second round 4 teams out of the 8 teams can become the Asian Qualifiers, where the first and second place teams in the two leagues can become the Asian Qualifiers (in June 2006, Iran and Japan from one league, and Saudi Arabia and Korea from another league became the Asian Qualifiers for 2006 FIFA World Cup).

3.1. *The probabilities of winning in the first round*

First we will give the probabilities of winning in the first round. Of course, for winning in each league, the candidate teams are from the first-best to the 29th-best teams and the 30th-, 31st-, 32nd-teams have no chance to win in the first round. After calculating all the probabilities, we will give the probabilities of winning the Asian Qualifiers in the next subsection.

The probabilities of winning in the first round are obtained by using the formulae in Eq. (5). We here define the probability $P(m, \pi')$ that the m-best qualifies for the second round as the π'-best in the second round.

It is trivial that $P(1, 1') = 1$ since the 1st-best team can never be defeated by the remaining teams. However, the 2nd-, 3rd-, 4th-, 5th-best teams have a chance to become the 2′nd best team in the second round. That is, the probabilities $P(m, 2')$ for 2, 3, 4, and 5, are positive based on the calculation using Eq. (5) as shown in Table 1. We note that the sum of these probabilities in Table 1 is a unity as expected. We also calculate the probabilities $P(m, \pi')$ for the m-best team to become the 3′rd-, 4′th-, 5′th-, 6′th-, 7′th-, and 8′th-best in the second round as shown in Tables 2, 3, 4, 5, 6, and 7, respectively.

Table 1. The probabilities $P(m, 2')$ for m = 2, 3, 4, 5

2	3	4	5	Total
$\frac{28}{31}$	$\frac{14}{155}$	$\frac{28}{4495}$	$\frac{1}{4495}$	1

Table 2. The probabilities $P(m, 3')$ for m = 3-9

3	4	5	6	7	8	9	Total
$\frac{112}{155}$	$\frac{1008}{4495}$	$\frac{204}{4495}$	$\frac{56}{8091}$	$\frac{28}{35061}$	$\frac{56}{876525}$	$\frac{7}{2629575}$	1

Table 3. The probabilities $P(m, 4')$ for m = 4-13

4	5	6	7	8	9
$\frac{448}{899}$	$\frac{288}{899}$	$\frac{1040}{8091}$	$\frac{1400}{35061}$	$\frac{1792}{175305}$	$\frac{1148}{525915}$
10	11	12	13	Total	
$\frac{308}{806403}$	$\frac{14}{268801}$	$\frac{4}{806403}$	$\frac{1}{4032015}$	1	

Table 4. The probabilities $P(m, 5')$ for m = 5-17

5	6	7	8	9	10	11
$\frac{256}{899}$	$\frac{2560}{8091}$	$\frac{22400}{105183}$	$\frac{896}{8091}$	$\frac{25424}{525915}$	$\frac{4928}{268801}$	$\frac{4928}{806403}$
12	13	14	15	16	17	Total
$\frac{480}{268801}$	$\frac{1816}{4032015}$	$\frac{112}{1178589}$	$\frac{56}{3535767}$	$\frac{112}{60108039}$	$\frac{7}{60108039}$	1

Table 5. The probabilities $P(m, 6')$ for m = 6-21

6	7	8	9	10	11
$\frac{1024}{8091}$	$\frac{2560}{11687}$	$\frac{39424}{175305}$	$\frac{7168}{40455}$	$\frac{158144}{1344005}$	$\frac{92512}{1344005}$
12	13	14	15	16	17
$\frac{29248}{806403}$	$\frac{23376}{1344005}$	$\frac{14944}{1964315}$	$\frac{5936}{1964315}$	$\frac{107968}{100180065}$	$\frac{6776}{20036013}$
18	19	20	21	Total	
$\frac{28}{310155}$	$\frac{2}{103385}$	$\frac{4}{1344005}$	$\frac{1}{4032015}$	1	

Table 6. The probabilities $P(m, 7')$ for m = 7-25

7	8	9	10	11	12	13
$\frac{4096}{105183}$	$\frac{28672}{292175}$	$\frac{379904}{2629575}$	$\frac{1089536}{6720025}$	$\frac{1792}{11687}$	$\frac{517888}{4032015}$	$\frac{1975616}{20160075}$
14	15	16	17	18	19	20
$\frac{679808}{9821575}$	$\frac{4037824}{88394175}$	$\frac{1700608}{60108039}$	$\frac{1648304}{100180065}$	$\frac{448}{50025}$	$\frac{7024}{1550775}$	$\frac{14128}{6720025}$
21	22	23	24	25	Total	
$\frac{236}{268801}$	$\frac{56}{175305}$	$\frac{28}{292175}$	$\frac{56}{2629575}$	$\frac{7}{2629575}$	1	

Table 7. The probabilities $P(m, 8')$ for m = 8-29

8	9	10	11	12	13
$\frac{16384}{2629575}$	$\frac{57344}{2629575}$	$\frac{28672}{650325}$	$\frac{272384}{4032015}$	$\frac{351232}{4032015}$	$\frac{670208}{6720025}$
14	15	16	17	18	19
$\frac{3082496}{29464725}$	$\frac{3021952}{29464725}$	$\frac{28614656}{300540195}$	$\frac{25375168}{300540195}$	$\frac{37184}{516925}$	$\frac{91616}{1550775}$
20	21	22	23	24	25
$\frac{314752}{6720025}$	$\frac{144464}{4032015}$	$\frac{4624}{175305}$	$\frac{5432}{292175}$	$\frac{32704}{2629575}$	$\frac{20468}{2629575}$
26	27	28	29	Total	
$\frac{4}{899}$	$\frac{2}{899}$	$\frac{4}{4495}$	$\frac{1}{4495}$	1	

3.2. *The probabilities of winning in the second round*

Let us consider the second round. All the combinations for the potential matches are given by (refer to Eq. (2))

$$|\Omega'| = \frac{8!}{(4!)^2 \cdot 2!} \tag{6}$$

Of course, the 7′th- and 8′th-teams have no chance to win. If we consider the 6′th-best team, the probability of the 6′th-best winning is given by

$$P_{6'} = \frac{\frac{5!2!}{4!2!}}{|\Omega'|} \tag{7}$$

where $P_{\pi'}$ denotes the probability of the π'-best team winning in the second round. Just similar to Eq. (7), we have

$$P_{5'} = \frac{\frac{4!3!}{4!3!} + \frac{4!3!}{3!2!}}{|\Omega'|} = \frac{13}{35} \tag{8}$$

$$P_{4'} = \frac{\frac{3!4!}{3!3!} + \frac{3!4!}{2!2!2!}}{|\Omega'|} = \frac{22}{35} \tag{9}$$

$$P_{3'} = \frac{\frac{2!5!}{2!3!2!} + \frac{2!5!}{2!3!}}{|\Omega'|} = \frac{6}{7} \tag{10}$$

It is clear that $P_{1'} = P_{2'} = 1$ since the first- and second-place teams can become the Asian Qualifiers with probability 1. Thus, we summarize the above probabilities in Table 8, where the sum of the probabilities is 4.

Table 8. The probabilities of winning for 1′-6′ in the second round

1′	2′	3′	4′	5′	6′	Total
1	1	$\frac{6}{7}$	$\frac{22}{35}$	$\frac{13}{35}$	$\frac{1}{7}$	4

We are now ready to give the probabilities of winning the Asian Qualifiers from the first round by using the above tables. It is quite trivial that the 1st-best team can become the Asian Qualifiers with probability 1. However, for the 2nd-best team, there is the probability $\frac{28}{31}$ to qualify for the second round, and once it is qualified for the second round, the 2nd-best team can become the Asian Qualifiers. That is, the probability of winning is $\frac{28}{31}$. For the 3rd-best team, the probability of winning is

$$\frac{14}{155} \cdot 1 + \frac{112}{155} \cdot \frac{6}{7} = \frac{22}{31}$$

For the 4th-best team, the probability of winning is

$$\frac{28}{4495} \cdot 1 + \frac{1008}{4495} \cdot \frac{6}{7} + \frac{448}{899} \cdot \frac{22}{31} = \frac{460}{899}$$

In a similar fashion, we can calculate the probabilities of winning in Table 9. Of course, the total sum of probabilities is 4 since 4 teams can become the Asian Qualifiers. In Fig. 4, we show the plot of the probabilities of winning from the 1st- to 21st-best teams. For instance, for the 21st-best team, there is the very low probability of winning with about 1 over 28 millions, which is equivalent to the probability of winning the First Prize for the typical Lottery.

Table 9. The probabilities of winning the Asian Qualifiers for 1-21 best teams in the first round

1	2	3	4	5	6	7	8
1	$\frac{28}{31}$	$\frac{22}{31}$	$\frac{460}{899}$	$\frac{2179}{6293}$	$\frac{12592}{56637}$	$\frac{100264}{736281}$	$\frac{41936}{525915}$
9	10	11	12	13	14	15	
$\frac{602}{13485}$	$\frac{1480}{62031}$	$\frac{3764}{310155}$	$\frac{184}{31465}$	$\frac{5758}{2171085}$	$\frac{46288}{41250615}$	$\frac{7736}{17678835}$	
16	17	18	19	20	21	Total	
$\frac{9296}{60108039}$	$\frac{14533}{300540195}$	$\frac{4}{310155}$	$\frac{2}{723695}$	$\frac{4}{9408035}$	$\frac{1}{28224105}$	4	

4. Concluding Remarks

We have calculated the probabilities of winning for the sports contest based on the assumption that the strength among the contest teams can be described by a linearly ordered set. We have developed our general formulae for the calculation of the probabilities of winning, and we demonstrated our calculation on the Asian Qualifiers for 2006 FIFA World Cup. Table 9 and Fig. 4 show the numerical results, where we can understand that there is a chance even if the rank of the team is quite low.

For 2006 FIFA World Cup, 4.5 countries can qualify for the Asian Qualifiers, where 0.5 countries mean that the winner of third place teams in the second round has the right to challenge against the similar team from North

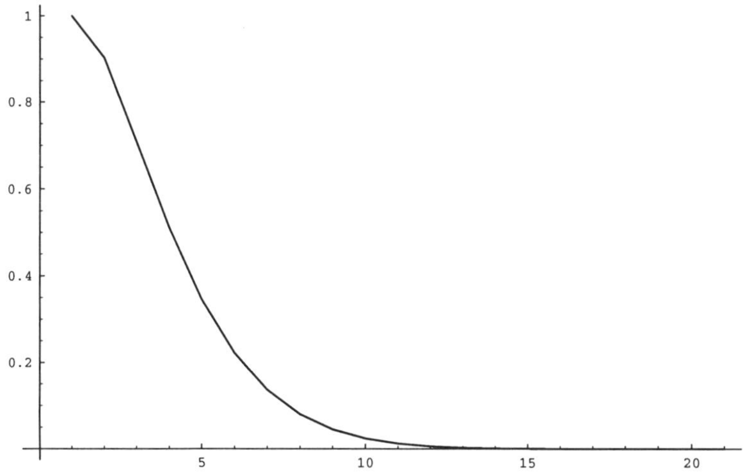

Figure 4. The plot of the probabilities of winning the Asian Qualifiers for 1-21 best teams in the first round

American and Caribbean Qualifiers. The winner has a last chance to qualify for 2006 FIFA World Cup. For the 0.5 countries issue, we believe that the similar calculations can be carried out under the suitable assumption.

Acknowledgments

This work was partially supported by a Grant-in-Aid for Scientific Research from the Ministry of Education, Sports, Science and Culture of Japan under Grant 16201035 and 16510128, and the Nanzan University Pache Research Subsidy I-A-2 for 2005.

References

1. IMA Journal of Management Mathematics, Vol. 16 ("OR in Sport" special issue), Oxford University Press, Oxford, 2005. (http://imaman.oxfordjournals.org/content/vol16/issue2/).
2. S. Butenko, J. Gil-Lafuente and P. M. Pardalos (Eds.), *Economics, Management and Optimization in Sports*, Springer-Verlag, Berlin, 2004.
3. M. Tamaki, *An Introductory Course in Probability*, (in Japanese), Makino-Shoten, Tokyo, 1992.
4. D. E. Knuth, *The Art of Computer Programming*, Vol.3, 2nd ed., Addison-Wesley, Boston, 1998.
5. F. Mosteller, *Fifty Challenging Problems in Probability with Solutions*, Dover, New York, 1965.

INDEX